Springer Handbook of Auditory Research

For further volumes:
http://www.springer.com/series/2506

Laurence O. Trussell · Arthur N. Popper
Richard R. Fay

Editors

Synaptic Mechanisms
in the Auditory System

 Springer

Editors

Laurence O. Trussell
Vollum Institute
Oregon Health & Science University
Portland, OR 97239, USA
trussell@ohsu.edu

Arthur N. Popper
Department of Biology
University of Maryland
College Park, MD 20742, USA
apopper@umd.edu

Richard R. Fay
Marine Biological Laboratory
Woods Hole, MA 02543, USA
rfay@luc.edu

ISBN 978-1-4419-9516-2   e-ISBN 978-1-4419-9517-9
DOI 10.1007/978-1-4419-9517-9
Springer New York Dordrecht Heidelberg London

Library of Congress Control Number: 2011935541

Printed on acid-free paper

Springer is part of Springer Science+Business Media (www.springer.com)

This volume is dedicated to Professor Alan D. Grinnell, whose career has embodied the twin themes that run through the entire book, the auditory system and the physiology of synapses. Alan, along with his advisor Donald Griffin, was the first to make electrophysiological studies in the bat auditory pathway, work he continued as a faculty member at UCLA. Having also trained with Bernard Katz and Ricardo Miledi, Alan has a genuine love for the synapse, which he has expressed in a parallel research career, producing many creative and elegant studies of the physiology of the neuromuscular junction. This was all accomplished by his enormous energy and amazing breadth of knowledge. Indeed, those fortunate enough to work with him know that Alan is a consummate scholar, a deeply inquisitive scientist, and an excellent friend.

Series Preface

The Springer Handbook of Auditory Research presents a series of comprehensive and synthetic reviews of the fundamental topics in modern auditory research. The volumes are aimed at all individuals with interest in hearing research, including advanced graduate students, postdoctoral researchers, and clinical investigators. The volumes are intended to introduce new investigators to important aspects of hearing science and to help established investigators to better understand the fundamental theories and data in fields of hearing that they may not normally follow closely.

Each volume presents a particular topic comprehensively, and each serves as a synthetic overview and guide to the literature. As such, the chapters present neither exhaustive data reviews nor original research that has not yet appeared in peer-reviewed journals. The volumes focus on topics that have developed a solid data and conceptual foundation rather than on those for which a literature is only beginning to develop. New research areas will be covered on a timely basis in the series as they begin to mature.

Each volume in the series consists of a few substantial chapters on a particular topic. In some cases, the topics will be ones of traditional interest for which there is a substantial body of data and theory, such as auditory neuroanatomy (Vol. 1) and neurophysiology (Vol. 2). Other volumes in the series deal with topics that have begun to mature more recently, such as development, plasticity, and computational models of neural processing. In many cases, the series editors are joined by a co-editor having special expertise in the topic of the volume.

Richard R. Fay, Falmouth, MA
Arthur N. Popper, College Park, MD

Volume Preface

This volume illustrates how two time-honored areas of research, auditory systems physiology and synaptic physiology, have come together to generate a new subfield of research, the synaptic mechanisms of auditory coding. That union has generated new insight into systems function, and its success is providing the stimulus for development or application of new techniques and ideas in our field. The topics primarily focus on synapses and ion channels in neurons of the central nervous system, with emphasis on the brainstem, but they also offer an informative look at the first auditory synapse in cochlear hair cells.

Chapter 1 by Trussell provides an overview and guide to the volume and shares thoughts about future research directions. Chapter 2 by Golding examines the voltage-gated ion channels of auditory neurons and how these determine the kind of computation that can be performed on acoustically driven inputs. With this as background, we turn to synapses in Chapter 3, wherein Nicolson shows how the molecular and physiological components of the hair cell synapse initiates coding.

The giant synapses of the auditory system have attracted attention of researchers both within and outside the auditory field, to great advantage. These terminals, the endbulbs and calyces of Held, are described in Chapter 4 by Manis, Xie, Wang, Marrs, and Spirou and also in Chapter 5 by Borst and Rusu. In Chapter 6, MacLeod and Carr describe the bases of synaptic coincidence detection and its role in sound localization, while in Chapter 7, Trussell examines how synaptic inhibition operates, with examples from the cochlear nucleus and superior olive.

Chapter 8 by Metherate and Chapter 9 by Tzounopoulos and Leão address the short- and long-term modifiability of auditory synapses and how this plasticity may be used in auditory processing. Metherate examines auditory neuromodulation and gives an example of its potential roles in attention. Tzounopoulos and Leão present the case for experience-dependent plasticity as a well-established component of auditory function from brainstem to cortex.

As in all previous SHAR volumes, there are chapters in other books of the series that have relevance to the general theme discussed in this volume. For example, the circuitry and computation in the auditory system, so related to synapse function, is discussed in chapters of Volume 15 (*Integrative Functions in the Mammalian*

Auditory Pathway), while synapses in the inner ear are considered in detail in Volume 8 (*The Cochlea*) and Volume 27 (*Vertebrate Hair Cells*). Finally, computational models of the auditory system, the topic of many chapters in this volume, are discussed in detail in Volume 6 (*Auditory Computation*) and Volume 35 (*Computational Models of the Auditory System*).

Laurence O. Trussell, Portland, OR
Arthur N. Popper, College Park, MD
Richard R. Fay, Falmouth, MA

Contents

Contributors

J.G.G. Borst Department of Neuroscience, Erasmus MC,
University Medical Center Rotterdam, Dr. Molewaterplein 50,
3015 GE, Rotterdam, The Netherlands
g.borst@erasmusmc.nl

Catherine E. Carr Department of Biology, University of Maryland,
College Park, MD 20742–4415, USA
cecarr@umd.edu

Nace L. Golding Section of Neurobiology, Institute for Neuroscience
and Center for Perceptual Systems, University of Texas at Austin, Austin,
TX 78712–0248, USA
golding@mail.utexas.edu

Ricardo M. Leão Department of Physiology, University of São Paulo,
Ribeirão Preto, SP, Brazil
rml34@pitt.edu

Paul B. Manis Department of Otolaryngology/Head and Neck Surgery,
UNC Chapel Hill, G127 Physician's Office Bldg., CB#7070, Chapel Hill,
NC 27599–7070, USA
pmanis@med.unc.edu

Glen S. Marrs Department of Otolaryngology, West Virginia University School
of Medicine, One Medical Center Drive, PO Box 9304, Health Sciences Center,
Morgantown, WV 26506–9304, USA
gmarrs@hsc.wvu.edu

Katrina M. MacLeod Department of Biology, University of Maryland,
College Park, MD 20742–4415, USA
macleod@umd.edu

Raju Metherate Department of Neurobiology and Behavior and Center for
Hearing Research, University of California, 2205 McGaugh Hall, Irvine,
CA 92697-4550, USA
rmethera@uci.edu

Teresa Nicolson Howard Hughes Medical Institute, Oregon Hearing Research
Center, and Vollum Institute, Oregon Health & Science University, 3181 SW Sam
Jackson Park Road, Portland, OR 97239, USA
nicolson@ohsu.edu

S.I. Rusu Department of Neuroscience, Erasmus MC, University Medical Center
Rotterdam, Dr. Molewaterplein 50, GE Rotterdam 3015, The Netherlands

George A. Spirou Department of Otolaryngology, West Virginia University
School of Medicine, One Medical Center Drive, PO Box 9304, Health Sciences
Center, Morgantown, WV 26506–9304, USA
gspirou@hsc.wvu.edu

Laurence O. Trussell Vollum Institute, Oregon Health & Science University,
3181 SW Sam Jackson Park Road, Portland, OR, USA
trussell@ohsu.edu

Thanos Tzounopoulos Department of Otolaryngology, University of Pittsburgh,
3501 Fifth Avenue, BSTW 10021, Pittsburgh, PA 15261, USA
thanos@pitt.edu

Yong Wang Otolaryngology/Neuroscience Program, 3C120 School of Medicine,
University of Utah, 30 North 1900 East, Salt Lake City, UT 84132, USA
yong.wang@hsc.utah.edu

Ruili Xie Department of Otolaryngology/Head and Neck Surgery,
UNC Chapel Hill, G127 Physician's Office Bldg., CB#7070, Chapel Hill,
NC 27599–7070, USA
ruili_xie@med.unc.edu

Chapter 1
Sound and Synapse

Laurence O. Trussell

1 Introduction

This volume is an expression of the ongoing application of the concepts and techniques of cellular neurophysiology and cell biology to understanding auditory function. Embedded in this application is a story of the fruits of cross fertilization among scientific fields. Rather than apply traditional methods of neuroanatomy, in vivo extracellular recordings, or spike frequency analysis, many labs began asking questions such as what ion channels are expressed in auditory neurons? How do these channels determine the cellular response to sound? Beyond simply identifying which transmitters were expressed in different neurons, scientists explored the biophysical responses to those transmitters and related them to the response times of synapses. Often, these were labs with background and training outside the auditory system. Among the pioneers in this effort were Donata Oertel, who first developed a viable brain slice preparation of the cochlear nucleus and characterized cellular response properties in identified cells (Oertel 1983), and Paul Manis, who first voltage clamped isolated auditory neurons (Manis and Marx 1991). Moreover, some of the most challenging projects in electrophysiology were first applied to the auditory system, such as the application of patch-clamp methods to presynaptic structures like the calyx of Held by Ian Forsythe, Gerard Borst, and colleagues (Forsythe 1994; Borst et al. 1995) or to tiny postsynaptic spiral ganglion cell dendrites by Elisabeth Glowatzki and Paul Fuchs (Glowatzki and Fuchs 2002). This had unexpected benefits: because the language of cellular physiology was common to many neural systems; this effort produced results understandable and of interest to diverse non-auditory neuroscientists and thus helped popularize the field.

L.O. Trussell (✉)
Vollum Institute, Oregon Hearing Research Center,
Oregon Health and Science University, 3181 Southwest
Sam Jackson Park Road, L335A, Portland, OR 97239, USA
e-mail: trussell@ohsu.edu

L.O. Trussell et al. (eds.), *Synaptic Mechanisms in the Auditory System*,
Springer Handbook of Auditory Research 41, DOI 10.1007/978-1-4419-9517-9_1,
© Springer Science+Business Media, LLC 2012

The results of these efforts have triggered a new appreciation of the synapse as a key to understanding auditory mechanisms. Synapses are more than just switches for excitation versus inhibition. Synapses vary in their strength and their ability to sustain activity over time. They vary in their temporal precision, their time course of action, and their ability to change in response to different patterns of activity. A central thesis, even an article of faith, is that this variation occurs in accordance with the particular demands for information processing in a given circuit. What is the physiological advantage conferred by expression of Kv1 K$^+$ channels in dendrites of coincidence detector cells? Why have a giant calyceal synapse, the largest in the brain, with a low probability for vesicle release? Questions like these motivate one to make teleological sense of "details." As important as these questions are, there is a danger in designing experiments that are yoked to such considerations. Focusing on what "makes sense" to the system presumes a rather complete understanding of the system, and this focus may lead to ignoring information that could eventually be of great consequence. For example, why do synapses in the dorsal cochlear nucleus, the lowest level of the auditory central nervous system (CNS), exhibit such amazingly rich and varied forms of plasticity? Our systems-level understanding of multisensory integration in this region is simply too rudimentary to answer this question. Thus a corollary to the central thesis presented earlier is that unbiased studies of cellular properties may lead to a novel or revised understanding of the system. As a result, it is not always bad practice to consider the circuit as a pretext for doing fascinating, and (dare I say it?) fun, experiments in cellular neuroscience!

2 Overview

The topics in this volume were chosen to highlight areas in which there is abundant insight into cellular function or in which the cellular properties are clearly essential to understanding how a circuit computes. Chapter 2 by Golding provides insight into the intrinsic response properties of neurons, how cells take their synaptic input and turn it into a particular pattern of action potential firing. This is a field that provides an excellent example in which studies of auditory cellular neuroscience must draw constantly from an ever-increasing body of multidisciplinary information. What ion channels are expressed in a given cell? What is their molecular composition and what is the consequence of this structure to their biophysical properties? How are these proteins distributed over the cell surface? How are they regulated and how do they change during development or in disease states? The study of ion channel properties also provides vital information for the construction of computational models that both are valid and have strong predictive power. From this will surely come deep insight into the function of auditory circuits.

Chapter 3 by Nicolson reveals the synaptic genesis of auditory processing in the hair cell. Hair cells transduce mechanical vibration to a voltage change that embodies key temporal and intensity features of sound. The synapse must then convert this voltage change into a neural code in two phases. First, voltage must translate to vesicle exocytosis in a manner that preserves these aspects of temporal and intensity information. Next, the postsynaptic dendrite must respond to the transmitter, generate an action potential, and then restore itself to be ready to respond again. Recent work has revealed that this is no "garden-variety" synapse; rather, it has the capacity to sustain continued exocytosis and to respond to voltage changes with exquisite temporal precision. Novel proteins are expressed at the synapse, and a remarkable exocytic mechanism called multivesicular release is prominent – presumably these and other novel features somehow figure into the specialized function of the hair cell synapse.

The auditory CNS features some of the largest synapses in the mammalian brain, the endbulbs and calyces of Held, explored in Chap. 4 by Manis, Xie, Wang, Marrs, and Spirou and in Chap. 5 by Borst and Rusu. These giant terminals, each making hundreds of synaptic release sites, have been an attractive preparation for study for a variety of reasons. Because they are so large, endbulbs and calyces are practically begging to be labeled as auditory relays and thus have all their physiological properties interpreted in that context. However, analysis of their detailed properties have revealed many surprises, such as short-term plasticity and presynaptic modulation, giving rise to speculation that such terminals do more than act as relays. Some laboratories approached these terminals with little interest in auditory function, instead taking the opportunity to study an accessible central synapse. Many significant advances have been made in this effort, which have informed a general understanding of how brain synapses work. However, as the results are compared to data in other preparations, it has become clear that endbulbs and calyces are not merely large generic synapses, but also structures highly specialized to specific components of auditory processing.

In Chap. 6, MacLeod and Carr explore how synapses mediate the amazingly precise coincidence detection that mediates some forms of sound localization. Basic components of coincidence detection are the innervation of distinct sets of dendrites, fast-acting transmitter receptors, and ultra-responsive membrane properties. These are features fundamental to function that appear be common to all vertebrates. Some aspects of sound localization, however, differ between birds and mammals, and perhaps even among some mammals. These may have resulted from animals' different frequency ranges of hearing and different head sizes, which determine what physical properties of sound are relevant and limit how circuits can extract information. For example, synaptic inhibition has been employed in different ways by mammals and birds, a topic of intense current debate. Although it is believed that inhibition is needed for refining coincidence detection, in fact inhibitory transmission of diverse types appears at every level of auditory processing and must therefore serve many functions. In Chap. 7, Trussell overviews mechanisms of synaptic inhibition and gives examples from two very different inhibitory pathways in the cochlear nucleus and superior olive. However, although there are some well-known examples of inhibition in the auditory system, the field is very young in terms of defining what are the variety of inhibitory cells, how each cell type modifies excitation at its different target cells, and how experience-dependent plasticity, drugs, or disease affects hearing through alterations in inhibition. Moreover, it is likely that in the world of intelligent design of prosthetic devices, construction of brainstem implants that mediate hearing in patients with damaged auditory nerves will have to account for refinements in processing imposed by inhibitory neurons.

A common misconception about auditory processing, especially in the lower auditory pathways, is that it needs to be invariant, to respond the same way at all times. Otherwise, preservation of fine temporal differences in the information contained in sound signals might be disturbed, thus compromising perception. Chapter 8 by Metherate and Chap. 9 by Tzounopoulos and Leão show that this view is not valid. Metherate defines for us the rather slippery term "neuromodulation" and discusses how it makes sense as a vital function for an auditory system that must operate in different situations with different states of attention. Tzounopoulos and Leão explore in detail how experience-dependent plasticity is a well-established part of auditory function, in the cortex, where it might be expected, but also in the lowest levels of auditory processing.

3 New Horizons

This introduction has tried to convey that our understanding of synaptic mechanisms in audition has required bringing in new skills sets and new outlooks. What new areas of research must come into the field to deepen our understanding of auditory function? Many of the chapters herein conclude with a look to the future. To the many insightful points they have made can be added the need to look at the functional significance of the complex array of descending connections within the auditory system. Being able to label vitally, and preferably to activate single axons, perhaps optogenetically, in identified descending pathways will bring clarity to a major area of research. Testing the role of single cells or single synapses by acute inactivate with modern genetic and molecular biological tools will be essential. Network-level computational models must take into account the kinds of work outlined in this volume.

Finally, it may be noted that many of the chapters in this book address exclusively synaptic mechanisms in auditory brainstem. Although there is much excellent work in cortex, by and large the studies of synaptic function are in their infancy for levels higher than the superior olive, including the lemniscal nuclei, the inferior colliculus, and the thalamus. One reason for this is the great complexity of their inputs. Even when recordings are made from identified cell types, it is difficult to identify the source of a particular excitatory or inhibitory input, especially when studied in vitro. New approaches to recording and stimulation, as well as new preparations, must be developed to extend the work outlined here through the full extent of the auditory system.

Acknowledgments I wish to thank the authors of these chapters for their hard work and scholarship. My support was provided by the NIH (grants NS028901 and DC004450).

References

Borst, J. G., Helmchen, F., & Sakmann, B. (1995). Pre- and postsynaptic whole-cell recordings in the medial nucleus of the trapezoid body of the rat. *Journal of Physiology, 489 (Pt 3)*, 825–840.

Forsythe, I. D. (1994). Direct patch recording from identified presynaptic terminals mediating glutamatergic EPSCs in the rat CNS, in vitro. *Journal of Physiology, 479 (Pt 3)*, 381–387.

Glowatzki, E., & Fuchs, P. A. (2002). Transmitter release at the hair cell ribbon synapse. *Nature Neuroscience, 5*(2), 147–154. doi: 10.1038/nn796.

Manis, P. B., & Marx, S. O. (1991). Outward currents in isolated ventral cochlear nucleus neurons. *Journal of Neuroscience, 11*(9), 2865–2880.

Oertel, D. (1983). Synaptic responses and electrical properties of cells in brain slices of the mouse anteroventral cochlear nucleus. *Journal of Neuroscience, 3*(10), 2043–2053.

Chapter 2
Neuronal Response Properties and Voltage-Gated Ion Channels in the Auditory System

Nace L. Golding

1   Introduction

One of the central challenges to auditory neuroscience is to understand how sound information is processed and transformed as it ascends to different levels in the brain. One way that the central auditory system is distinct from other sensory areas of the brain is the extent to which sound information is segregated at the earliest subcortical areas into different ascending pathways encoding different aspects of sound. For example, in the visual system, the first stage of information processing in the brain takes place in the lateral geniculate nucleus of the thalamus before proceeding directly to the primary visual cortex, where many of the major transformations in visual receptive fields occur. In olfaction, although extensive processing occurs in the olfactory bulb prior to the cortex, it is not apparent that there are topographic differences in how olfactory information is processed.

   The aim of this chapter is to review how the coding of auditory information in different ascending pathways is influenced by synaptic integration, the process by which excitatory and inhibitory inputs sum together and trigger patterns of action potentials that reflect salient features of sounds. It will be made clear in this chapter that synaptic integration is strongly influenced, and in some cases dominated, by interactions between synaptic inputs and different classes of voltage-gated ion channels. Although mammalian systems are the primary focus, work from the avian auditory system will be discussed in specific instances. Particular attention will be on neurons of the cochlear nucleus and superior olivary complex, where the role of

N.L. Golding (✉)
Section of Neurobiology, Institute for Neuroscience, and Center for Perceptual Systems, University of Texas at Austin, Austin, TX 78712-0248, USA
e-mail: golding@mail.utexas.edu

L.O. Trussell et al. (eds.), *Synaptic Mechanisms in the Auditory System*,
Springer Handbook of Auditory Research 41, DOI 10.1007/978-1-4419-9517-9_2,
© Springer Science+Business Media, LLC 2012

voltage-gated ion channels can be more easily understood within the context of well-defined circuit computations and functional roles. Two broad classes of neurons emerge: those with electrical properties that precisely maintain the temporal features encoded in their auditory inputs and those with electrical properties that transform synaptic input patterns into new patterns.

2 The Spatial and Temporal Structure of Auditory Input to the Brain

In order to understand the nature of different auditory neurons' responses to sound stimuli, it is important to review two fundamental concepts in auditory neuroscience: tonotopy and phase locking. Sounds of different frequencies vibrate the basilar membrane of the cochlea in a topographic manner, with low frequencies vibrating more apical locations and high frequencies vibrating more basal locations. These vibrations are transduced into graded electrical signals by the cochlear hair cells, which in turn trigger patterns of action potentials in the spiral ganglion neurons (Nicolson, Chap. 3). Because neurons in the spiral ganglion innervate a limited number of hair cells, each ganglion cell carries information about a limited range of frequencies. Deflections of the stereocilia embedded in the basilar membrane cause a depolarization that leads to activation of voltage-gated calcium channels, causing calcium influx and the release of the excitatory neurotransmitter glutamate onto the endings of the spiral ganglion neurons. The activity of hair cells is converted into trains of action potentials by the spiral ganglion cells whose axons project to the brain via the eighth cranial, or auditory, nerve. All auditory information to the brain is carried by the auditory nerve fibers, which in turn synapse onto diverse cell targets in the cochlear nucleus, the first and obligatory integrative station in the brain. The cochlear nucleus possesses at least six classes of projecting neurons. Each of these pathways conveys different kinds of information, despite the fact that the presynaptic pattern of action potentials to these neurons is the same. There is an orderly representation of frequencies imposed by the paths of these auditory afferents in the brain. Their parallel orientation to one another in the cochlear nucleus creates a series of "iso-frequency" slabs, imposing frequency selectivity on the different cell types present in the cochlear nucleus. This organization is maintained through the projection patterns of neurons in the cochlear nucleus, thus creating tonotopic maps at successively higher levels in the auditory pathway.

Auditory afferents also convey critical information about sounds due to their ability to precisely represent periodic information in the patterns of their action potential output. This is commonly referred to as temporal coding. In hair cells, timed neurotransmitter release is brought about by the fact that hair cell signaling is directionally selective, with positive deflections of the stereocilia (toward the tallest stereocilia) triggering membrane depolarization and negative deflections resulting in membrane hyperpolarization. Thus, during acoustic deflections of the basilar membrane, hair cells respond with cyclical depolarizing and hyperpolarizing voltage changes that reflect the frequency content of the stimulus. The corresponding cyclical release of neurotransmitter onto spiral ganglion cells imposes a restricted interval over which firing occurs, a phenomenon known as phase locking. In auditory nerve fibers, the axons of spiral ganglion neurons, phase locking occurs at frequencies generally below 2–4 kHz in mammals but extends up to 9 kHz in barn owls (Johnson 1980; Köppl 1997; Taberner and Liberman 2005). It is important to note that the precise phase locking of an individual auditory neuron does not require perfect one-for-one firing with each cycle of the acoustic stimulus. As many neurons encode a given frequency, the interval of the stimulus is encoded by the overall firing responses of the neural population as a whole.

3   Synaptic and Voltage-Gated Ion Channel Properties for Precise Temporal Coding

Given the importance of timing information in the auditory system, a major focus of this chapter is on how interactions between synaptic inputs and voltage-gated ion channels maintain, and in some cases improve, the precision of the firing of action potentials. Some of the most intensely studied circuits that utilize timing information are introduced here.

3.1   Circuits That Utilize Timing Information

3.1.1   Coincidence Detection Across Frequencies in Octopus Cells of the Ventral Cochlear Nucleus

Octopus cells are located in a distinct area of the posteroventral cochlear nucleus called the octopus cell area (Osen 1969). Their axons form a major ascending projection, giving rise to large calyceal endings in the contralateral ventral nucleus of the lateral lemniscus as well as the superior paraolivary nuclei (reviewed in Oertel 1999). These neurons are named after their distinctive dendritic architecture, which consists of large-caliber dendrites emanating from one pole of the soma. Octopus cells exhibit a distinct orientation with respect to the paths of the auditory nerve fibers, which provide their primary excitation. The cell body tends to be oriented toward the posterior octopus cell area, which receives inputs from lower-frequency afferents, and the dendrites extend roughly perpendicularly to the paths of the auditory nerve fibers toward higher-frequency regions (Fig. 2.1a) (Osen 1969; Kane 1973). Accordingly, octopus cells in vivo exhibit broad tuning curves and are effectively driven by transient broadband stimuli such as clicks (Godfrey et al. 1975;

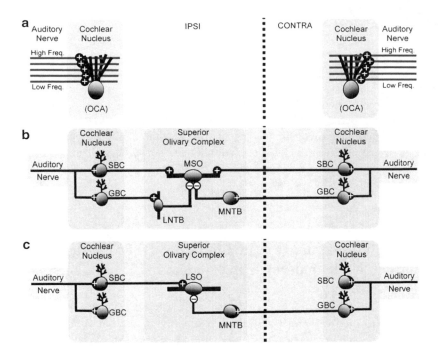

Fig. 2.1 Three time-coding pathways in the auditory brainstem. (**a**) Octopus cells are clustered in a distinctive area of the posteroventral cochlear nucleus, the octopus cell area (OCA). Excitatory, glutamatergic inputs from auditory nerve fibers are organized tonotopically, with low-frequency fibers forming synapses on more proximal dendrites and higher-frequency fibers contacting progressively more distal dendrites. (**b**) Excitatory synaptic coincidence detection of binaural inputs in the medial superior olive (MSO). MSO principal neurons present in the superior olivary complex receive glutamatergic excitation from both ipsilateral and contralateral spherical bushy cells (SBCs) in the anteroventral cochlear nucleus, which in turn are driven by large calyceal synapses of auditory nerve fibers, the endbulbs of Held. MSO neurons are driven by two feedforward inhibitory nuclei, the medial and lateral nuclei of the trapezoid body (MNTB and LNTB). Both neuron types are primarily glycinergic and are driven by excitation from globular bushy cells (GBCs) of the posteroventral cochlear nucleus. Glycinergic inhibition in MSO principal neurons is targeted to the soma and proximal dendrites, whereas excitation is primarily dendritic and segregated to one side of a bipolar arbor. (**c**) Binaural processing in the lateral superior olive (LSO). LSO principal neurons receive ipsilateral excitation from ipsilateral spherical bushy cells and contralateral inhibition from MNTB principal cells. Similar to MSO principal neurons, LSO principal neurons receive somatic/proximal dendritic inhibition and dendritic excitation within a bipolar dendritic structure

Rhode and Smith 1986; Oertel et al. 2000). In response to tones and noise stimuli, octopus cells respond with an "onset" firing pattern, with a single well-timed spike followed by nearly no subsequent firing for the duration of the sound stimulus. Octopus cells likely integrate the convergence of at least 50 auditory nerve fibers (Golding et al. 1995). Because each input contributes only a small submillivolt depolarization to octopus cells' postsynaptic responses, the initiation of action potentials requires strong synchronous activation of many auditory nerve fibers tuned to a broad range of frequencies. In this way, octopus cell dendrites detect the coincident activity of a large population of auditory nerve fibers encoding a broad range of frequencies.

3.1.2   Computation of Interaural Time Differences in the Medial Superior Olive

Neurons of the medial superior olive (MSO) are one of the major cell groups in the superior olivary complex and will be described, along with their avian homologs, in Chap. 6 by McLeod and Carr. The MSO is one of the first sites for integrating auditory activity from the two ears. MSO neurons are innervated by the spherical bushy cells that reside in the ipsilateral and contralateral ventral cochlear nucleus (Fig. 2.1b) (Cant and Casseday 1986; Smith et al. 1993; Beckius et al. 1999). Spherical bushy cells themselves are driven by only a few (1–3) powerful specialized endings from the auditory nerve, the endbulbs of Held (Manis et al., Chap. 4). The spherical bushy cells then provide conventional bouton-type excitatory synapses to MSO principal cells. The dendritic architecture of MSO principal cells is bipolar, with ipsilateral bushy cell input segregated to the lateral dendrites and contralateral bushy cells inputs restricted to the medial dendrites (Stotler 1953; Lindsey 1975). MSO neuron responses are also shaped by two feed-forward inhibitory nuclei, the medial and lateral nucleus of the trapezoid body (MNTB and LNTB, respectively). The principal neurons of the MNTB and LNTB are driven by contralateral and ipsilateral globular bushy cells in the cochlear nucleus, respectively (Borst and Rusu, Chap. 5).

As low-frequency sound sources move along the horizontal plane, the relative timing of bushy cell inputs to the superior olivary complex changes systematically, thereby changing the relative timing of excitatory and inhibitory inputs to MSO neurons. MSO neurons respond to these synaptic alterations by changing their rate of action potential firing. This activity is conveyed via the axonal projections of MSO neurons to the central nucleus of the inferior colliculus (Henkel and Spangler 1983; Nordeen et al. 1983; Loftus et al. 2004). In this way, MSO neurons detect the relative coincidence of synaptic inputs driven by the two ears and translate these differences into a rate code. Ultimately, this activity is utilized for the localization of sounds along the horizontal plane. Thus, the temporal resolution of the detection of binaural coincidence in the MSO has a clear relationship to the spatial acuity of horizontal sound localization.

3.1.3   Computation of Interaural Level Difference in the Lateral Superior Olive

Neurons in the lateral superior olive (LSO) comprise the second major integrative stage for processing binaural cues in the auditory brainstem. The principal neurons of the LSO vary their rate of action potential firing according to differences in the level of sound intensity between the two ears. These level differences are most acute at high frequencies because these frequencies are more susceptible to the effects of head shadowing. Consistent with this role, the frequency representation in the population of LSO neurons appears biased toward high frequencies (Guinan et al. 1972; Tsuchitani 1977). The circuitry of the LSO contains many of the same components as that of the MSO: principal neurons of the LSO receive phase-locked excitatory activity from ipsilateral spherical bushy cells of the cochlear nucleus and inhibitory glycinergic inputs from the contralateral principal neurons of the MNTB (Fig. 2.1c). As in the MSO, timing is critical in the LSO. As sounds move along the horizontal plane, the relative timing and hence the balance between excitatory and inhibitory synaptic inputs is altered, systematically changing the rate at which action potentials are generated. Whereas MSO neurons detect correlations between binaural excitatory inputs, effective signaling by LSO principal neurons relies on decorrelations between excitation and inhibition. Despite the different nature of their respective circuit computations, the ability of both MSO and LSO principal neurons to signal changes in sound location relies on the temporal precision of presynaptic excitatory and inhibitory inputs, as well as precision in the postsynaptic neurons themselves.

3.2   Glutamate Receptor Properties for Fast Synaptic Excitation in Time-Coding Auditory Neurons

In order to encode the temporal structure of sounds, neurons in auditory pathways concerned with timing information must reduce the time window over which they integrate auditory information. An obvious specialization required for this to occur is a reduction in the time course of excitatory synaptic currents. In birds and mammals, auditory neurons encoding fine timing information exhibit excitatory currents mediated primarily by α-amino-3-hydroxy-5-methyl-4-isoxazolepropionic acid (AMPA)-type glutamate receptors. These receptors are characterized by fast rise times and short durations (generally <0.2 and 0.5 ms, respectively, at 34°C) (Raman et al. 1994; Isaacson and Walmsley 1995; Gardner et al. 1999). These properties reflect the molecular composition of the receptors. In the cochlear nucleus, bushy cells express primarily GluR3 and GluR4 subunits and express GluR2 subunits in less abundance (Hunter et al. 1993; Wang et al. 1998b). Studies in cell expression systems and auditory neurons have shown that these subunits yield receptors with fast gating and high calcium permeability (Geiger et al. 1995; Otis et al. 1995). Although the molecular composition of AMPA receptors has not been as thoroughly studied in other auditory nuclei, EPSP responses reflecting fast AMPA receptor gating is seen as a repeated motif in neurons in auditory nuclei involved in temporal coding, including the ventral cochlear nucleus (bushy cells), superior olivary complex (principal neurons of the MSO, LSO, and MNTB), and the ventral nucleus of the lateral lemniscus.

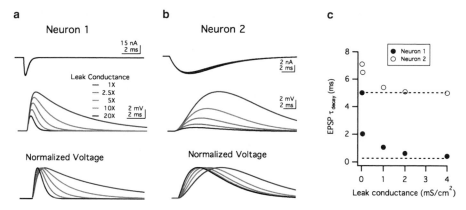

Fig. 2.2 Effects of resting conductances on synaptic timing. Point-neuron models of a neuron exhibiting fast synaptic currents (**a**, "Neuron 1") and synaptic currents 25-fold slower (**b**, "Neuron 2"). EPSCs and EPSPs (*upper* and *middle traces*, respectively) are shown for models exhibiting a 1- to 20-fold increase in membrane leak conductance. As membrane leak conductance increases (from 0.0002 to 0.004 S/cm^2), the amplitude and duration of EPSPs decrease in both models. In both cases, the time course of the EPSC is best reflected by models with the greatest leak conductance. However, EPSP duration in Neuron 2 is limited by the time course of the longer underlying synaptic current, whereas EPSP duration in Neuron 1 is sensitive to a larger range of leak conductance values. Synaptic conductances: 0.5 and 0.065 μS for Neurons 1 and 2, respectively (**c**). Graphical representation of the decay time constant of the EPSP (EPSP τ_{decay}) as a function of membrane leak conductance for the examples in A and B. *Dotted lines* indicate the single-exponential time constant of decay for EPSCs in Neurons 1 and 2 (0.25 and 5 ms, respectively)

3.3 Resting Membrane Properties Establish the Time Course and Sensitivity of Synaptic Integration

3.3.1 Contribution of Passive Leak Channels to Resting Membrane Properties

The fast kinetic properties of glutamate receptors would not translate into brief postsynaptic responses without corresponding specializations in the postsynaptic intrinsic membrane properties. Indeed, the speed of postsynaptic voltage responses is limited by the number of ion channels open at rest (which contributes to the membrane conductance). A simple illustration of this concept is shown in Fig. 2.2, which draws a comparison between two neurons with different integrative properties. The two model cells in this example have only a single compartment (i.e., soma only, no dendrites or axon) but Neuron 1 receives glutamatergic excitation that is 25 times faster than Neuron 2 (Fig. 2.2a, b, respectively). In these simulations, the number of leak channels is incrementally increased up to 20-fold from an initial value and the amplitude and time course of excitatory postsynaptic potentials (EPSPs) are compared. In both neurons, the amplitude of EPSPs drops in proportion to the increased conductance according to Ohm's law:

$$\text{Ohm's law}: V_m = I_m / G_m \tag{2.1}$$

In Eq. 2.1, V_m is the membrane voltage, I_m is the synaptic current, and G_m is the resting membrane conductance. Note that while Ohm's law is most typically expressed in terms of membrane resistance (the inverse of membrane conductance), conductance is used here because of its direct and more intuitive relationship to ion channel expression. The speed of membrane voltage changes (quantified by the membrane time constant, or τ_m), is also inversely proportional to the membrane conductance, as:

$$\tau_m = C_m / G_m \tag{2.2}$$

In Eq. 2.2, C_m represents the membrane capacitance, which is proportional to the cell's surface area. Here the two model neurons differ in their responses: at the lowest membrane conductance level (1×), the response in Neuron 1 far exceeds the underlying synaptic current, because the slow membrane time constant limits the speed of the repolarization. As the resting leak conductance is increased, the time course of membrane repolarization more closely approaches the time course of the underlying synaptic current. In Neuron 2, the same effects are observed, but they are not nearly as dramatic because the longer time course of synaptic current imposes a lower limit on the time course of EPSPs. Thus, there are two important concepts illustrated by this simulation. First, there is a trade-off between membrane sensitivity and temporal precision, as increasing leak conductance simultaneously reduces voltage changes and the membrane time constant. Second, in order to encode synaptic timing information with precision, it is not sufficient for a neuron to exhibit neurotransmitter-gated receptors with rapid kinetics. The cell must also exhibit a resting conductance level that confers a membrane time constant low enough to avoid limiting the intrinsic time course of excitation.

At the molecular level, a family of genes has been identified and cloned that give rise to potassium leak currents. These are the so-called two-pore domain potassium channels, named after the two pore-forming loops in their predicted membrane topology. Of the eight family members that form functional channels and are expressed in the mammalian CNS, three (TASK-1, TASK-2, and TWIK-1) have been shown to be expressed in neurons of the cochlear nucleus (Talley et al. 2001; Berntson and Walmsley 2008). However, it is not yet known how these channel proteins are distributed in different cell types in auditory pathways. More recently, sodium-dependent leak channels have also been identified (NALCN, or "sodium leak channel, nonselective") (Lu et al. 2007). While this channel shows widespread distribution throughout the brain, its expression in auditory neurons has not yet been investigated. While leak channels are an important component of the passive membrane properties of all neurons, it will be seen that in many auditory neurons membrane properties may be dominated by voltage-gated ion channels active around the resting potential.

3.3.2   Contribution of Voltage-Gated Ion Channels to Resting Membrane Properties

While voltage-insensitive leak channels provide an important contribution to the passive electrical properties of neurons, there are few, if any neurons in the central nervous system that are purely passive. Along with leak channels, voltage-gated ion channels may also make important contributions to the resting properties of neurons, as long as they have a non-inactivating component of current that is active at the cell's resting potential. Two types of channels have been documented to be especially important: low voltage-activated (LVA) potassium channels, and the hyperpolarization and cyclic nucleotide-gated cation channels (HCN channels). Both channel types are discussed below.

Low Voltage-Activated Potassium Channels

A recurring channel motif in auditory brainstem neurons concerned with temporal coding is the presence of potassium currents that are activated at relatively hyperpolarized voltages. Current through these low voltage-activated potassium channels will be denoted as I_{K-LVA}, but in different studies this current has also been abbreviated as I_L, I_{KLT}, and I_{KL}. The LVA potassium channels are expressed widely in many neurons in the brain and are often found enriched in the axon initial segment as well as in the nodes of Ranvier (Hopkins et al. 1994; Sheng et al. 1994; Kole et al. 2007). However, these channels are expressed in particularly high density in the cell bodies and dendrites of auditory brainstem neurons concerned with temporal coding (Manis and Marx 1991; Brew and Forsythe 1995; Bal and Oertel 2001). While there are differences in the details of the LVA potassium channel characteristics across different neuron types, they share several general features. First, the activation of these channels typically occurs at voltages slightly negative to the resting potential (~−65 to −70 mV), and given that LVA potassium channels have a non-inactivating component, the channels can provide a tonic conductance at rest. Second, LVA potassium channels across auditory neurons exhibit a similar pharmacological profile, being blocked by submillimolar concentrations of 4-aminopyridine as well as different fractions of dendrotoxin (dendrotoxin-I, α-dendrotoxin, and dendrotoxin-K), a high-affinity toxin isolated from the African black or green mamba snake. Finally, LVA potassium channels exhibit fast (typically submillisecond) activation kinetics, allowing these channels to influence the timing, amplitude, and shape of subthreshold EPSPs and action potentials. These interactions will be discussed in Sect. 3.4.

The precise molecular composition of LVA potassium channels expressed in time-coding auditory neurons is not fully understood, but it is clear that they consist of members of the K_v1 subfamily (also referred to as Shaker-type potassium channels in *Drosophila*). Progress in understanding LVA potassium channel composition has been aided by the availability of toxins that target different K_v1 subunits with high specificity. LVA potassium currents in time-coding auditory neurons are consistently blocked by α-dendrotoxin and dendrotoxin-I, toxins that block channels containing $K_v1.1$, 1.2, and 1.6 subunits (Brew and Forsythe 1995;

Ferragamo and Oertel 2002; Leão et al. 2006). Indeed, immunocytochemical and in situ hybridization studies indicate that the Kv1.1 subunit is expressed consistently in these areas, an observation supported by the extensive block of LVA potassium currents by dendrotoxin-K, which specifically targets potassium channels containing at least one $K_v1.1$ alpha subunit in the tetramer (Bal and Oertel 2001; Dodson et al. 2002; Mathews et al. 2010). However, K_v1 subunits can co-assemble with other members of the same family to form a functional channel (Hopkins et al. 1994), and several lines of evidence support the idea that LVA potassium channels are heteromeric combinations of different K_v1 family subunits. For example, in principal neurons of the MNTB, LVA potassium currents have been subdivided into two approximately equal components based on their sensitivity to tityustoxin-Kα, a blocker of $K_v1.2$ subunits (Dodson et al. 2002). However, ~90% of the LVA potassium current is blocked by dendrotoxin-K, indicating that most channels contain $K_v1.1$ subunits. These findings support the conclusion that LVA potassium channels in the MNTB consist of at least two classes of heteromeric channels, one containing $K_v1.1$ and $K_v1.2$ subunits, and the other containing $K_v1.1$ and another K_v1 family member, possibly $K_v1.6$. Similarly, in octopus cells LVA potassium current is largely (~80%) dendrotoxin-K sensitive, but only about 50% of the LVA potassium current is blocked by tityustoxin-Kα (Bal and Oertel 2001). The overall conclusion from these and other studies is that LVA potassium current is mediated by a heterogeneous population of channels containing different K_v1 family subunits, with $K_v1.1$ being present in the majority of channels.

What might be the functional consequences of having potassium channels comprised of different combinations of K_v1 subunits? In expression systems, homomeric combinations of $K_v1.1$ exhibit more negative activation ranges and faster kinetics than channels composed of only $K_v1.2$ subunits, and heteromeric channels containing both subunits display intermediate properties (Hopkins et al. 1994). Thus the expression of channels with different subunit stoichiometry would provide a means for auditory neurons to, in a sense, "tune" channels to display the correct voltage sensitivity and kinetics in order to carry out specific computational roles. This is not the only way through which channel properties can be adjusted: channel kinetics and voltage-sensitivity can also be modified by the presence of accessory (nonpore-forming) subunits and neuromodulators. Thus at the molecular level there are multiple mechanisms by which channel properties can be modified, increasing the difficulty of interpreting differences in channel properties across different neuron types and even differences observed within the same neuron class.

Hyperpolarization-Activated Cation Channels

HCN channels are comprised of a family of four gene products (HCN1–4). These channels are non-inactivating and typically have negative activation ranges that overlap to some extent with the resting potential. These features make HCN channels important determinants of the resting conductance of neurons, which in turn influences the speed of membrane voltage changes, as discussed in Sect. 3.3.1. Thus, the degree to which these channels shape the resting conductance and membrane time constant depends critically on the density of channels, the voltage-dependence of activation, and the value of the resting potential. These factors vary widely across neuron types, even among time-coding auditory neurons. For example, octopus cells express HCN channels in high density (~150 nS whole-cell conductances), with 50% of the channels activated at −65 mV. At the average resting potential of these cells of −60 mV, I_h is a major determinant of the extraordinarily fast membrane time constant of these cells, which is ~200 µs (Golding et al. 1995; Bal and Oertel 2000). By contrast, bushy cells recorded under similar conditions exhibit lower whole-cell conductances (~30 nS), but perhaps more importantly the half-activation voltage for HCN channels is −83 mV, possibly reflecting differences in subunit composition and/or modulatory state (Cao et al. 2007). As this value is far more negative relative to the resting potential of bushy cells, HCN channels make a relatively smaller contribution to the resting conductance, and the temporal precision found in these cells is more strongly influenced by other voltage-gated ion channels, particularly LVA potassium channels.

The interplay between HCN channels and LVA potassium channels deserves comment. The activation range of both channels overlaps in the voltage range near rest. Since the reversal potential of HCN channels typically resides between −20 and −40 mV, tonic activation of these channels near rest produces an inward current. By contrast, the reversal potential of LVA potassium currents is ~−90 mV, and thus activation of these channels at rest produces an outward current. When both channel types are expressed together in significant densities, the resting potential reflects the influence of both channels (Bal and Oertel 2000, 2001; Hassfurth et al. 2009). Each channel's activation moves the membrane potential away from its own activation range and toward that of the other. Thus the co-expression of HCN channels and LVA potassium channels comprises a homeostatic system that promotes the stability of the resting potential.

3.4   Control of Synaptic Integration and Action Potential Timing by Low Voltage-Activated Potassium Channels

The LVA potassium channels possess several properties that render them especially effective in controlling the timing of synaptic activity. First, as discussed in the previous section, LVA potassium channels typically activate at voltages similar or negative to the typical resting potentials of neurons (between −80 and −60 mV). Second, the activation of K_v1 channels exhibits steep voltage dependence over the subthreshold voltage range (generally between ~−65 and −45 mV). Third, K_v1 channels exhibit activation kinetics that are fast enough to allow the channel to contribute to the repolarization of both subthreshold EPSPs and action potential waveforms.

3.4.1   Influence of Low Voltage-Activated Potassium Channels on Synaptic Potentials

In time-coding neurons of the auditory brainstem, the strength of the activation of K_v1 potassium channels by EPSPs depends strongly on the rate of rise of excitation. Differences in rise time can arise in several ways. First, in neurons that integrate excitation from many inputs, a slower depolarizing envelope would be produced by acoustic stimuli that generate lower synchrony in presynaptic inputs. Second, concurrent inhibition also has the potential to shape the overall rate of rise of the depolarization. A striking demonstration of this concept comes from the octopus cells of the cochlear nucleus (Ferragamo and Oertel 2002). These authors observed that in octopus cells action potential initiation did not exhibit an absolute voltage threshold, but rather one that was highly dependent on the rate of rise of excitation. Indeed, spike initiation required a slope of greater than 12 mV/ms: with slower rising depolarizations, spikes could not be generated at any stimulus amplitude. The explanation for this result has to do with the interplay between K_v1 channel kinetics and the rising phase of excitation. With sharply rising EPSPs, peak excitation is reached prior to strong activation of LVA potassium channels, allowing for the regenerative activation of sodium channels and the initiation of spikes to occur prior to strong activation of outward potassium current. With increasingly slower events, sodium channel activation is less synchronous and more extensively overlaps with the activation of potassium current, thus giving rise to ratios of inward to outward current that are increasingly less favorable for action potential initiation. In this way, K_v1 channels narrow the time window over which synaptic inputs may sum effectively to produce action potential signaling.

Another way of thinking about the effects of K_v1 channels on synaptic integration is as a high-pass filter that attenuates more slowly rising EPSPs. In MSO principal neurons, where action potential generation requires the submillisecond coincidence of binaural synaptic inputs, K_v1 channels have been shown to reduce the number of "false-positive" spikes generated from synaptic activity that is uncorrelated with the synaptic activity driven by the stimulus (Svirskis et al. 2002, 2004).

In MSO principal neurons, excitation is targeted to the dendrites and then propagates "forward" to the soma and axon. This introduces a potential complication for temporal coding, as synapses located at different dendritic locations would be subject to variable degrees of electrotonic cable filtering. Cable filtering arises from the combined resistive and capacitive properties of the dendritic membrane, the effect of which is to delay the rise time of EPSPs, prolong their duration, and attenuate their amplitude. (Mathews et al. 2010) used paired somatic and dendritic recordings as well as modeling to examine how LVA potassium channels shape dendritic EPSPs as they propagate along the dendrites (Fig. 2.3). They found that while EPSPs undergo strong attenuation in their propagation to the soma, as would be expected, EPSPs become narrower in duration at the same time (Fig. 2.3a, b). This effect is due to the activation of dendrotoxin-sensitive K_v1 channels, which are expressed at highest density in the perisomatic region of the cell, providing strong active repolarization of EPSPs, particularly near firing threshold. Based on this

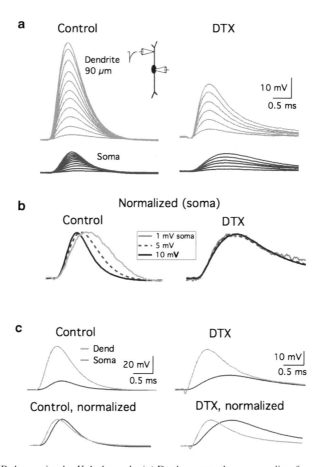

Fig. 2.3 EPSP sharpening by K_v1 channels. (**a**) Dual current-clamp recording from an MSO principal neuron from the soma and lateral dendrite 90 μm away. Responses are shown to a family of simulated EPSCs (sEPSCs) injected into the dendrite. sEPSPs traversing the entire subthreshold voltage were recorded in physiological saline (*left traces*; 0.2–2.2, 0.2 nA step) and also in the presence of 100 nM α–dendrotoxin (DTX), a blocker of K_v1 potassium channels (right traces; 0.2–1.2, 0.2 nA step). (**b**) Normalized sEPSPs from three selected traces from the somatic recording show that larger voltage responses narrowed by ~40% at 10 mV relative to 1 mV. This voltage-dependent sharpening was blocked in the presence of DTX. (**c**) The largest subthreshold sEPSPs show that cable distortions in EPSP rise time and duration as the sEPSP propagated from the dendrite to the soma were greatly enhanced when K_v1 channels were blocked by DTX (Figure modified from (Mathews et al. 2010). With permission)

data, computational modeling of MSO principal neurons demonstrated that LVA potassium channels act as a high-pass filter. The stronger the broadening of EPSPs during dendritic propagation, the more robust the truncation of EPSPs by potassium channel activation. In this way, LVA potassium channels impose a relatively uniform duration for EPSPs arising from different dendritic regions, allowing EPSPs to more accurately reflect the relative timing of sounds to the two ears.

3.4.2 Effects of Low Voltage-Activated Potassium Channels on Action Potentials

Given the rapid activation of LVA potassium channels, it might also be expected that these channels influence the action potential waveform. This has been examined closely in the principal neurons of the MNTB, which fire trains of brief action potentials with high temporal fidelity at rates of up to ~600–800 Hz in vitro and in vivo (Taschenberger and von Gersdorff 2000; McLaughlin et al. 2008; Lorteije et al. 2009). Analyses of $I_{K\text{-}LVA}$ in MNTB neurons have shown that K_v1-containing LVA potassium channels are indeed activated during the course of the action potential, narrowing the action potential by up to ~40%, as well as increasing the depth of the spike afterhypolarization (Brew and Forsythe 1995; Klug and Trussell 2006). The large afterhyperpolarization in turn improves temporal coding by preventing multiple spikes from firing in response to a single synaptic volley (Fig. 2.4a), thereby reducing the variability in the timing of action potentials (Dodson et al. 2002; Gittelman and Tempel 2006; Klug and Trussell 2006). These results are in accord with the increase in action potential jitter observed in MNTB recordings in vivo from $K_v1.1$ knockout mice (Kopp-Scheinpflug et al. 2003). While different K_v1 channel subtypes are found to be localized to the soma, these channels also appear to be present in high density in the initial segment of the axon (Dodson et al. 2002), and are thus in a position to provide tight control over the threshold and timing of action potentials near their presumed site of initiation.

In octopus cells and MSO principal neurons, LVA potassium channels strongly regulate not only the shape of the action potential, but its amplitude as well. In these neurons, somatic action potentials appear unusually small (~5–20 mV). Blockade of LVA potassium channels results in roughly a threefold increase in action potential amplitude at the soma (Ferragamo and Oertel 2002; Scott et al. 2005) and improves the efficacy with which the larger action potential invades the distal dendrites (Scott et al. 2010) (Fig. 2.4b). Thus, in MSO principal neurons and octopus cells, an additional role for LVA potassium channels is to restrict access of the action potential to the soma and dendrites. Scott et al. (2007) have hypothesized that minimizing the amplitude of the action potential in the soma and dendrites might improve the temporal fidelity of synaptic integration by reducing the influence of the spike afterhyperpolarization on future cycles of synaptic integration during high-frequency synaptic input.

3.5 Control of Action Potential Signaling by High Voltage-Activated Potassium Channels

In addition to LVA potassium channels, many neurons contain potassium channels activated at voltages more positive to −30 mV, the high voltage-activated (HVA) potassium channels. A substantial proportion of these channels are typically blocked by millimolar concentrations of tetraethyl ammonium (TEA), but in many cases the molecular identity of the channel subunit configuration remains poorly defined. Members of the K_v3 family of voltage-gated potassium channels have been shown to

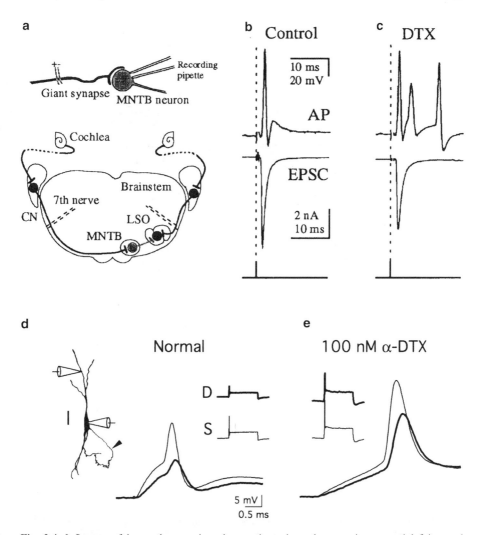

Fig. 2.4 Influence of low voltage-activated potassium channels on action potential firing and backpropagation. Response of an MNTB principal neuron to presynaptic stimulation of the calyx of Held. (**a**) Experimental configuration. (Modified from Brew and Forsythe 1995, Figs. 1 and 7). (**b**) Under control conditions, calyceal stimulation evoked a single well-timed postsynaptic action potential. (Modified from Scott et al. 2007, Fig. 4. With permission). (**c**) However, when K$_v$1 channels were blocked by 100 nM dendrotoxin-I, calyceal stimulation triggered repetitive firing. (**d**) Dual patch recordings from an MSO principal neuron from the soma (*thin traces*) and dendrite (*thick traces*; 75 μm dendritic distance). A 1.6-nA somatic current pulse elicited transient firing (inset). In normal saline, the relatively small action potential was observed first at the soma and then at the dendrite, showing strong amplitude attenuation at the latter site. Anatomical scale bar: 20 μm. Blockade of K$_v$1 channels with 100 μM α–dendrotoxin (DTX) increased action potential amplitude threefold and strongly reduced the relative attenuation of the spike between the soma and dendrite. Current pulse 0.4 nA

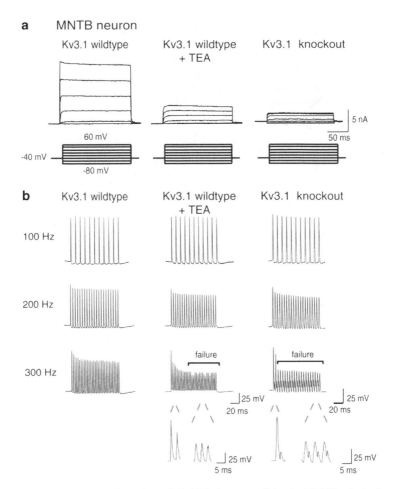

Fig. 2.5 Role of $K_v3.1$ potassium channels in high-frequency firing in MNTB principal neurons. (**a**) Whole-cell voltage clamp recordings of outward potassium currents in mouse MNTB neurons. Left: A family of potassium currents in an MNTB neuron from a wildtype mouse elicited by a family of voltage steps from −80 to +60 mV in 10 mV steps (*left*; "$K_v3.1$ wildtype"). Middle: Currents were extensively blocked in the presence of 1 mM tetraethylammonium chloride (TEA). Right: Outward currents in MNTB neurons from $K_v3.1$ knockout mice showed little high voltage-activated outward current, resembling the TEA condition from wildtype mice. (**b**) Action potential firing responses in current clamp under the same conditions as in A. Action potential failures were apparent at 300 Hz when $K_v3.1$ currents were eliminated by TEA blockade or by genetic deletion in knockout mice (Figure taken from Macica et al. (2003) Fig. 1. With permission)

be expressed in several nuclei in the auditory pathway, including the cochlear nucleus, superior olive, and inferior colliculus (Perney et al. 1992; Perney and Kaczmarek 1997; Grigg et al. 2000). The expression of channels containing the $K_v3.1$ subunit is particularly strong in the principal neurons of the MNTB. In these neurons, HVA potassium currents ($I_{K\text{-}HVA}$) consist of a fast-activating current with little inactivation, a biophysical profile that bears high similarity to that exhibited by homomeric channels consisting of $K_v3.1$ subunits (Fig. 2.5a) (Wang et al. 1998a; Macica et al. 2003).

While pharmacological block or genetic deletion of these channels has little effect on the ability of MNTB neurons to fire at frequencies lower than 200 Hz, at higher frequencies the absence of the channel leads to a greater incidence of action potential failures (Fig. 2.5b). An increase in action potential failures arising from a *reduction* in potassium current may seem counterintuitive, given that HVA potassium channels have been shown to narrow the width of MNTB spikes and increase the depth of spike afterhyperpolarizations (Klug and Trussell 2006). However, these results can be understood by considering how spike shape and repolarization influence voltage-gated sodium channels during repetitive firing. In the absence of $I_{K\text{-HVA}}$, an increase in the duration of the action potential will cause a larger population of voltage-gated sodium channels to inactivate, whereas the smaller afterhyperpolarization will reduce the proportion of voltage-gated sodium channels that can recover from inactivation. Thus, an essential function of $I_{K\text{-HVA}}$ in MNTB neurons is to limit sodium channel inactivation by "resetting" the membrane potential to a relatively more hyperpolarized level.

3.6 Ion Channel Gradients Across Tonotopic Maps and Within Cells

3.6.1 Potassium Channel Gradients Across Tonotopic Maps

Previous sections have described neurons in several time-coding auditory circuits as biophysically homogenous populations. However, there is a growing body of evidence indicating that in some auditory brainstem nuclei, the electrical properties of neurons vary according to tonotopic location. Just as the expression of different subtypes of voltage-gated ion channels critically influences the excitability of auditory neurons, differences in the expression levels of channels also strongly affect how neurons integrate synaptic information. Some of the best examples of these influences come from studies of time-coding principal neurons in the LSO and MNTB. Neurons in both the LSO and MNTB are arranged tonotopically, with low-frequency neurons positioned more laterally and high-frequency neurons located progressively more medially (Guinan et al. 1972; Tsuchitani 1977). In the LSO, Barnes-Davies and colleagues found that principal neurons could be divided into two electrophysiological types: transient-firing cells with relatively low input resistance (~70 MΩ) and repetitively firing cells with more than two times the input resistance of transient cells (Barnes-Davies et al. 2004). Furthermore, the distribution of these neurons varied according to tonotopic location, with transient-firing neurons more prevalent in lateral, low-frequency areas and repetitive-firing neurons in more medial, higher-frequency areas. The relative expression levels of dendrotoxin-sensitive LVA potassium channels play a key role in establishing this dichotomy. In voltage-clamp recordings, LVA potassium currents were significantly larger in transient versus repetitive-firing neurons, and in current-clamp recordings, blockade of these channels could convert transient firing neurons into repetitive-firing neurons. An idea presented by these authors is that the role of LVA potassium channels in preventing multiple spikes per stimulus cycle is most important at acoustic frequencies below the phase-locking limit, where temporal precision is more relevant.

In the MNTB, the tonotopic expression pattern of HVA potassium channels appears reversed relative to that of LVA potassium channels in the LSO. Peak HVA potassium currents appear larger in more medial, high-frequency areas of the MNTB, in agreement with the relatively higher intensity of immunolabeled $K_v3.1$ subunits in these regions (Li et al. 2001; von Hehn et al. 2004; Leão et al. 2006). Interestingly, similar findings have been observed in the avian cochlear nucleus and MSO (called the nucleus magnocellularis and laminaris, respectively) (Parameshwaran et al. 2001). Leão and colleagues have documented that MNTB principal neurons exhibit an increased ability to entrain to high-frequency depolarizations, consistent with the idea that K_v3 channels improve high-frequency firing (Leão et al. 2006).

The modulatory state of $K_v3.1$ channels also contributes to the functional gradient of $K_v3.1$ currents in MNTB principal neurons. MNTB principal neurons express two splice variants of the $K_v3.1$ gene, $K_v3.1a$ and $K_v3.1b$. However, $K_v3.1b$ is expressed more highly in mature neurons (Perney et al. 1992; Liu and Kaczmarek 1998). Only $K_v3.1b$ is subject to phosphorylation by protein kinase C, which reduces current through the channel by decreasing its open probability (Macica et al. 2003). Song and colleagues have shown that quiescent MNTB neurons exhibit higher levels of phosphorylated $K_v3.1b$, but trains of synaptic stimulation in vitro or auditory activity in vivo dephosphorylate the channel, thereby increasing the amplitude of $K_v3.1$ current. These activity-dependent changes in channel function ultimately improve the ability of MTNB neurons to follow higher-frequency stimuli, though apparently at the cost of temporal precision (Song et al. 2005). In slices, the ratio of dephosphorylated to phosphorylated Kv3.1b differs topographically, being higher in medial high-frequency regions.

While many questions remain regarding the nature and functional significance of potassium channel gradients in auditory nuclei, the preceding studies indicate that the expression of ion channels is not rigidly programmed within a cell type but has the capability of being fine-tuned to better respond to varying demands imposed by differing patterns of input activity. The mechanisms by which neurons achieve this tuning remain an important avenue of investigation.

3.6.2 Gradients of Voltage-Gated Potassium and Sodium Channel Gradients in the Dendrites of Single Cells

Just as differential expression of voltage-gated ion channels affects the integrative properties of neurons located in tonotopic regions, recent findings in MSO principal neurons support the idea that ion channels can be nonuniformly expressed along the dendrites of individual neurons. In time-coding neurons of the MSO, both voltage-gated sodium channels and LVA potassium channels are expressed in higher density at the soma than in the dendrites (Mathews et al. 2010; Scott et al. 2010). The lack of strong sodium channel expression in MSO dendrites is a major factor underlying the weak backpropagation of action potentials in these neurons (Scott et al. 2007). Electrophysiological mapping of the relative density of LVA potassium currents along MSO dendrites revealed that potassium channels are expressed in a gradient, with somatic channels expressed at an approximately fourfold higher density than in the distal half of the dendrites. As a result, LVA potassium currents are strongest at the soma and axon, close to the site of action potential initiation. As discussed in Sect. 3.4.1, this pattern of channel expression has been shown to be important for maintaining a uniform duration of EPSPs as they propagate along the dendrites to the soma and axon.

4 Circuits that Transform Action Potential Patterns: The Influence of Voltage-Gated Ion Channels

The previous sections of this chapter have focused on how different ion channels enable the precise encoding of the timing of presynaptic inputs. However, in many circuits of the auditory system neurons do not exhibit the kinds of synaptic and biophysical specializations that are prerequisites for fast temporal computations. Instead, the firing pattern of presynaptic inputs is transformed into new patterns. As is the case in most neurons, the type, density, and distribution of voltage-gated channels plays an integral part in generating these patterns.

4.1 Chopping Responses in Neurons of the Ventral Cochlear Nucleus and Superior Olive

Chopping refers to the repetitive firing behavior exhibited by certain auditory neurons in response to acoustic stimulation. A chopping response consists of a well-timed initial spike followed by a train of spikes generated at a regular interval that is unassociated with that of the acoustic stimulus (Fig. 2.6). Chopping responses can be found in many neurons at several different levels of the auditory system, including the dorsal and ventral cochlear nuclei, LSO, periolivary nuclei, and inferior colliculus (Tsuchitani 1977; Blackburn and Sachs 1989; Rees et al. 1997). However, chopping has been most extensively studied in the ventral cochlear nucleus, where two broad categories of chopping responses have been correlated with two different types of stellate cells (reviewed in Oertel et al. 2010). Sustained chopping responses ("C_S" units in vivo) exhibit regular firing for the duration of the acoustic stimulus (Rhode et al. 1983; Young et al. 1988; Smith and Rhode 1989). These cells correspond to "T stellate cells," multipolar neurons that provide excitatory input to several higher-order auditory nuclei (Oertel 1983; Smith and Rhode 1989; Oertel et al. 1990). Onset choppers ("O_C" units in vivo) display broad frequency tuning, and in contrast to sustained choppers, they provide chopping responses only for a limited number of cycles during acoustic stimulation. Onset choppers have been associated

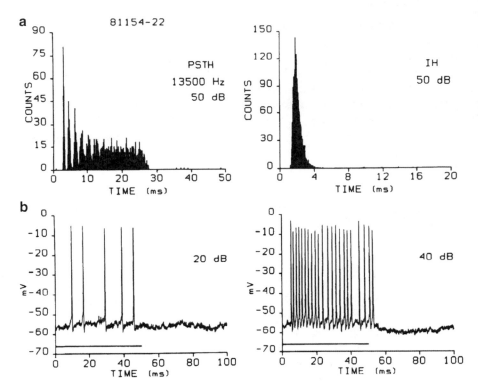

Fig. 2.6 Extracellular and intracellular recording of chopping responses in presumed VCN stellate neurons in vivo. (**a**) Peristimulus time histogram ("PSTH," left) showing the temporal response pattern of a VCN chopper unit in response to a 25-ms pure tone stimulus presented at the cell's characteristic frequency of 13.5 kHz. The interspike interval histogram ("IH," *right*) illustrates the cell's preferential interspike interval over a narrow range of frequencies. (**b**) Intracellular records from the same cell in (**a**), showing action potential trains, with increased variability in spike timing apparent later in the response to the higher intensity stimulus (*right*) (Figure taken from Rhode and Smith 1986, Fig. 17. With permission)

with the "D stellate cells" (see Trussell, Chap. 7), which provide inhibitory input to targets in the ipsilateral dorsal cochlear nucleus and contralateral cochlear nuclei (Smith and Rhode 1989; Oertel et al. 1990; Doucet and Ryugo 1997).

Sustained firing in stellate cells requires appropriate properties of both synapses and voltage-gated ion channels. At synapses between auditory nerve fibers and stellate cells, excitation through AMPA-type glutamate receptors is fast and brief, similar to other auditory nerve synapses in the VCN (Gardner et al. 1999, 2001). However, T stellate cells express a relatively higher NMDA receptor density than other VCN neurons, which might broaden EPSPs at depolarized membrane potentials (Cao and Oertel 2010). The longer membrane time constant (on the order of milliseconds) of stellate cells combined with polysynaptic excitation from neighboring t-stellate cells (Ferragamo et al. 1998) improves the ability of these cells to maintain prolonged excitation throughout trains of synaptic stimuli.

In contrast to the time-coding neurons discussed in Sect. 3, in stellate cells the voltage-gated conductances that shape subthreshold synaptic responses are notably absent. Accordingly, blockade of voltage-gated potassium conductances in time-coding neurons often generates repetitive, chopping-like responses (Banks and Smith 1992; Barnes-Davies et al. 2004; Svirskis et al. 2002). While stellate cells do express HCN channels, the activation range of these channels is far more hyperpolarized than time-coding neurons such as octopus cells and principal neurons of the MSO and LSO, and thus HCN channels make only a small contribution to the resting potential and input resistance (Rodrigues 2005). In dissociated VCN cells presumed to be stellate cells, little or no LVA potassium current was recorded (Manis and Marx 1991). Together these results explain the highly linear voltage-current relationship that has been described in VCN stellate cells (Wu and Oertel 1984; Manis and Marx 1991; Rodrigues 2005). Such properties allow for extensive temporal summation of even weak excitatory synaptic inputs from the auditory nerve, an important aspect of maintaining sustained firing throughout the course of a stimulus.

The observation that chopping responses occur at an intrinsic frequency independent of the frequency of the acoustic stimulus highlights the critical role played by voltage-gated ion channels, particularly in regard to the repolarization of the action potential. Manis and Marx (1991) showed in dissociated VCN neurons that "type I" cells (presumed to be stellate cells) possessed two types of HVA potassium currents with similar voltage sensitivity. Both currents activated near $-30\,\mathrm{mV}$ but exhibited differing kinetics, with fast and slow activation time constants of ~1 and 13 ms. These authors noted the importance of the slower component in controlling the interspike interval of firing, as cell-to-cell differences in the magnitude and kinetics of the slower component would be expected to influence the depth and rate of recovery from the spike afterhyperpolarization. They further proposed that the cell-to-cell variability in kinetics of the slow component might underlie in part the heterogeneity in chopping frequency and regularity that has been reported in single-unit recordings in vivo. While many factors can influence the frequency and regularity of chopping responses, including the number and dendritic location of excitatory and inhibitory synaptic inputs (Banks and Sachs 1991; Paolini et al. 2005), the strength and kinetics of action potential afterhyperpolarizations are important considerations. In studies of regular-firing neurons in the LSO in vitro, calcium-activated potassium channels have been shown to be a prominent component of the spike afterhyperpolarization (Adam et al. 2001), and modeling studies have predicted that these channels are likely to be important in establishing the frequency of chopping (Zacksenhouse et al. 1998; Zhou and Colburn 2010). Thus while the synaptic and ion channel mechanisms underlying chopping responses may differ across neuron types in different nuclei, regulation of the action potential afterhyperpolarization is undoubtedly a common factor that shapes chopping characteristics in many neuron types, even if there is diversity in the underlying ion channels.

4.2 The Influence of A-Type Potassium Channels on Pauser/Buildup Responses in Fusiform Cells of the Dorsal Cochlear Nucleus

Fusiform cells of the dorsal cochlear nucleus (DCN) receive direct excitation from auditory nerve fibers and provide the major excitatory output from the dorsal cochlear nucleus. In vivo, fusiform cells often generate a characteristic temporal response known as a "pauser/buildup" pattern, consisting of either a transient response followed by a ramped increase in firing rate or simply a ramped increase in firing in the absence of an initial transient response (i.e., a pause).

Fusiform cell responses are influenced by the interaction between synaptic responses and "A-type" potassium channels so-named from their original description in the marine gastropod *anisodoris* (Connor and Stevens 1971). The critical features of these channels include activation at voltages below action potential threshold and fast inactivation that occurs over a voltage range near and depolarized to the resting potential. Kanold and Manis (1999) showed that in DCN fusiform cells, one of two types of inactivating potassium currents fits this profile, which they termed I_{KIF}, for "fast-inactivating potassium current." In fusiform cells, I_{KIF} activates rapidly, with a time constant less than 1 ms, and inactivates on the order of ~11 ms (Kanold and Manis 1999). Since this A-current inactivates extensively in the voltage range near rest, its magnitude depends sharply on the electrophysiological context in which it is activated: A-current is large in magnitude when activated from negative voltages but almost completely absent when activated from a voltage range near or just above the typical resting potential of fusiform cells (Fig.2.7a). It should be noted that while the A-current terminology is a useful functional description of currents found in many neurons from many types of organisms, at the molecular level A-current can be generated from alpha-subunits of different potassium channel subfamilies, most notably $K_v1.4$ or $K_v4.2$. Fusiform cells have been shown to express the mRNA for $K_v4.2$ channels (Fitzakerley et al. 2000), but the expression level of $K_v1.4$ mRNA or protein is not known.

There are two important functional consequences of these channel characteristics for synaptic integration in fusiform cells. First, at membrane potentials just negative to rest, rapid activation of A-current by EPSPs transiently suppresses repetitive firing, inducing a "buildup" of firing rate with time. If strong enough, A-current can suppress initial firing entirely, generating a long "pause." However, once activated, A-current inactivates over a time scale of tens of milliseconds, releasing the membrane from its hyperpolarizing influence and allowing repetitive firing to occur later during the course of the stimulus (Fig. 2.7b).

Fusiform cells have complex receptive fields that are strongly shaped by two broad types of inhibitory inputs: a fast frequency-specific input that exhibits similar tuning as excitation and a far more broadly tuned input (Young et al. 1992; Oertel and Young 2004). Because A-type potassium channels are sensitive to small changes in membrane potential near rest, the prediction is that the firing pattern of fusiform cells depends on both the level and duration of synaptic inhibition immediately prior

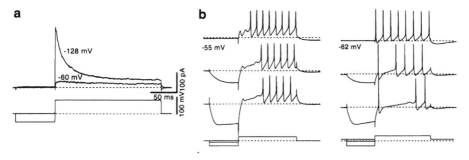

Fig. 2.7 Role of A-type potassium channels in shaping pauser/buildup firing patterns in DCN fusiform cells. (**a**) Isolation of A-type potassium currents in whole-cell voltage clamp recordings from a fusiform cell. Voltage steps to 0 mV were preceded by a 70-ms prepulse to −60 mV or −128 mV. A large, transient A-type potassium current is apparent in the −128 mV prepulse condition but not the −60 mV condition, reflecting steady-state inactivation of the channel. (**b**) A-current imposes sensitivity of firing pattern to the prior level of membrane hyperpolarization. Two different cells are shown. In voltage traces, dotted lines indicate the resting potential for firing responses. In current injection protocols (*bottom*) *dotted lines* demarcate the absence of steady injected current (Figure adapted from Kanold and Manis 1999, Figs. 2 and 4. With permission)

to a stimulus. Thus, inhibition has not only an immediate effect on firing through its integration with concurrent excitation, but through its relief of A-type potassium channels it modifies the pattern of firing activity that occurs immediately thereafter.

4.3   Complex Spikes in Cartwheel Cells of the DCN

The cartwheel cells of the DCN comprise an interconnected network of local glycinergic inhibitory interneurons that provide feedforward inhibition to fusiform cells, thus shaping the major excitatory output pathway of the nucleus. Cartwheel cells are not driven by direct input from primary auditory nerve fibers, but rather by multisensory inputs conveyed by the parallel fibers of cochlear nucleus granule cells (reviewed in Oertel and Young 2004). Additionally, granule cells provide glycinergic inputs to other cartwheel cells, which are weakly excitatory due to their higher concentration of intracellular chloride (Golding and Oertel 1996; Kim and Trussell 2009). The timing of multisensory inputs is not strictly maintained in cartwheel cells. Synaptic excitation from granule cell activation rises slowly and triggers highly variable trains of both single spikes and bursts of action potentials, defined here as 3–5 spikes at high frequency, generally up to ~300 Hz, with later spikes in the burst showing a decrement in amplitude (Fig. 2.8A) (Zhang and Oertel 1993; Manis et al. 1994; Parham and Kim 1995). In cartwheel cells, these bursts of spikes are referred to as complex spikes, and single action potentials are called simple spikes. Thus, in contrast to the fast temporal integration carried out in auditory timing pathways, complex spikes in cartwheel cells confer a highly nonlinear relationship between the action potential output and the strength of synaptic input.

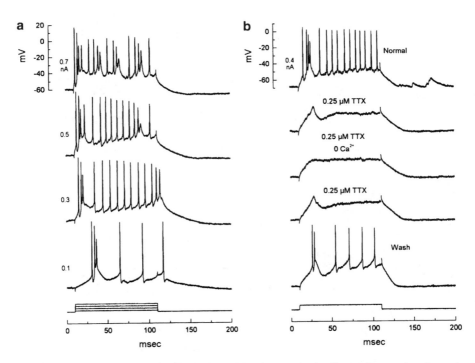

Fig. 2.8 Pattern and ionic basis of complex spiking in cartwheel cells. (**a**) Sharp microelectrode recording from a cartwheel cell recorded in the mouse DCN slice. Current pulses of increasing magnitude trigger a mixture of simple and complex spikes. (**b**) Pharmacological dissection of cartwheel cell spikes. Bath application of 0.25 μM TTX, a blocker of voltage-gated sodium channels, reversibly eliminated the fast overshooting spikes, leaving a smaller more slowly rising spike. This slower spike was mediated by voltage-gated calcium channels based on its reversible elimination when calcium was omitted from the external solution (Figure taken from Golding and Oertel 1997, Fig. 1. With permission)

Intracellular recordings from cartwheel cells have identified a critical role for voltage-gated calcium channels in producing action potential bursting. Complex spikes consist of a series of fast TTX-sensitive spikes superimposed on the rising phase of a slower smaller spike that is in turn sensitive to blockers of voltage-gated calcium channels (Fig. 2.8b) (Zhang and Oertel 1993; Golding and Oertel 1997; Kim and Trussell 2007). Cartwheel cells bear considerable homology to cerebellar Purkinje neurons, whose complex spikes are driven extensively by the activation of both T-type and P/Q-type calcium channels in the dendrites (Usowicz et al. 1992; Swensen and Bean 2003). However, calcium spikes in cartwheel cells appear more complicated, with the channel contribution dependent on the electrophysiological context. Kim and Trussell showed that there were two subsets of calcium spikes in cartwheel cells. Onset calcium spikes could be initiated from relatively negative voltages and were eliminated by blockers of T- and possibly R-type low voltage-activated calcium channels. By contrast, calcium spikes initiated later in responses at more depolarized voltages were sensitive to blockers of P/Q- and L-type high voltage-activated calcium channels (Kim and Trussell 2007). Thus, no single calcium channel subtype appears solely responsible for driving complex spiking. It has sometimes been assumed that calcium spikes are initiated in the dendrites on the basis of homology to Purkinje neurons as well as the nonlinear increase in calcium influx that has been observed in the dendrites during complex spikes (Molitor and Manis 2003; Roberts et al. 2008). However, both T- and R-type calcium channels are also localized at the initial segment of the axon of cartwheel cells (Bender and Trussell 2009). Focal blockade of either channel type at the initial segment reduces the number of spikes in a complex spike and causes stimuli that trigger simple spikes to fall below threshold. Taken together, the evidence from electrophysiological and imaging studies have shown that in cartwheel cells, complex and simple spiking involves the interplay between voltage-gated sodium channels and multiple classes of voltage-gated calcium channels. Furthermore, this interplay is sensitive to channel localization within different compartments of the cell as well as to the recent history of synaptic activity and spiking. These factors provide cartwheel cells with many potential ways in which they can fine-tune the time course and magnitude of their inhibitory influences on the output of the DCN.

What is the role of complex spikes in the processing of information in the DCN, and what computational advantage does this mode of firing afford relative to simple spiking? While a definitive answer is not known, bursts of action potentials in cartwheel cells drive summating clusters of inhibitory synaptic potentials and thus comprise a powerful form of transient inhibition that can arise even from relatively weak excitation. Furthermore, the high-frequency burst of action potentials triggered by calcium spikes induces strong calcium influx in presynaptic terminals, thereby increasing the probability of glycine release at synapses between cartwheel cells and fusiform cells as well as other cartwheel cells. Finally, the timing of complex spikes relative to their excitatory synaptic inputs from granule cells strongly influences the induction of synaptic plasticity (Tzounopoulos and Leão, Chap. 9). Thus, the timing and frequency of complex spikes exerts long-term control over the ability of cartwheel cells to modify the output of the DCN through fusiform cells.

5 Influence of Action Potential Initiation and Backpropagation on Auditory Coding

The action potential represents the final stage of synaptic integration, in which the information from an appropriate pattern of synaptic excitation is relayed to the synaptic terminals, potentially triggering neurotransmitter release and contributing to synaptic integration in the cell's network targets. While some of the ion channel mechanisms regulating the size and shape of the action potential in auditory neurons concerned with temporal coding were discussed in Sect. 3.4.2, the impact of the action potential on synaptic integration also depends critically on where it is initiated in the cell and the efficacy with which it subsequently propagates. These important factors are discussed below.

One of the most striking aspects of different auditory brainstem neurons is the diversity in the size and shape of their action potentials. At one end of a continuum of properties, MSO principal neurons and octopus cells exhibit action potentials that are extraordinarily small and narrow (5–15 mV and 200–400 μs duration at half-maximal amplitude). Contrary to the textbook idea that the action potential is a stereotyped "all-or-none" waveform, action potentials in MSO principal neurons and octopus cells can exhibit graded amplitudes, depending on the rate of rise of excitation (Golding et al. 1995; Scott et al. 2007). In MSO principal neurons, simultaneous intracellular patch-clamp recordings from the soma and dendrites have revealed that the action potential appears first at the soma, near the axon, and then subsequently in the dendrites, consistent with classical studies showing that the action potential is initiated in the axon and then propagates orthodromically down the axon as well as antidromically back into the soma and dendrites (the latter process is often referred to as "backpropagation"). Bushy cells and principal neurons of the LSO show larger-spike waveforms but rarely overshoot 0 mV. Other neurons that have been discussed in this chapter, such as the principal neurons of the MNTB, stellate cells in the VCN, and fusiform cells in the DCN, exhibit more conventional action potentials that overshoot 0 mV and display a far more consistent waveform.

Does the diversity in action potential shape among different types of neurons matter functionally? The technical difficulty of obtaining intracellular recordings from myelinated axons has made this a challenging question to address experimentally. In MSO principal neurons, where somatically recorded action potentials are graded according to the intensity of synaptic excitation, paired whole-cell and loose-patch recordings from the soma and axon, respectively, have shown that graded action potentials in the soma are in fact uniform in the axon itself and can be reliably initiated during trains of brief depolarizations up to ~1 kHz (Scott et al. 2007). Thus, in MSO principal neurons, all spikes at the soma, even those as small as a few millivolts, represent fully regenerative action potential propagation down the axon. The small size of the action potential thus reflects a rather extraordinary electrical isolation of the spike-generating region in the axon from the somatodendritic compartment of the cell, where summation of binaural synaptic inputs takes place. While the necessity of isolating the spike from the somatodendritic compartment of the cell is not known definitively, an attractive idea is that minimizing the amplitude of spikes at the soma allows action potential signaling to occur in MSO neurons while reducing distortion of future binaural synaptic comparisons, which may occur at submillisecond intervals.

The preceding results raise the question of how the mechanistic details of spike initiation influence the encoding of auditory information. Kuba et al. (2006) have shown in neurons of nucleus laminaris (NL), the avian MSO, that somatic action potential amplitude and the physical dimensions of the initial segment show a striking correlation with the tonotopic location of the neuron. Neurons in the high-frequency regions of NL exhibit smaller action potentials than those in low-frequency regions. Further, the axon initial segment of high-frequency neurons, defined by the clustering of immunolabeled voltage-gated sodium channels and ankyrin-G (an associated anchoring protein), is longer and located further from the soma (50 versus 20 μm on average). Using a combination of modeling and experiments, these authors showed that the more distal location of the initial segment in high-frequency neurons allows neurons to maintain higher firing rates during trains of high-frequency synaptic excitation, because the stronger attenuation of the

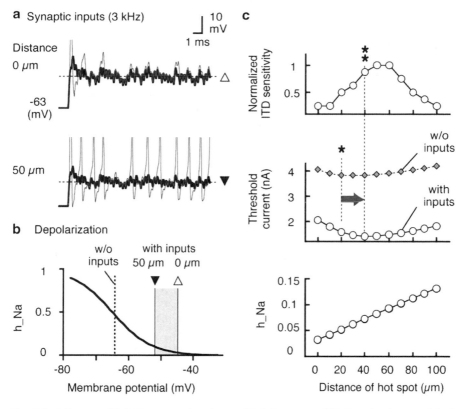

Fig. 2.9 Influence of initial segment location on high-frequency ITD detection in a model of nucleus laminaris neurons. (**a**) Simulated coincident binaural excitatory synaptic input trains at 3 kHz for a "hot spot" of voltage-gated sodium channels located next to the soma (0 μm) or in the axon 50 μm away (synaptic conductance 90 nS). Repetitive firing is far more apparent for the axonal hot spot location. Dark traces indicate membrane potential with spiking blocked. *Dashed lines* and *triangles* indicate average membrane depolarization. (**b**) Inactivation gating (h_{Na}) of voltage-gated sodium channels as a function of the average membrane potential during high-frequency stimulation. Channel inactivation was reduced in the axonal versus somatic location (closed vs open triangle, respectively) due to the greater cable attenuation of axonal voltage. *Dotted line*: resting potential. *Gray box* indicates the range of average membrane depolarizations demarcated by the examples shown in (**a**) (*triangles*). (**c**) From *top* to *bottom*: simulated interaural time difference (ITD) sensitivity, threshold current, and fractional sodium channel inactivation as a function of axonal hot spot location. In the presence of 3 kHz synaptic trains (*open circles* in all three graphs), near-maximal ITD sensitivity occurred when the hot spot was located at a distal axonal location (~50 μm; *top panel*). The minimum current threshold (*middle panel*) shifted to more distal axonal locations in the stimulated (*circles*, *double asterisk*) versus unstimulated condition (*diamonds*, *single asterisk*). This behavior reflects the decreasing steady-state inactivation of axonal voltage-gated sodium channels in more distal axonal regions during summating excitation (*bottom panel*) (Figure adapted from Kuba et al. 2006, Fig. 4. With permission

summed synaptic response in the longer axon section prior to the initial segment induces less sodium channel inactivation (Fig. 2.9) (Kuba et al. 2006). A similar relationship of action potential amplitude and initial segment geometry on tonotopic location has also been demonstrated in the avian cochlear nucleus (Kuba and

Ohmori 2009). Recently, it has been shown that auditory deprivation in NL neurons induces an increase in the length of the initial segment as well as an increase in whole-cell sodium currents, resulting in a decrease in spike threshold and the incidence of spontaneous firing (Kuba et al. 2010). These findings demonstrate that the structural and functional features of the initial segment are not rigidly programmed, but rather are shaped homeostatically by early auditory activity. Thus, the physical structure of the initial segment and the control of action potential initiation are important regulated determinants of auditory coding. The detailed molecular understanding of the initial segment and the mechanisms underlying its plasticity remains an exciting frontier for future investigations.

The action potential represents the final outcome of synaptic integration, whereupon the neuron has made a "decision" to signal its network targets. However, the backpropagation of the action potential from the axon into the soma and dendrites also has important short- and long-term consequences for neuronal excitability and auditory coding. First, the backpropagating action potential provides a way for the cell to communicate its firing state to synapses located in distal dendritic regions. It does so indirectly, through the activation of voltage-gated calcium channels by the action potential waveform. The relative timing between action potential-triggered calcium influx and the timing of excitatory synaptic potentials has been shown to be a critical factor in the induction of synaptic plasticity in both cartwheel and fusiform cells of the dorsal cochlear nucleus (Tzounopoulos and Leão, Chap. 9). Calcium influx triggered by backpropagating action potentials has also been detected in the dendrites of neonatal LSO principal neurons and in dendrites of nucleus laminaris (Kullmann and Kandler 2008; Blackmer et al. 2009), suggesting that backpropagation of the action potential is either naturally efficient or propagated actively by voltage-gated sodium channels located in the dendrites in these cell types. However, direct electrophysiological recordings from the dendrites of these cells have not been made.

6   Summary

In the brain, auditory information coded in the firing patterns of auditory nerve fibers is immediately channeled into different ascending pathways, each of which extracts and conveys certain features embedded in the neural code. This chapter has introduced some of the ways in which ion channels interact with synaptic input to shape the encoding process. Given the salience of temporal information in the auditory system for interpreting speech and communication signals as well as localizing sound sources, neurons in the cochlear nucleus and superior olivary complex that process and transmit this information have attracted considerable attention. These neurons and their related circuitry are elegant systems in which researchers can begin to understand how cell morphology, ion channel biophysics, and synapses work together to perform well-defined computations.

This discussion has emphasized that rapid kinetics of excitatory neurotransmitter-gated receptors and large resting conductances are prerequisites for submillisecond precision in temporal coding. Both passive leak channels and voltage-gated ion channels contribute to these resting conductance and thus play critical roles in providing tight control over synaptic timing. Superimposed on the resting conductance are several mechanisms by which time-coding auditory neurons can further refine the amplitude and time course of synaptic excitation. The properties of LVA potassium channels are well adapted for interacting with subthreshold synaptic input. These channels are partially activated at rest and are further recruited by synaptic excitation, shaping both subthreshold synaptic activity as well as the action potential itself.

6.1   Two Broad Classes of Time-Coding Auditory Neurons

When considering the roles played by voltage-gated ion channels in controlling the fidelity of temporal coding in different brainstem auditory circuits, it is informative to think of the neurons in these circuits as comprising two broad categories. In the first category, exemplified by octopus cells and MSO principal neurons, synaptic integration involves the summation of numerous but individually weak excitatory inputs in the dendritic arbor. These neurons exhibit high levels of resting conductances, mediated both by LVA potassium channels and HCN channels. In these neurons, LVA potassium channels play critical roles in shaping subthreshold synaptic integration: once activated, these channels act as a high-pass filter, suppressing slowly rising, less temporally coincident synaptic input, and causing a high current threshold for action potential initiation. LVA potassium channels are also partly responsible for limiting the invasion of the action potential into the soma and dendrites. In the other category, exemplified by spherical bushy cells and MNTB principal neurons, action potentials are driven with high probability by one or a few powerful synapses. In general, these neurons are less influenced by HCN channels, display lower levels of resting conductances, and exhibit action potentials of larger amplitudes (though still smaller than conventional action potentials). In these neurons, the role of both LVA and HVA potassium channels appears more oriented to controlling the timing and duration of action potentials, as well as conferring the ability to respond with a consistent latency at high frequency.

6.2   Plasticity and Regulation of Ion Channel Function

This chapter has highlighted some prominent examples of how the pattern of action potential firing is controlled by the density, biophysical properties, and even spatial distribution of ion channels in neurons. An important future challenge is to understand how each neuron achieves its own unique complement of voltage-gated ion channels and electrical properties. The finding that LVA potassium channels increase fourfold in the principal neurons of the MSO over the first week of hearing (Scott et al. 2005) is but one example among many that the electrical properties of most auditory neurons are not established immediately at birth but are established progressively during early auditory experience according to developmental programs and possibly also by neural activity. However, the mechanisms that regulate the density and distribution of voltage-gated ion channels in auditory neurons are only beginning to be elucidated and remain an exciting future direction. The answers to these questions are prerequisites not only for understanding how the central auditory system is refined during early sensory experience, but also for understanding how the central auditory system is affected by presbycusis, deafness and cochlear prostheses, all of which introduce substantial long-term changes in the firing activity of auditory pathways.

Acknowledgements The author would like to thank Drs. S. Cherry and M. Roberts for their comments on the manuscript. The author was supported by a grant from the National Institutes of Health (R01 DC 0006877).

References

Adam, T. J., Finlayson, P. G., & Schwarz, D. W. (2001). Membrane properties of principal neurons of the lateral superior olive. *Journal of Neurophysiology, 86*(2), 922–934.

Bal, R., & Oertel, D. (2000). Hyperpolarization-activated, mixed-cation current (I(h)) in octopus cells of the mammalian cochlear nucleus. *Journal of Neurophysiology, 84*(2), 806–817. doi:10938307.

Bal, R., & Oertel, D. (2001). Potassium currents in octopus cells of the mammalian cochlear nucleus. *Journal of Neurophysiology, 86*(5), 2299–2311.

Banks, M. I., & Sachs, M. B. (1991). Regularity analysis in a compartmental model of chopper units in the anteroventral cochlear nucleus. *Journal of Neurophysiology, 65*(3), 606–629.

Banks, M. I., & Smith, P. H. (1992). Intracellular recordings from neurobiotin-labeled cells in brain slices of the rat medial nucleus of the trapezoid body. *Journal of Neuroscience, 12*, 2819–2837.

Barnes-Davies, M., Barker, M. C., Osmani, F., & Forsythe, I. D. (2004). Kv1 currents mediate a gradient of principal neuron excitability across the tonotopic axis in the rat lateral superior olive. *European Journal of Neuroscience, 19*(2), 325–333.

Beckius, G. E., Batra, R., & Oliver, D. L. (1999). Axons from anteroventral cochlear nucleus that terminate in medial superior olive of cat: Observations related to delay lines. *Journal of Neuroscience, 19*(8), 3146–3161.

Bender, K. J., & Trussell, L. O. (2009). Axon initial segment Ca²⁺ channels influence action potential generation and timing. *Neuron, 61*(2), 259–271. doi:10.1016/j.neuron.2008.12.004.

Berntson, A. K., & Walmsley, B. (2008). Characterization of a potassium-based leak conductance in the medial nucleus of the trapezoid body. *Hearing Research, 244*(1–2), 98–106. doi:10.1016/j.heares.2008.08.003.

Blackburn, C. C., & Sachs, M. B. (1989). Classification of unit types in the anteroventral cochlear nucleus: PST histograms and regularity analysis. *Journal of Neurophysiology, 62*(6), 1303–1329.

Blackmer, T., Kuo, S. P., Bender, K. J., Apostolides, P. F., & Trussell, L. O. (2009). Dendritic calcium channels and their activation by synaptic signals in auditory coincidence detector neurons. *Journal of Neurophysiology, 102*(2), 1218–1226. doi:10.1152/jn.90513.2008.

Brew, H. M., & Forsythe, I. D. (1995). Two voltage-dependent K+ conductances with complementary functions in postsynaptic integration at a central auditory synapse. *Journal of Neuroscience*, *15*(12), 8011–8022.

Cant, N. B., & Casseday, J. H. (1986). Projections from the anteroventral cochlear nucleus to the lateral and medial superior olivary nuclei. *Journal of Comparative Neurology*, *247*(4), 457–476. doi:10.1002/cne.902470406.

Cao, X. J., & Oertel, D. (2010). Auditory nerve fibers excite targets through synapses that vary in convergence, strength, and short-term plasticity. *Journal of Neurophysiology*, *104*(5), 2308–2320. doi:10.1152/jn.00451.2010.

Cao, X. J., Shatadal, S., & Oertel, D. (2007). Voltage-sensitive conductances of bushy cells of the mammalian ventral cochlear nucleus. *Journal of Neurophysiology*, *97*(6), 3961–3975. doi:10.1152/jn.00052.2007.

Connor, J. A., & Stevens, C. F. (1971). Voltage clamp studies of a transient outward membrane current in gastropod neural somata. *Journal of Physiology*, *213*(1), 21–30.

Dodson, P. D., Barker, M. C., & Forsythe, I. D. (2002). Two heteromeric kv1 potassium channels differentially regulate action potential firing. *Journal of Neuroscience*, *22*(16), 6953–6961. doi:20026709.

Doucet, J. R., & Ryugo, D. K. (1997). Projections from the ventral cochlear nucleus to the dorsal cochlear nucleus in rats. *Journal of Comparative Neurology*, *385*(2), 245–264.

Ferragamo, M. J., & Oertel, D. (2002). Octopus cells of the mammalian ventral cochlear nucleus sense the rate of depolarization. *Journal of Neurophysiology*, *87*(5), 2262–2270. doi:10.1152/jn.00587.2001.

Ferragamo, M. J., Golding, N. L., & Oertel, D. (1998). Synaptic inputs to stellate cells in the ventral cochlear nucleus. *Journal of Neurophysiology*, *79*(1), 51–63.

Fitzakerley, J. L., Star, K. V., Rinn, J. L., & Elmquist, B. J. (2000). Expression of shal potassium channel subunits in the adult and developing cochlear nucleus of the mouse. *Hearing Research*, *147*(1–2), 31–45.

Gardner, S. M., Trussell, L. O., & Oertel, D. (1999). Time course and permeation of synaptic AMPA receptors in cochlear nuclear neurons correlate with input. *Journal of Neuroscience*, *19*(20), 8721–8729.

Gardner, S. M., Trussell, L. O., & Oertel, D. (2001). Correlation of AMPA receptor subunit composition with synaptic input in the mammalian cochlear nuclei. *Journal of Neuroscience*, *21*(18), 7428–7437.

Geiger, J. R., Melcher, T., Koh, D. S., Sakmann, B., Seeburg, P. H., Jonas, P., & Monyer, H. (1995). Relative abundance of subunit mrnas determines gating and Ca^{2+} permeability of AMPA receptors in principal neurons and interneurons in rat CNS. *Neuron*, *15*(1), 193–204.

Gittelman, J. X., & Tempel, B. L. (2006). Kv1.1-containing channels are critical for temporal precision during spike initiation. *Journal of Neurophysiology*, *96*(3), 1203–1214. doi:10.1152/jn.00092.2005.

Godfrey, D. A., Kiang, N. Y., & Norris, B. E. (1975). Single unit activity in the posteroventral cochlear nucleus of the cat. *Journal of Comparative Neurology*, *162*(2), 247–268. doi:10.1002/cne.901620206.

Golding, N. L., & Oertel, D. (1996). Context-dependent synaptic action of glycinergic and gabaergic inputs in the dorsal cochlear nucleus. *Journal of Neuroscience*, *16*(7), 2208–2219.

Golding, N. L., & Oertel, D. (1997). Physiological identification of the targets of cartwheel cells in the dorsal cochlear nucleus. *Journal of Neurophysiology*, *78*(1), 248.

Golding, N. L., Robertson, D., & Oertel, D. (1995). Recordings from slices indicate that octopus cells of the cochlear nucleus detect coincident firing of auditory nerve fibers with temporal precision. *Journal of Neuroscience*, *15*(4), 3138–3153.

Grigg, J. J., Brew, H. M., & Tempel, B. L. (2000). Differential expression of voltage-gated potassium channel genes in auditory nuclei of the mouse brainstem. *Hearing Research*, *140*(1–2), 77–90.

Guinan, J. J., Norris, B. E., & Guinan, S. S. (1972). Single auditory units in the superior olivary complex. II: Locations of unit categories and tonotopic organization. *International Journal of Neuroscience*, *4*, 147–166.

Hassfurth, B., Magnusson, A. K., Grothe, B., & Koch, U. (2009). Sensory deprivation regulates the development of the hyperpolarization-activated current in auditory brainstem neurons. *European Journal of Neuroscience, 30*(7), 1227–1238. doi:10.1111/j.1460-9568.2009. 06925.x.

Henkel, C. K., & Spangler, K. M. (1983). Organization of the efferent projections of the medial superior olivary nucleus in the cat as revealed by HRP and autoradiographic tracing methods. *Journal of Comparative Neurology, 221*(4), 416–428. doi:10.1002/cne.902210405.

Hopkins, W. F., Allen, M. L., Houamed, K. M., & Tempel, B. L. (1994). Properties of voltage-gated K+ currents expressed in xenopus oocytes by mkv1.1, mkv1.2 and their heteromultimers as revealed by mutagenesis of the dendrotoxin-binding site in mkv1.1. *Pflügers Archiv: European Journal of Physiology, 428*(3–4), 382–390.

Hunter, C., Petralia, R. S., Vu, T., & Wenthold, R. J. (1993). Expression of AMPA-selective glutamate receptor subunits in morphologically defined neurons of the mammalian cochlear nucleus. *Journal of Neuroscience, 13*(5), 1932–1946.

Isaacson, J. S., & Walmsley, B. (1995). Counting quanta: Direct measurements of transmitter release at a central synapse. *Neuron, 15*(4), 875–884.

Johnson, D. H. (1980). The relationship between spike rate and synchrony in responses of auditory-nerve fibers to single tones. *Journal of the Acoustic Society of America, 68*, 1115–1122.

Kane, E. C. (1973). Octopus cells in the cochlear nucleus of the cat: Heterotypic synapses upon homeotypic neurons. *International Journal of Neuroscience, 5*(6), 251–279.

Kanold, P. O., & Manis, P. B. (1999). Transient potassium currents regulate the discharge patterns of dorsal cochlear nucleus pyramidal cells. *Journal of Neuroscience, 19*(6), 2195–2208.

Kim, Y., & Trussell, L. O. (2007). Ion channels generating complex spikes in cartwheel cells of the dorsal cochlear nucleus. *Journal of Neurophysiology, 97*(2), 1705–1725. doi:10.1152/jn.00536.2006.

Kim, Y., & Trussell, L. O. (2009). Negative shift in the glycine reversal potential mediated by a ca2+− and ph-dependent mechanism in interneurons. *Journal of Neuroscience, 29*(37), 11495–11510. doi:10.1523/JNEUROSCI.1086-09.2009.

Klug, A., & Trussell, L. O. (2006). Activation and deactivation of voltage-dependent K+ channels during synaptically driven action potentials in the MNTB. *Journal of Neurophysiology, 96*(3), 1547–1555. doi:10.1152/jn.01381.2005.

Kole, M. H., Letzkus, J. J., & Stuart, G. J. (2007). Axon initial segment kv1 channels control axonal action potential waveform and synaptic efficacy. *Neuron, 55*(4), 633–647. doi:10.1016/j. neuron.2007.07.031.

Köppl, C. (1997). Phase locking to high frequencies in the auditory nerve and cochlear nucleus magnocellularis of the barn owl, *tyto alba. Journal of Neuroscience, 17*(9), 3312.

Kopp-Scheinpflug, C., Fuchs, K., Lippe, W. R., Tempel, B. L., & Rübsamen, R. (2003). Decreased temporal precision of auditory signaling in kcna1-null mice: An electrophysiological study in vivo. *Journal of Neuroscience, 23*(27), 9199–9207.

Kuba, H., & Ohmori, H. (2009). Roles of axonal sodium channels in precise auditory time coding at nucleus magnocellularis of the chick. *Journal of Physiology, 587*(Pt 1), 87–100. doi:10.1113/jphysiol.2008.162651.

Kuba, H., Ishii, T., & Ohmori, H. (2006). Axonal site of spike initiation enhances auditory coincidence detection. *Nature, 444*(7122), 1069–1072. doi:10.1038/nature05347.

Kuba, H., Oichi, Y., & Ohmori, H. (2010). Presynaptic activity regulates na(+) channel distribution at the axon initial segment. *Nature, 465*(7301), 1075–1078. doi:10.1038/nature09087.

Kullmann, P. H., & Kandler, K. (2008). Dendritic Ca^{2+} responses in neonatal lateral superior olive neurons elicited by glycinergic/gabaergic synapses and action potentials. *Neuroscience, 154*(1), 338–345. doi:10.1016/j.neuroscience.2008.02.026.

Leão, R. N., Sun, H., Svahn, K., Berntson, A., Youssoufian, M., Paolini, A. G., Fyffe, R. E., & Walmsley, B. (2006). Topographic organization in the auditory brainstem of juvenile mice is disrupted in congenital deafness. *Journal of Physiology, 571*(Pt 3), 563–578. doi:10.1113/jphysiol.2005.098780.

Li, W., Kaczmarek, L. K., & Perney, T. M. (2001). Localization of two high-threshold potassium channel subunits in the rat central auditory system. *Journal of Comparative Neurology, 437*(2), 196–218.

Lindsey, B. G. (1975). Fine structure and distribution of axon terminals from the cochlear nucleus on neurons in the medial superior olivary nucleus of the cat. *Journal of Comparative Neurology, 160*(1), 81–103. doi:10.1002/cne.901600106.

Liu, S. J., & Kaczmarek, L. K. (1998). The expression of two splice variants of the kv3.1 potassium channel gene is regulated by different signaling pathways. *Journal of Neuroscience, 18*(8), 2881–2890.

Loftus, W. C., Bishop, D. C., Saint Marie, R. L., & Oliver, D. L. (2004). Organization of binaural excitatory and inhibitory inputs to the inferior colliculus from the superior olive. *Journal of Comparative Neurology, 472*(3), 330–344. doi:10.1002/cne.20070.

Lorteije, J. A., Rusu, S. I., Kushmerick, C., & Borst, J. G. (2009). Reliability and precision of the mouse calyx of held synapse. *Journal of Neuroscience, 29*(44), 13770–13784. doi:10.1523/JNEUROSCI.3285-09.2009.

Lu, B., Su, Y., Das, S., Liu, J., Xia, J., & Ren, D. (2007). The neuronal channel NALCN contributes resting sodium permeability and is required for normal respiratory rhythm. *Cell, 129*(2), 371–383. doi:10.1016/j.cell.2007.02.041.

Macica, C. M., von Hehn, C. A., Wang, L. Y., Ho, C. S., Yokoyama, S., Joho, R. H., & Kaczmarek, L. K. (2003). Modulation of the kv3.1B potassium channel isoform adjusts the fidelity of the firing pattern of auditory neurons. *Journal of Neuroscience, 23*(4), 1133–1141.

Manis, P. B., & Marx, S. O. (1991). Outward currents in isolated ventral cochlear nucleus neurons. *Journal of Neuroscience, 11*(9), 2865–2880.

Manis, P. B., Spirou, G. A., Wright, D. D., Paydar, S., & Ryugo, D. K. (1994). Physiology and morphology of complex spiking neurons in the guinea pig dorsal cochlear nucleus. *Journal of Comparative Neurology, 348*(2), 261–276. doi:10.1002/cne.903480208.

Mathews, P. J., Jercog, P. E., Rinzel, J., Scott, L. L., & Golding, N. L. (2010). Control of submillisecond synaptic timing in binaural coincidence detectors by $K(v)1$ channels. *Nature Neuroscience, 13*, 601–609. doi:10.1038/nn.2530.

McLaughlin, M., van der Heijden, M., & Joris, P. X. (2008). How secure is in vivo synaptic transmission at the calyx of held? *Journal of Neuroscience, 28*(41), 10206–10219. doi:10.1523/JNEUROSCI.2735-08.2008.

Molitor, S. C., & Manis, P. B. (2003). Dendritic Ca^{2+} transients evoked by action potentials in rat dorsal cochlear nucleus pyramidal and cartwheel neurons. *Journal of Neurophysiology, 89*(4), 2225–2237. doi:10.1152/jn.00709.2002.

Nordeen, K. W., Killackey, H. P., & Kitzes, L. M. (1983). Ascending auditory projections to the inferior colliculus in the adult gerbil, meriones unguiculatus. *Journal of Comparative Neurology, 214*(2), 131–143. doi:10.1002/cne.902140203.

Oertel, D. (1983). Synaptic responses and electrical properties of cells in brain slices of the mouse anteroventral cochlear nucleus. *Journal of Neuroscience, 3*(10), 2043–2053.

Oertel, D. (1999). The role of timing in the brain stem auditory nuclei of vertebrates. *Annual Review of Physiology, 61*, 497–519. doi:10.1146/annurev.physiol.61.1.497.

Oertel, D., & Young, E. D. (2004). What's a cerebellar circuit doing in the auditory system? *Trends in Neurosciences, 27*(2), 104–110. doi:10.1016/j.tins.2003.12.001.

Oertel, D., Wu, S. H., Garb, M. W., & Dizack, C. (1990). Morphology and physiology of cells in slice preparations of the posteroventral cochlear nucleus of mice. *Journal of Comparative Neurology, 295*(1), 136–154. doi:10.1002/cne.902950112.

Oertel, D., Bal, R., Gardner, S. M., Smith, P. H., & Joris, P. X. (2000). Detection of synchrony in the activity of auditory nerve fibers by octopus cells of the mammalian cochlear nucleus. *Proceedings of the National Academy of Sciences of the United States of America, 97*(22), 11773–11779. doi:10.1073/pnas.97.22.11773.

Oertel, D., Wright, S., Cao, X. J., Ferragamo, M., & Bal, R. (2010). The multiple functions of T stellate/multipolar/chopper cells in the ventral cochlear nucleus. *Hearing Research*. doi:10.1016/j.heares.2010.10.018.

Osen, K. K. (1969). The intrinsic organization of the cochlear nuclei. *Acta Otolaryngol, 67*(2), 352–359.

Otis, T. S., Raman, I. M., & Trussell, L. O. (1995). AMPA receptors with high Ca^{2+} permeability mediate synaptic transmission in the avian auditory pathway. *Journal of Physiology, 482 (Pt 2)*, 309–315.

Paolini, A. G., Clarey, J. C., Needham, K., & Clark, G. M. (2005). Balanced inhibition and excitation underlies spike firing regularity in ventral cochlear nucleus chopper neurons. *European Journal of Neuroscience, 21*(5), 1236–1248. doi:10.1111/j.1460-9568.2005.03958.x.

Parameshwaran, S., Carr, C. E., & Perney, T. M. (2001). Expression of the kv3.1 potassium channel in the avian auditory brainstem. *Journal of Neuroscience, 21*(2), 485–494.

Parham, K., & Kim, D. O. (1995). Spontaneous and sound-evoked discharge characteristics of complex-spiking neurons in the dorsal cochlear nucleus of the unanesthetized decerebrate cat. *Journal of Neurophysiology, 73*(2), 550–561.

Perney, T. M., & Kaczmarek, L. K. (1997). Localization of a high threshold potassium channel in the rat cochlear nucleus. *Journal of Comparative Neurology, 386*(2), 178–202.

Perney, T. M., Marshall, J., Martin, K. A., Hockfield, S., & Kaczmarek, L. K. (1992). Expression of the mrnas for the kv3.1 potassium channel gene in the adult and developing rat brain. *Journal of Neurophysiology, 68*(3), 756–766.

Raman, I. M., Zhang, S., & Trussell, L. O. (1994). Pathway-specific variants of AMPA receptors and their contribution to neuronal signaling. *Journal of Neuroscience, 14*(8), 4998–5010.

Rees, A., Sarbaz, A., Malmierca, M. S., & Le Beau, F. E. (1997). Regularity of firing of neurons in the inferior colliculus. *Journal of Neurophysiology, 77*(6), 2945–2965.

Rhode, W. S., & Smith, P. H. (1986). Encoding timing and intensity in the ventral cochlear nucleus of the cat. *Journal of Neurophysiology, 56*(2), 261–286.

Rhode, W. S., Oertel, D., & Smith, P. H. (1983). Physiological response properties of cells labeled intracellularly with horseradish peroxidase in cat ventral cochlear nucleus. *Journal of Comparative Neurology, 213*(4), 448–463. doi:10.1002/cne.902130408.

Roberts, M. T., Bender, K. J., & Trussell, L. O. (2008). Fidelity of complex spike-mediated synaptic transmission between inhibitory interneurons. *Journal of Neuroscience, 28*(38), 9440–9450. doi:10.1523/JNEUROSCI.2226-08.2008.

Rodrigues, A. (2005). Hyperpolarization-activated currents regulate excitability in stellate cells of the mammalian ventral cochlear nucleus. *Journal of Neurophysiology, 95*(1), 76–87. doi:10.1152/jn.00624.2005.

Scott, L. L., Mathews, P. J., & Golding, N. L. (2005). Posthearing developmental refinement of temporal processing in principal neurons of the medial superior olive. *Journal of Neuroscience, 25*(35), 7887–7895. doi:10.1523/JNEUROSCI.1016-05.2005.

Scott, L. L., Hage, T. A., & Golding, N. L. (2007). Weak action potential backpropagation is associated with high-frequency axonal firing capability in principal neurons of the gerbil medial superior olive. *Journal of Physiology, 583*(Pt 2), 647–661. doi:10.1113/jphysiol.2007.136366.

Scott, L. L., Mathews, P. J., & Golding, N. L. (2010). Perisomatic voltage-gated sodium channels actively maintain linear synaptic integration in principal neurons of the medial superior olive. *Journal of Neuroscience, 30*(6), 2051–2062.

Sheng, M., Tsaur, M. L., Jan, Y. N., & Jan, L. Y. (1994). Contrasting subcellular localization of the kv1.2 K+ channel subunit in different neurons of rat brain. *Journal of Neuroscience, 14*(4), 2408–2417.

Smith, P. H., & Rhode, W. S. (1989). Structural and functional properties distinguish two types of multipolar cells in the ventral cochlear nucleus. *Journal of Comparative Neurology, 282*(4), 595–616. doi:10.1002/cne.902820410.

Smith, P. H., Joris, P. X., & Yin, T. C. (1993). Projections of physiologically characterized spherical bushy cell axons from the cochlear nucleus of the cat: Evidence for delay lines to the medial superior olive. *Journal of Comparative Neurology, 331*(2), 245–260. doi:10.1002/cne.903310208.

Song, P., Yang, Y., Barnes-Davies, M., Bhattacharjee, A., Hamann, M., Forsythe, I. D., & Kaczmarek, L. K. (2005). Acoustic environment determines phosphorylation state of the kv3.1

potassium channel in auditory neurons. *Nature Neuroscience*, *8*(10), 1335–1342. doi:10.1038/ nn1533.

Stotler, W. A. (1953). An experimental study of the cells and connections of the superior olivary complex of the cat. *Journal of Comparative Neurology*, *98*(3), 401–431.

Svirskis, G., Kotak, V., Sanes, D. H., & Rinzel, J. (2002). Enhancement of signal-to-noise ratio and phase locking for small inputs by a low-threshold outward current in auditory neurons. *Journal of Neuroscience*, *22*(24), 11019–11025.

Svirskis, G., Kotak, V., Sanes, D. H., & Rinzel, J. (2004). Sodium along with low-threshold potassium currents enhance coincidence detection of subthreshold noisy signals in MSO neurons. *Journal of Neurophysiology*, *91*(6), 2465–2473. doi:10.1152/jn.00717.2003.

Swensen, A. M., & Bean, B. P. (2003). Ionic mechanisms of burst firing in dissociated purkinje neurons. *Journal of Neuroscience*, *23*(29), 9650–9663.

Taberner, A. M., & Liberman, M. C. (2005). Response properties of single auditory nerve fibers in the mouse. *Journal of Neurophysiology*, *93*(1), 557–569. doi:10.1152/jn.00574.2004.

Talley, E. M., Solorzano, G., Lei, Q., Kim, D., & Bayliss, D. A. (2001). CNS distribution of members of the two-pore-domain (KCNK) potassium channel family. *Journal of Neuroscience*, *21*(19), 7491–7505.

Taschenberger, H., & von Gersdorff, H. (2000). Fine-tuning an auditory synapse for speed and fidelity: Developmental changes in presynaptic waveform, EPSC kinetics, and synaptic plasticity. *Journal of Neuroscience*, *20*(24), 9162–9173.

Tsuchitani, C. (1977). Functional organization of lateral cell groups of cat superior olivary complex. *Journal of Neurophysiology*, *40*(2), 296–318.

Usowicz, M. M., Sugimori, M., Cherksey, B., & Llinás, R. (1992). P-Type calcium channels in the somata and dendrites of adult cerebellar purkinje cells. *Neuron*, *9*(6), 1185–1199.

von Hehn, C. A., Bhattacharjee, A., & Kaczmarek, L. K. (2004). Loss of kv3.1 tonotopicity and alterations in camp response element-binding protein signaling in central auditory neurons of hearing impaired mice. *Journal of Neuroscience*, *24*(8), 1936–1940. doi:10.1523/ JNEUROSCI.4554-03.2004.

Wang, L. Y., Gan, L., Forsythe, I. D., & Kaczmarek, L. K. (1998). Contribution of the kv3.1 potassium channel to high-frequency firing in mouse auditory neurones. *Journal of Physiology*, *509* (Pt 1), 183–194.

Wang, Y. X., Wenthold, R. J., Ottersen, O. P., & Petralia, R. S. (1998). Endbulb synapses in the anteroventral cochlear nucleus express a specific subset of ampa-type glutamate receptor subunits. *Journal of Neuroscience*, *18*(3), 1148–1160.

Wu, S. H., & Oertel, D. (1984). Intracellular injection with horseradish peroxidase of physiologically characterized stellate and bushy cells in slices of mouse anteroventral cochlear nucleus. *Journal of Neuroscience*, *4*(6), 1577–1588.

Young, E. D., Robert, J. M., & Shofner, W. P. (1988). Regularity and latency of units in ventral cochlear nucleus: Implications for unit classification and generation of response properties. *Journal of Neurophysiology*, *60*(1), 1–29.

Young, E. D., Spirou, G. A., Rice, J. J., & Voigt, H. F. (1992). Neural organization and responses to complex stimuli in the dorsal cochlear nucleus. *Philosophical Transactions of the Royal Society of London. Series B, Biological Sciences*, *336*(1278), 407–413. doi:10.1098/ rstb.1992.0076.

Zacksenhouse, M., Johnson, D. H., Williams, J., & Tsuchitani, C. (1998). Single-neuron modeling of LSO unit responses. *Journal of Neurophysiology*, *79*(6), 3098–3110.

Zhang, S., & Oertel, D. (1993). Cartwheel and superficial stellate cells of the dorsal cochlear nucleus of mice: Intracellular recordings in slices. *Journal of Neurophysiology*, *69*(5), 1384–1397.

Zhou, Y., & Colburn, H. S. (2010). A modeling study of the effects of membrane afterhyperpolarization on spike interval statistics and on ILD encoding in the lateral superior olive. *Journal of Neurophysiology*, *103*(5), 2355–2371. doi:10.1152/jn.00385.2009.

Chapter 3
The Hair Cell Synapse

Teresa Nicolson

1 Introduction

The sensory cells of the auditory/vestibular system rely on two highly specialized
structures to transmit information to the brain: hair bundles and ribbon synapses.
Hair bundles transduce sound and head movements into changes in membrane
potential, which in turn stimulate ribbon synapses to release neurotransmitter. As
opposed to the all-or-none firing of neurons, both mechanotransduction and release
of neurotransmitter occur in a graded fashion. When hair bundles are deflected,
cations flow through transduction channels, initiating a change in membrane potential. Stronger deflections result in greater changes in receptor potential, which in
turn lead to graded gating of L-type calcium channels. These basally located channels mediate calcium influx and subsequent fusion of synaptic vesicles. After vesicle
release, sufficient amounts of neurotransmitter cause an action potential to fire in
afferent neurons.

Graded transmission passes on information about the intensity of the mechanical
stimulus during signaling to the acousticovestibular nerves. In addition to encoding
the intensity of a stimulus, another challenge for the ribbon synapse is to accurately
transmit information about the frequency of the stimulus. Ideally, the timing of the
release of neurotransmitter must be in phase with the mechanical stimulus, otherwise information about frequency would be lost. Moreover, the ribbon synapse has
to keep up with the demands of nearly constant stimulation provided by ordinary
activities such as walking or conversation. Intense stimulation of hair cells results in

T. Nicolson (✉)
Howard Hughes Medical Institute, Oregon Hearing Research Center, and Vollum Institute,
Oregon Health and Science University, 3181 SW Sam Jackson Park Road,
Portland, OR 97239, USA
e-mail: nicolson@ohsu.edu

L.O. Trussell et al. (eds.), *Synaptic Mechanisms in the Auditory System*,
Springer Handbook of Auditory Research 41, DOI 10.1007/978-1-4419-9517-9_3,

high rates of evoked release of neurotransmitter. Even in the absence of stimuli, hair cell synapses have relatively high rates of spontaneous release of synaptic vesicles, creating the need for a large supply of transmitter-filled vesicles. The necessity for high rates of spontaneous release of synaptic vesicles is not clear but presumably leads to greater responsiveness of the hair cell synapse to small changes in membrane potential. Packed with hundreds of synaptic vesicles and capable of high rates of evoked or spontaneous vesicle release, the hair cell synapse is essentially poised to relay any signals from the hair bundle, however small or large, and however fast or prolonged.

The following sections cover our current knowledge of the anatomy, physiology, and molecular components that mediate transmission at the first synapse of the auditory/vestibular system. As mentioned earlier, mature hair cells act as receptors and do not fire action potentials upon stimulation. Changes in membrane depolarization are induced by mechanotransduction, and are proportional to the probability of exocytosis. Synaptic transmission at hair cell synapses, therefore, involves graded release of neurotransmitter. Exocytosis of synaptic vesicles depends on local concentrations of calcium that are primarily determined by inward calcium currents through voltage-gated calcium channels. Accumulating evidence suggests that release of vesicles at this first synapse of the auditory/vestibular system occurs in a linear fashion, at least in mature synapses. For example, twice as much calcium influx results in twice as much vesicle fusion. Saturation of vesicle fusion is not apparent in mature hair cells. At lower frequencies, the graded release of vesicles occurs in phase with the mechanical stimulus (i.e., synchronization or phase-locking). Some hair cells can release in phase at rates in the kilohertz range. Higher-frequency stimuli result in steady depolarization of hair cells. In the case in which phase locking is no longer physically possible, the physical position of the hair cell along the tonotopic gradient of the neuroepithelium conveys information about frequency to the brain.

For synaptic transmission, afferent synapses of the auditory/vestibular system are glutamatergic and rely on components found at conventional synapses, but they also utilize specialized structures known as synaptic ribbons. These ribbon structures are also found in other neuronal sensory receptors and are thought to mediate fusion of multiple vesicles at a time. Multiple fusions of vesicles in turn release multiple quanta or packets of neurotransmitter. Analyses of multiquantal release along with other functional and genetic studies are beginning to reveal the physiological and molecular bases of synaptic transmission at afferent synapses in hair cells. These studies are the main focus of this chapter.

2 Anatomy of the Hair Cell Synapse

Hair cells are innervated by afferent neurons that have bipolar morphology. The axons of afferent neurons project to multiple auditory/vestibular nuclei of the hindbrain where the signals are processed and communicated among the nuclei and to

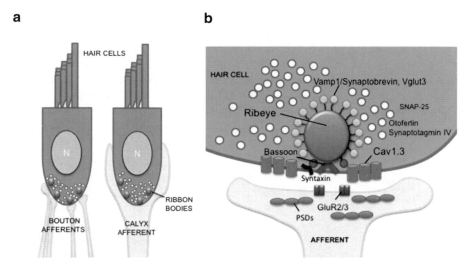

Fig. 3.1 Cartoon of afferent synapses in the auditory/vestibular system. (**a**) Examples of bulb-like boutons seen in both auditory and vestibular end organs, and a calyx bouton surrounding a vestibular hair cell. (**b**) Molecules present at an afferent hair cell synapse. Ribeye is the main constituent of the ribbon body, which sits above clusters of $Ca_v1.3$ calcium channels. Some of the conventional exocytic machinery of traditional synapses has been detected in hair cells, such as synaptobrevin, SNAP25, and syntaxin. Candidates for calcium sensors include otoferlin and synaptotagmin IV. Glutamatergic signaling is achieved via Vglut3 transporters on the presynaptic side and GluR2/3 channels in postsynaptic dendrites

other areas of the brain. The dendritic boutons of afferent neurons vary enormously in morphology, ranging from fully ensheathing calyx structures to simple, bulb-like formations (Fernández et al. 1988, 1995; Goldberg et al. 1990; see Fig. 3.1a). Apposed to the afferent boutons are the synaptic ribbon bodies, which are electron-dense entities located at the active zone of the plasma membrane. In some cases, each ribbon body is innervated by a single afferent neuron as seen in the cochlea (Liberman 1980, 1982). Within the neuroepithelium of the cochlea, there are two types of hair cells: inner hair cells that play a sensory role and outer hair cells that are thought to amplify sound. The majority of auditory afferents innervate the inner hair cells whereas only 5% make contact with outer hair cells. In contrast to the auditory system, a single calyx ending of a vestibular neuron can form multiple synapses with all of the available ribbon bodies in a vestibular hair cell (Lysakowski and Goldberg 1997). Moreover, a single vestibular fiber may form several calyxes or bouton contacts with multiple hair cells, thus receiving input from dozens of ribbons (Desai et al. 2005). Hair cells may also receive innervation from descending efferent neurons (see Sect. 3.5).

Like the afferent boutons, the proteinaceous structure of the ribbon body varies greatly in morphology (Lysakowski and Goldberg 2008). Some hair cell ribbons resemble the long, rod-shaped structures seen in photoreceptors. These synaptic structures were first seen using transmission electron microscopy in the 1950s

(Sjostrand 1953). Although they derived their name from the string-like appearance in photoreceptors, many ribbon bodies are ellipsoid or spherically shaped in hair cells (Sobkowicz et al. 1982; Lysakowski and Goldberg 2008). Their diameters range from 200 to 400 nm. Spherical ribbon bodies are also found in bipolar neurons of the retina. One other cell type, pinealocytes, also possesses these specialized synaptic structures. Variation of ribbon number and shape within a single end organ has been observed (Johnson and Marcotti 2008; Meyer et al. 2009). Some hair cells may have as few as three ribbon bodies, whereas others may have up to 60 ribbon bodies (Schnee et al. 2005; Obholzer et al. 2008). The functional significance of the shape of a ribbon body is not known.

Although variant in size and shape, all ribbon bodies are surrounded by synaptic vesicles, many of which appear to be tethered to the surface of the body via fine filaments (Fig. 3.1b). Estimates range from tens to hundreds of vesicles aligned to the surface of the ribbon body (Lenzi et al. 1999; Schnee et al. 2005). On electron micrographs, those vesicles facing the active zone between the ribbon body and plasma membrane often appear to be much darker than the other tethered vesicles, suggesting that additional molecules are bound to them. The typical darkened appearance of the active zone membrane, indicative of neurotransmission, often includes this subset of active-zone synaptic vesicles. Some micrographs include "pedicles" of electron-dense material extending from the ribbon body to the plasma membrane (Lysakowski and Goldberg 2008). Such pedicles are thought to be the equivalent of the attachment structure seen in photoreceptor synapses. Ribbon-tethered vesicles are typically small, clear, and uncoated, resembling conventional synaptic vesicles. However, there are some examples of amorphous vesicular/tubular structures associated with ribbon bodies (Matthews and Sterling 2008). It has been suggested that these amorphous structures may represent several synaptic vesicles caught in the act of compound fusion.

Because ribbon bodies are connected to active zones and tether numerous synaptic vesicles, they are thought to play a major role in facilitating synaptic transmission. How they facilitate transmission is not clear. One possibility is that they increase the readily releasable pool by tethering and directing vesicles to the active zone. Thus, the large structure of the ribbon body may create a larger active zone, clustering tens of vesicles underneath its structure, potentially increasing the number of vesicle release sites.

3 Physiology of the Hair Cell Synapse

Neurotransmission at hair cell synapses has been probed in various ways including measurement of capacitance changes in hair cells, postsynaptic currents of afferent neurons, and paired voltage-clamp recordings. Such recordings have provided evidence that hair cell synapses operate mostly in a linear fashion and can release multiple vesicles in response to membrane depolarization; this release follows the phase of the stimulus. The intensity of stimuli is conveyed via transmitter-evoked discharge or firing rates of afferent neurons. It is not clear why hair cells release multiple vesicles at each active zone or why these synapses work in a linear fashion, but phase locking of transmitter release and modulation of discharge rate are thought to be crucial for transmitting information about the frequency and intensity of a stimulus.

Capable of releasing vesicles over very long periods, hair cell synapses are seemingly indefatigable. Capacitance data indicate that hair cells are capable of releasing thousands of vesicles per second and that each individual ribbon synapse can release hundreds per second (Parsons et al. 1994; Moser and Beutner 2000). Remarkably, hair cells can release neurotransmitter at such high rates for seconds, whereas conventional synapses or endocrine cells can perform at this level for 100 ms at best (Parsons et al. 1994; Goutman and Glowatzki 2007). The existence of large pools of vesicles available for release, beyond those tethered to the body, suggests that the ribbon is reloaded with free-floating synaptic vesicles of the so-called reserve pool (Parsons et al. 1994; Moser and Beutner 2000; Eisen et al. 2004; Griesinger et al. 2005). In micrographs of stimulated hair cells, the number of docked and tethered vesicles is reduced and hallmarks of endocytosis of plasma membrane are apparent (Lenzi et al. 2002). Endocytosis is, however, quite slow ($\tau > 7$ s; Moser and Beutner 2000; Schnee et al. 2005) and cannot account for rapid reloading of ribbon bodies. Photobleaching experiments of fluorescently labeled docked vesicles at hair cell ribbons resulted in quick repopulation of ribbons ($\tau < 60$ ms) with preexisting labeled vesicles in the nearby cytoplasm (Griesinger et al. 2005). Thus, the ribbon synapse appears to draw from a large pool of free-floating synaptic vesicles for sustained transmission.

3.1 Linear Processing

With respect to physiologically relevant stimuli, the calcium dependence of neurotransmitter release at hair cell synapses is linear. There is a strict proportional relationship between vesicle fusion and calcium currents, with greater influx causing more fusion (Johnson et al. 2005; Schnee et al. 2005). A linear calcium dependence of fusion coupled with the apparent absence of calcium sensors such as synaptotagmin 1, has sparked curiosity about the calcium sensor present at this synapse (see Sect. 3.4.3). Various experiments to identify the properties of the calcium sensor at the active zone have revealed cooperative calcium binding as seen at conventional synapses. Uncaging of calcium in murine cochlear hair cells held at positive potentials producing small calcium currents suggests that five calcium-binding steps precede vesicle fusion (Beutner et al. 2001). Under these conditions, an intracellular concentration of 30 μM calcium was sufficient to drive release of docked vesicles (Beutner et al. 2001). Measurements of activity at single synapses held at positive potentials yielded a similar result of a higher (third- to fourth-) order calcium dependence of release (Goutman and Glowatzki 2007).

In contrast to mature hair cells, the calcium dependence of exocytosis in immature hair cells was found to be nonlinear (Johnson et al. 2005). Immature inner hair cells at P6–7 display a fourth-order calcium dependence as opposed to the linear relation seen at P16. During this period in development, the number of calcium channels decreases, yet exocytotic calcium efficiency increases (Johnson et al. 2005). How efficiency increases and the calcium dependence decreases from a fourth-order process to a first-order process during maturation is not clear. Possible scenarios include a tighter coupling between calcium channels and release sites, or a change in expression of calcium sensors (see Sect. 3.4.3).

Many characteristics of the transfer function of the hair cell synapse have been inferred by measuring activity on one side of the synapse. Recently, paired voltage-clamp recordings of bullfrog (*Rana catesbeiana*) papillar or rat mammalian hair cells and afferent fibers allowed direct measurement of the input–output or transfer characteristics of hair cell synapses. These measurements verified that hair cell synapses operate in the negative or physiological voltage range in a linear fashion (Keen and Hudspeth 2006; Goutman and Glowatzki 2007). Greater influx of calcium through voltage-gated calcium channels resulted in a proportional increase of synaptic vesicle release as inferred by an increase in excitatory postsynaptic currents (EPSCs; Keen and Hudspeth 2006).

3.2   Vesicle Pools and Multivesicular Release

In a study of spontaneous release from auditory hair cells in vivo, the notion of multiquantal or synchronous release of several vesicles at afferent synapses was invoked to explain the appearance of multicomponent EPSPs and ineffective EPSPs that did not result in action potentials (Siegel 1992). Further evidence for multivesicular release at hair cell synapses was obtained from rat cochlear explants (Glowatzki and Fuchs 2002). In this landmark study, afferent boutons smaller than a micron across were targeted for whole-cell recordings. It was noted that the amplitude and waveform of EPSCs varied greatly, suggesting that multiple vesicles were released in a more or less coordinated fashion. Assuming linear summation of neurotransmitter release, the average currents in boutons were evoked by a release of three to six vesicles, whereas the largest currents were produced by the release of 22 vesicles. Later studies using paired recordings of hair cells and afferent fibers confirmed the occurrence of multiquantal events at hair cell synapses (Keen and Hudspeth 2006; Li et al. 2009). In bullfrog papillar hair cells, waveforms of both small- and large-amplitude EPSCs were smooth and had identical kinetics, indicating that nonsynchronous independent fusion of synaptic vesicles did not occur (Li et al. 2009). Interestingly, a recent report suggests that multivesicular release does not occur at outer hair cell synapses (Weisz et al. 2009). Ribbon synapses of outer hair cells are not very active, and the resulting EPSCs are slower and smaller in amplitude than those in inner hair cell boutons.

In the case of multivesicular release, the tight coordination of vesicle fusion may be accomplished by cooperative, highly synchronous fusion of individual vesicles or compound fusion of several vesicles either before or during fusion with the plasma membrane. In favor of compound fusion, large exocytic bursts in leopard frog (*Rana pipiens*) saccular hair cells suggest that fast fusion involves an order of magnitude more vesicles than the subset docked at the active zone (Edmonds et al. 2004). It is assumed that vesicles tethered to the surface of the ribbon body cannot travel quickly enough to participate in fast fusion. Instead of traveling down to the active zone, free-floating vesicles may attach anywhere along the surface of the ribbon body. Alternatively, endocytosis could replenish vesicles, but not on such a short time scale (Parsons et al. 1994; Moser and Beutner 2000; Schnee et al. 2005).

To date, there is ample evidence for multiple pools of vesicles that release at different rates. Depending on the preparation and the temporal resolution of the recording technique, the fast pool of vesicles is released with time constants ranging between 2.7 and 53 ms (reviewed by Nouvian et al. 2006). Comparison between low- and high-frequency papillar hair cells indicates that high-frequency hair cells release at a slower rate ($\tau = 43$ ms for high frequency versus $\tau = 18$ ms for low frequency; Schnee et al. 2005). Recordings of single synapses suggest that the rate of vesicle fusion depends on intracellular calcium concentration with release probability proportional to calcium influx (Goutman and Glowatzki 2007). The finding that presynaptic calcium governs release probability but not the number of vesicles released per event was deduced by measuring the activity of single synapses, although these recordings were performed using immature rat cochlear hair cells (Goutman and Glowatzki 2007). Other studies extrapolate their findings from recordings that measure the activity of dozens of ribbon synapses simultaneously. The caveat with measurements from multiple ribbons is that it is not known whether each synapse is equally active within a hair cell. In vivo spontaneous rates of afferent firing certainly vary from fiber to fiber in mature animals, suggesting that individual ribbons operate at different rates (Taberner and Liberman 2005; Heil et al. 2007). Despite the differences in recording techniques, measurements at single ribbon synapses of inner hair cells yielded a similar time constant of 3 ms for the fast pool (Goutman and Glowatzki 2007). Synaptic vesicles tethered beyond the active zone appear to constitute the slower pool of vesicles, releasing with a time constant greater than 100 ms (Schnee et al. 2005). Capacitance data suggest that a slower kinetic component correlates well with the number of tethered, yet not docked, vesicles. A slower pool of vesicles fits with the notion of "conveyor belt" activity (Parsons et al. 1994; Lenzi et al. 1999; Schnee et al. 2005); the slowed kinetics of release occurs because tethered vesicles are not available for immediate release, that is, they must first travel to the active zone. For turtle papillar hair cells, refilling of the active zone is rate-limiting during the first 20 ms of stimulation (Schnee et al. 2005). After this initial period, it appears that synaptic vesicle fusion is the rate-limiting step during sustained release of neurotransmitter.

3.3 Adaptation

Although hair cell synapses can transmit precisely timed signals for prolonged periods of time, the synapse adapts over time. The time course of adaptation within the first second is on the order of milliseconds and includes a rapid component (a few milliseconds) and a slower component (tens of milliseconds; Westerman and Smith 1984). During this time the amount of neurotransmitter release appears to level off by more than half, as inferred by postsynaptic recordings (Furukawa and Matsuura 1978; Goutman and Glowatzki 2007). This decrease in postsynaptic response is independent of AMPA receptor sensitization (Goutman and Glowatzki 2007) and is likely due to reduced vesicle fusion. The reduction of vesicle fusion or fast adaptation can be readily explained by a rapid decrease in the number of docked or immediately releasable synaptic vesicles after depolarization. The slower pool of vesicles may not be available for release until they are shuttled to the active zone (Moser and Beutner 2000; Schnee et al. 2005; Goutman and Glowatzki 2007). However, a step in membrane fusion or some other process may be rate-limiting. Interestingly, in bassoon −/− hair cell synapses, the adaptation of afferent discharge rate was unaffected in in vivo recordings from the auditory nerve (Buran et al. 2010). Despite the absence of ribbon bodies and presumably the lack of tethered vesicle pools in bassoon mutants, sustained exocytosis was still possible (see Sect. 3.4.1). Both the unaltered adaptation of afferent discharge rate and the sustained exocytosis suggest that neither the kinetics nor the size of the slower pool of vesicles is determined by the ribbon body.

4 Molecular Components of the Hair Cell Synapse

4.1 The Ribbon Complex

For several decades after their initial discovery, not much was known about the composition of ribbon synapses. A few early studies using enzymatic digestion of thin-sectioned tissue revealed that photoreceptor ribbons are proteinaceous but did not appear to contain polysaccharides (Bunt 1971; Matsusaka 1967). Although the molecular nature of these electron-dense structures has not been fully explored, the first protein isolated from bovine retinal preparations was RIBEYE (Schmitz et al. 2000; Fig. 3.1b). It turns out that RIBEYE is the most abundant component of the dense structure, comprising two-thirds of the protein present (Zenisek et al. 2004). RIBEYE is an odd protein in that it consists of the transcriptional repressor, C-terminal binding protein 2 (CTBP2), spliced to an N-terminal domain with no homology to other proteins. The N-terminal domain can bind to itself; therefore it is referred to as the aggregation domain or A domain of RIBEYE. However, the CTBP2 domain or B domain can also bind to itself. In the absence of NADH, the B domain can associate with the A domain as there is a flexible linker between the two domains (Magupalli et al. 2008).

Other presynaptic components found at ribbon synapses include the large scaffolding proteins bassoon and piccolo, both of which are also present in conventional synapses. In photoreceptors, these two scaffold proteins appear to travel in transport packets that also contain RIBEYE protein. Knock-out of bassoon in mice reveals that bassoon plays a role in anchoring ribbon bodies to the plasma membrane in receptor cells (Khimich et al. 2005). However, bassoon is probably not the only protein capable of anchoring ribbons because ribbon bodies in bipolar neurons are unaffected (Dick et al. 2003), and not all ribbon bodies in hair cells are unattached (Khimich et al. 2005). Moreover, in photoreceptors of knock-out animals, ectopic synapses can form elsewhere, suggesting some other mode of attachment to the plasma membrane (Dick et al. 2003). It is not known whether ribbon bodies are initially attached in immature hair cells and then detach at later stages. Existing data suggest that bassoon is at least important for maintaining the attachment of ribbons to the active zone. Nevertheless, the bassoon knock-out mice offered an opportunity to examine the function of mainly ribbon-less hair cells. Khimich and colleagues found that the auditory brainstem response (ABR) was abnormal in bassoon knock-out mice. They observed a smaller initial peak of the ABR, indicating that activity of the auditory nerve was diminished. Using capacitance recordings, they also found that in mutant hair cells, the fast component of release, most likely representing the readily releasable pool, was compromised, whereas the slower component of release mostly approximated levels of exocytosis seen in wildtype mice. This result is somewhat surprising, as the ribbon body was thought to be vital for the resupply of vesicles during continuous synaptic transmission. A later study of single auditory afferent fibers in bassoon −/− mice found that spontaneous and sound-evoked firing of afferent neurons still occurred, but at lower rates (Buran et al. 2010). The reduction of discharge rates was more apparent when mice were presented with clicks in comparison to tone bursts. A lack of response to the onset of a tone burst stimulus was in fact very striking in bassoon −/− afferent fibers. These results suggest that the reliability of neurotransmitter release from hair cells is diminished in bassoon knock-out mice (Buran et al. 2010).

4.2   Calcium Channels

A truly critical component of the ribbon synapse is the presynaptic L-type calcium channel (Fig. 3.1b). These channels consist of dihydropyridine-sensitive $Ca_V1.3$ alpha subunits that are densely packed directly beneath ribbon bodies (Issa and Hudspeth 1994; Platzer et al. 2000; Spassova et al. 2001). In hair cells, $Ca_V1.3$ channels are rapidly gated by changes in membrane voltage and mostly noninactivating, two properties that are highly conducive to fast, tonic release of vesicles (Glowatzki et al. 2008; Johnson and Marcotti 2008). Freeze fracture experiments of frog saccular hair cells suggest that there are up to 125 channels per active zone (Roberts et al. 1990). Although many channels are present, there is evidence that only one or a few $Ca_V1.3$ channels need to open for neurotransmitter release (Brandt et al. 2005). Both knock-out mice and mutant $ca_V1.3$ zebrafish (*Danio rerio*) are deaf (Platzer et al. 2000; Sidi et al. 2004). In mice, capacitance changes in $Ca_V1.3$−/− hair cells are absent (Brandt et al. 2003). Together, these findings illustrate that the $Ca_V1.3$ calcium channel is critical for synaptic vesicle fusion in hair cells.

4.3 Exo- and Endocytosis Machinery and Calcium Sensors

Upon activation of $Ca_V1.3$ channels, intracellular calcium rises and these ions act locally and are otherwise strongly buffered beyond the ribbon body (Roberts 1993). An initial study of conventional exocytic or SNARE complex components reported that proteins such as syntaxin 1, SNAP25, and VAMP1 were detectable using RT-PCR of organ of Corti tissue (Safieddine and Wenthold 1999). However, some of the normally essential SNARE components were absent: synaptophysin, synapsin I and II, and synaptotagmins I–III and V were not present (Safieddine and Wenthold 1999). These results lead to the idea that hair cell synapses must rely in part on novel membrane fusion components. The protein otoferlin, with six predicted C2 calcium-binding domains, was put forth as a candidate for a novel calcium sensor at the ribbon synapse (Roux et al. 2006). Mutations in otoferlin cause deafness in humans, and knock-out of the gene in mice results in the drastic reduction of fast exocytosis in auditory hair cells (Yasunaga et al. 2000; Varga et al. 2003; Roux et al. 2006). Exocytosis is also reduced in otoferlin −/− vestibular hair cells and shows a less linear dependence on calcium (Dulon et al. 2009). Interestingly, pachanga mice carrying a mutation in otoferlin (aspartate to glycine at position 1767) are deaf but display normal docking and fusion of rapidly releasable vesicles (Pangrsic et al. 2010). Moreover, the number of synaptic vesicles present at the ribbon was comparable to that seen in wildtype, but mutant pachanga hair cells exhibited slower rates or fatigue of exocytosis in hair cells during longer stimulation (Pangrsic et al. 2010). In the case of pachanga mutant hair cells, the defect appears to be in vesicle replenishment rather than vesicle fusion.

How otoferlin might regulate vesicle replenishment remains to be investigated. One question is whether otoferlin is acting at the ribbon synapse. Immunolabel of otoferlin shows protein present throughout the hair cell body, and two studies have found that otoferlin interacts with the Golgi expressed Rab8b GTPase and the unconventional myosin VI motor protein (Schug et al. 2006; Heidrych et al. 2008, 2009). Although otoferlin binds to other active-zone SNARE complex members such as syntaxin 1A, SNAP25, and $Ca_V1.3$ (Ramakrishnan et al. 2009), it may act at multiple sites in the cell. Moreover, synaptic ribbons are morphologically normal in otoferlin−/− immature hair cells, but over time there is a decrease in the number of attached ribbons, not as striking as but similar to the phenotype seen in bassoon knock-out mutants (Roux et al. 2006). Also, the ribbon body and calcium channels are not as tightly coupled in otoferlin −/− hair cells (Heidrych et al. 2009). Perhaps fast exocytosis requires a tight association of ribbons and channels. In photoreceptors, bassoon may travel via a transport packet to the synapse, and because otoferlin can interact with Golgi or endosomal proteins, a trafficking defect in either mutant could impact ribbon function.

Trafficking of basolateral synapse components and synaptic vesicle recycling in hair cells is a relatively unexplored area. Clathrin-mediated endocytosis may play a role in vesicle recycling to a certain extent. Mutations in *synaptojanin 1* (*synj1*), a lipid phosphatase that uncoats clathrin-coated vesicles, lead to fatigue of neurotransmitter release in zebrafish hair cells (Trapani et al. 2009). Moreover, a defect in phase locking is present in mutant *synj1* hair cells, indicating that vesicle recycling is not only necessary to keep up with the demands of a prolonged stimulus but is also important for timing of release (Trapani et al. 2009).

Recently, another candidate has emerged as a calcium sensor in cochlear hair cells: synaptotagmin IV (SYT IV). SYT IV is unusual in that it has two C2 domains, but the C2A domain does not bind calcium. Knock-out mice display deficits in memory and learning (Ferguson et al. 2004a, b), but the role of SYT IV in conventional synapses is not clear as it has been reported to both support and inhibit vesicle fusion (Dean et al. 2009; Zhang et al. 2009). In cochlear hair cells, SYT IV is required for linear dependence of exocytosis (Johnson et al. 2010). Fusion of synaptic vesicles still occurs in *Syt IV −/−* hair cells, but not in proportion to calcium influx. SYT IV is not expressed at P7 in gerbil hair cells and is not required for the fourth-order dependence of exocytosis seen in immature mouse hair cells (Johnson et al. 2010). In contrast to the earlier study on SNARE components, Johnson and colleagues found that immature hair cells express SYT I and II. It is interesting to note that the single calcium-binding site of SYT IV, as opposed to the multiple binding sites in SYT I or II, would fit with a sensor that operates in a linear fashion. On the other hand, it is possible that a saturated, higher-order calcium sensor would work in a linear fashion as well.

4.4 Glutamatergic Components

Like other peripheral synapses, the auditory/vestibular system utilizes glutermatergic neurotransmission at its first chemical synapse (Fig. 3.1b). Packaging of neurotransmitter appears to be carried out by vesicular glutamate transporter 3 (VGLUT3). As with the calcium channel, loss of *VGLUT3* causes deafness in mice and fish (Obholzer et al. 2008; Seal et al. 2008). In vivo recordings of action potentials in afferent neurons evoked by mechanical stimulation of hair cells were absent in *vglut3* mutant fish (Obholzer et al. 2008). Likewise, in VGLUT3 null mice, recordings of cochlear explants revealed that synaptic activity was absent as well (Seal et al. 2008). A missense mutation in *VGLUT3* is also associated with deafness in humans (Ruel et al. 2008). Interestingly, Ruel and colleagues found that calcium-evoked exocytosis was not impaired in VGLUT3 null mice.

On the other side of the cleft, α-amino-3-hydroxy-5-methyl-4-isoxazolepropionic acid (AMPA) receptors bind glutamate and mediate subsequent depolarization of afferent fibers (Ruel et al. 2000; Glowatzki and Fuchs 2002). In the bullfrog papilla, it has been estimated that a single vesicle activates approximately 30 AMPA receptors on the postsynaptic membrane and that each ribbon synapse contains approximately 100 AMPA receptors (Li et al. 2009). Again, fitting for a synapse capable of tonic transmission, the abundant AMPA receptors do not saturate or quickly desensitize (Furukawa and Matsuura 1978; Starr and Sewell 1991; Li et al. 2009).

5   Efferent Synapses on Hair Cells

The configurations of efferent innervation of the peripheral auditory/vestibular system are numerous and vary quite substantially between species. It is likely that innervation varies according to needs. For example, a fish may want to reduce information to hair cells of the lateral-line system elicited from self-stimulation such as fin movements. Interestingly, the auditory papilla basilaris end organ receives efferent innervation in iguanas and crocodiles, but not in bullfrogs. Apparently, the need for efferent modulation of this particular end organ is dispensable for bullfrogs. Functional implications of efferent modulation of the auditory/vestibular system include feedback systems to enhance sensory input or to dampen self-stimulation and feedback that can lead to habituation or adaptation to a stimulus (Highstein 1991). In the inner ear, the efferent system appears to protect the cochlea from acoustic injury as animals without efferent feedback suffer damage of the auditory nerve (Darrow et al. 2006).

Efferent fibers either contact auditory/vestibular hair cells directly or they create synapses on afferent nerve endings. For the most part, inner hair cells do not receive any direct innervation from efferent neurons, whereas outer hair cells may be densely innervated with up to eight direct contacts (Klinke and Galley 1974). The number of contacts varies according to position within the cochlea. In the vestibular system, hair cells without afferent calyces may have one or more efferent contacts. In both the vestibular and auditory systems, the major neurotransmitter acting on hair cells is acetylcholine, although other transmitters have been implicated in efferent-to-afferent signaling such as gamma-amino butyric acid (GABA), opioid peptides, and dopamine (reviewed by Ruel et al. 2007). The alpha nine and ten subunits of the nicotinic acetylcholine receptors (nAChR), along with molecules associated with nAChR assembly such as rapsyn and RIC-3, are present in cochlear hair cells (Osman et al. 2008). The alpha nine subunit is also present in vestibular hair cells (Kong et al. 2006). These receptors mediate calcium influx, which in turn modulates calcium-activated SK potassium channels that are colocalized with nAChRs (Kong et al. 2008). Opening of SK channels results in hyperpolarization of the membrane potential and would consequently reduce hair cell activity. This rapid coupling of an excitatory receptor with a hyperpolarizing channel is not unique to hair cells as there are multiple examples of receptors coupled with SK channels in the CNS.

Another consequence of calcium influx through nAChRs is the activation of ryanodine receptors in postsynaptic cisternae that are closely associated with the hair cell basolateral membrane (Sridhar et al. 1997; Lioudyno et al. 2004; de San et al. 2007). Interestingly, calcium-induced calcium release increases the affinity of acetylcholine for the nAChRs in hair cells (de San et al. 2007). Release from the subsurface cisternae could account for the slow effects of cholinergic signaling on hair cell activity (Sridhar et al. 1997). The function of slow modulation is not clear but may be a mechanism to protect hair cells from overstimulation.

6   Summary

The first synapse of the auditory/vestibular system possesses a number of distinctive features that enable it to rapidly and faithfully pass on sensory information to neurons that innervate the hindbrain. One feature is graded neurotransmission that appears to be linear in nature. A second feature is the presence of specialized ribbon structures, surrounded by an inexhaustible pool of synaptic vesicles. Yet another specialized feature observed in inner hair cells and frog papillar hair cells is multivesicular release in response to calcium influx.

How the ribbon body facilitates neurotransmission is still unresolved. Does it simply act as a vesicle trap? Or does it participate in active transport of vesicles? How mobile are the vesicles once they are bound to the surface of the ribbon? If vesicles are transported, do the filaments tethering the vesicles participate in trafficking? Intuitively, one might guess that the filaments have the opposite effect of holding a vesicle in place, as is the case for synapsins, which apparently are not present in hair cells (Safieddine and Wenthold 1999).

How multivesicular release occurs is also not clear. Does compound fusion of docked vesicles occur, or is there coordinated co-release? If the release is coordinated, then it must be tightly cooperative, otherwise it is difficult to reconcile the recent results of dual recordings. These recordings suggest that multivesicular release could occur via compound fusion (Li et al. 2009). What is lacking is any evidence besides physiological recordings. Experiments designed to capture compound fusion at a fine structural level would be convincing but may be technically challenging. Does compound fusion require specialized exocytic machinery? Does the surface of the ribbon body provide some means of achieving compound fusion? These questions remain unanswered but are exciting topics of ongoing research.

The core SNARE fusion molecules (synaptobrevin, SNAP25, and syntaxin) are present in hair cells. But some of their accessory proteins are not found in hair cells. The absence of certain components is intriguing and has led to efforts directed toward identifying the missing parts, such as the calcium sensor. Typical calcium sensors, such as synaptagmin I, II, III or V, were not detectable in adult auditory hair cells. However, some of the lesser characterized synaptotagmins are present: synaptotagmin IV, VI–VIII, and IX. Otoferlin, with its multiple C2 calcium-binding domains appeared to be a good candidate, but recent data also suggest that otoferlin may be required for other critical roles such as trafficking of proteins and replenishment of vesicles. Synaptotagmin IV, on the other hand, is another candidate that may provide linear sensing of calcium. In either case, knock-out of each candidate sensor has different effects in the auditory or vestibular system, suggesting that neither is a universal calcium sensor in hair cells.

Investigation of how ribbons facilitate vesicle fusion continues to be an exciting but challenging process. The development of novel genetic or physiological methods may be necessary to reveal insights into how the ribbon body promotes or regulates vesicle fusion, and further genetic studies may define the molecular mechanisms that mediate multivesicular release.

Acknowledgments I wish to thank Elisabeth Glowatzki, Josef Trapani, and Laurence Trussell for their helpful suggestions and comments on this chapter.

References

Beutner, D., Voets, T., Neher, E., & Moser, T. (2001). Calcium dependence of exocytosis and endocytosis at the cochlear inner hair cell afferent synapse. *Neuron, 29*(3), 681–690.

Brandt, A., Striessnig, J., & Moser, T. (2003). CaV1.3 channels are essential for development and presynaptic activity of cochlear inner hair cells. *Journal of Neuroscience, 23*(34), 10832–10840.

Brandt, A., Khimich, D., & Moser, T. (2005). Few CaV1.3 channels regulate the exocytosis of a synaptic vesicle at the hair cell ribbon synapse. *Journal of Neuroscience, 25*(50), 11577–11585.

Bunt, A. H. (1971). Enzymatic digestion of synaptic ribbons in amphibian retinal photoreceptors. *Brain Research, 25*(3), 571–577.

Buran, B. N., Strenzke, N., Neef, A., Gundelfinger, E. D., Moser, T., & Liberman, M. C. (2010). Onset coding is degraded in auditory nerve fibers from mutant mice lacking synaptic ribbons. *Journal of Neuroscience, 30*(22), 7587–7597.

Darrow, K. N., Simons, E. J., Dodds, L., & Liberman, M. C. (2006). Dopaminergic innervation of the mouse inner ear: evidence for a separate cytochemical group of cochlear efferent fibers. *Journal of Comparative Neurology, 498*(3), 403–414.

Zorrilla de San Martín, J., Ballestero, J., Katz, E., Elgoyhen, A. B., & Fuchs, P. A. (2007). Ryanodine is a positive modulator of acetylcholine receptor gating in cochlear hair cells. *JARO: Journal of the Association for Research in Otolaryngology, 8*(4), 474–483.

Dean, C., Liu, H., Dunning, F. M., Chang, P. Y., Jackson, M. B., & Chapman, E. R. (2009). Synaptotagmin-IV modulates synaptic function and long-term potentiation by regulating BDNF release. *Nature Neuroscience, 12*(6), 767–776.

Desai, S. S., Zeh, C., & Lysakowski, A. (2005). Comparative morphology of rodent vestibular periphery. I: Saccular and utricular maculae. *Journal of Neurophysiology, 93*(1), 251–266.

Dick, O., Tom Dieck, S., Altrock, W. D., Ammermüller, J., Weiler, R., Garner, C. C., Gundelfinger, E. D., & Brandstätter, J. H. (2003). The presynaptic active zone protein bassoon is essential for photoreceptor ribbon synapse formation in the retina. *Neuron, 37*(5), 775–786.

Dulon, D., Safieddine, S., Jones, S. M., & Petit, C. (2009). Otoferlin is critical for a highly sensitive and linear calcium-dependent exocytosis at vestibular hair cell ribbon synapses. *Journal of Neuroscience, 29*(34), 10474–10487.

Edmonds, B. W., Gregory, F. D., & Schweizer, F. E. (2004). Evidence that fast exocytosis can be predominantly mediated by vesicles not docked at active zones in frog saccular hair cells. *Journal of Physiology, 560*(Pt 2), 439–450.

Eisen, M. D., Spassova, M., & Parsons, T. D. (2004). Large releasable pool of synaptic vesicles in chick cochlear hair cells. *Journal of Neurophysiology, 91*(6), 2422–242.

Ferguson, G. D., Herschman, H. R., & Storm, D. R. (2004a). Reduced anxiety and depression-like behavior in synaptotagmin IV (−/−) mice. *Neuropharmacology, 47*(4), 604–611.

Ferguson, G. D., Wang, H., Herschman, H. R., & Storm, D. R. (2004b). Altered hippocampal short-term plasticity and associative memory in synaptotagmin IV (−/−) mice. *Hippocampus, 14*(8), 964–974.

Fernández, C., Baird, R. A., & Goldberg, J. M. (1988). The vestibular nerve of the chinchilla. I. Peripheral innervation patterns in the horizontal and superior semicircular canals. *Journal of Neurophysiology, 60*(1), 167–181.

Fernández, C., Lysakowski, A., & Goldberg, J. M. (1995). Hair-cell counts and afferent innervation patterns in the cristae ampullares of the squirrel monkey with a comparison to the chinchilla. *Journal of Neurophysiology, 73*(3), 1253–1269.

Furukawa, T., & Matsuura, S. (1978). Adaptive rundown of excitatory post-synaptic potentials at synapses between hair cells and eight nerve fibres in the goldfish. *Journal of Physiology, 276*, 193–209.

Glowatzki, E., & Fuchs, P. A. (2002). Transmitter release at the hair cell ribbon synapse. *Nature Neuroscience, 5*(2), 147–154.

Glowatzki, E., Grant, L., & Fuchs, P. (2008). Hair cell afferent synapses. *Current Opinion in Neurobiology, 18*(4), 389–395.

Goldberg, J. M., Lysakowski, A., & Fernández, C. (1990). Morphophysiological and ultrastructural studies in the mammalian cristae ampullares. *Hearing Research, 49*(1–3), 89–102.

Goutman, J. D., & Glowatzki, E. (2007). Time course and calcium dependence of transmitter release at a single ribbon synapse. *Proceedings of the National Academy of Sciences of the United States of America, 104*(41), 16341–1634.

Griesinger, C. B., Richards, C. D., & Ashmore, J. F. (2005). Fast vesicle replenishment allows indefatigable signalling at the first auditory synapse. *Nature, 435*(7039), 212–215.

Heidrych, P., Zimmermann, U., Bress, A., Pusch, C. M., Ruth, P., Pfister, M., Knipper, M., & Blin, N. (2008). Rab8b GTPase, a protein transport regulator, is an interacting partner of otoferlin, defective in a human autosomal recessive deafness form. *Human Molecular Genetics, 17*(23), 3814–3821.

Heidrych, P., Zimmermann, U., Kuhn, S., Franz, C., Engel, J., Duncker, S. V., Hirt, B., Pusch, C. M., Ruth, P., Pfister, M., Marcotti, W., Blin, N., & Knipper, M. (2009). Otoferlin interacts with myosin VI: Implications for maintenance of the basolateral synaptic structure of the inner hair cell. *Human Molecular Genetics, 18*(15), 2779–2790.

Heil, P., Neubauer, H., Irvine, D. R. F., & Brown, M. (2007). Spontaneous activity of auditory-nerve fibers: insights into stochastic processes at ribbon synapses. *Journal of Neuroscience, 27*(31), 8457–8474.

Highstein, S. M. (1991). The central nervous system efferent control of the organs of balance and equilibrium. *Neuroscience Research, 12*(1), 13–30.

Issa, N. P. & Hudspeth, A. J. (1994). Clustering of Ca2+ channels and Ca(2+)-activated K+ channels at fluorescently labeled presynaptic active zones of hair cells. *Proceedings of the National Academy of Sciences of the United States of America, 91*(16), 7578–7582.

Johnson, S. L., & Marcotti, W. (2008). Biophysical properties of CaV1.3 calcium channels in gerbil inner hair cells. *Journal of Physiology, 586*(4), 1029–1042.

Johnson, S. L., Marcotti, W., & Kros, C. J. (2005). Increase in efficiency and reduction in Ca2+ dependence of exocytosis during development of mouse inner hair cells. *Journal of Physiology, 563*(Pt 1), 177–191.

Johnson, S. L., Franz, C., Kuhn, S., Furness, D. N., Rüttiger, L., Münkner, S., Rivolta, M. N., Seward, E. P., Herschman, H. R., Engel, J., Knipper, M., & Marcotti, W. (2010). Synaptotagmin IV determines the linear Ca2+ dependence of vesicle fusion at auditory ribbon synapses. *Nature Neuroscience, 13*(1), 45–52.

Keen, E. C., & Hudspeth, A. J. (2006). Transfer characteristics of the hair cell's afferent synapse. *Proceedings of the National Academy of Sciences of the United States of America, 103*(14), 5537–5542.

Khimich, D., Nouvian, R., Pujol, R., Tom Dieck, S., Egner, A., Gundelfinger, E. D., & Moser, T. (2005). Hair cell synaptic ribbons are essential for synchronous auditory signalling. *Nature, 434*(7035), 889–894.

Klinke, R., & Galley, N. (1974). Efferent innervation of vestibular and auditory receptors. *Physiological Reviews, 54*(2), 316–357.

Kong, W.-J., Cheng, H.-M., & van Cauwenberge, P. (2006). Expression of nicotinic acetylcholine receptor subunit alpha9 in type II vestibular hair cells of rats. *Acta Pharmacologica Sinica, 27*(11), 1509–1514.

Kong, J.-H., Adelman, J. P., & Fuchs, P. A. (2008). Expression of the SK2 calcium-activated potassium channel is required for cholinergic function in mouse cochlear hair cells. *Journal of Physiology, 586*(Pt 22), 5471–5485.

Lenzi, D., Runyeon, J. W., Crum, J., Ellisman, M. H., & Roberts, W. M. (1999). Synaptic vesicle populations in saccular hair cells reconstructed by electron tomography. *Journal of Neuroscience, 19*(1), 119–132.

Lenzi, D., Crum, J., Ellisman, M. H., & Roberts, W. M. (2002). Depolarization redistributes synaptic membrane and creates a gradient of vesicles on the synaptic body at a ribbon synapse. *Neuron, 36*(4), 649–659.

Li, G.-L., Keen, E., Andor-Ardó, D., Hudspeth, A. J., & von Gersdorff, H. (2009). The unitary event underlying multiquantal EPSCs at a hair cell's ribbon synapse. *Journal of Neuroscience, 29*(23), 7558–7568.

Liberman, M. C. (1980). Morphological differences among radial afferent fibers in the cat cochlea: An electron-microscopic study of serial sections. *Hearing Research, 3*(1), 45–63.

Liberman, M. C. (1982). Single-neuron labeling in the cat auditory nerve. *Science, 216*(4551), 1239–1241.

Lioudyno, M., Hiel, H., Kong, J.-H., Katz, E., Waldman, E., Parameshwaran-Iyer, S., Glowatzki, E., & Fuchs, P. A. (2004). A "synaptoplasmic cistern" mediates rapid inhibition of cochlear hair cells. *Journal of Neuroscience, 24*(49), 11160–11164.

Lysakowski, A., & Goldberg, J. M. (1997). A regional ultrastructural analysis of the cellular and synaptic architecture in the chinchilla cristae ampullares. *Journal of Comparative Neurology, 389*(3), 419–443.

Lysakowski, A., & Goldberg, J. M. (2008). Ultrastructural analysis of the cristae ampullares in the squirrel monkey (*Saimiri sciureus*). *Journal of Comparative Neurology, 511*(1), 47–64.

Magupalli, V. G., Schwarz, K., Alpadi, K., Natarajan, S., Seigel, G. M., & Schmitz, F. (2008). Multiple RIBEYE-RIBEYE interactions create a dynamic scaffold for the formation of synaptic ribbons. *Journal of Neuroscience, 28*(32), 7954–7967.

Matsusaka, T. (1967). Lamellar bodies in the synaptic cytoplasm of the accessory cone from the chick retina as revealed by electron microscopy. *Journal of Ultrastructure Research, 18*(1), 55–70.

Matthews, G., & Sterling, P. (2008). Evidence that vesicles undergo compound fusion on the synaptic ribbon. *Journal of Neuroscience, 28*(21), 5403–5411.

Meyer, A. C., Frank, T., Khimich, D., Hoch, G., Riedel, D., Chapochnikov, N. M., Yarin, Y. M., Harke, B., Hell, S. W., Egner, A., & Moser, T. (2009). Tuning of synapse number, structure and function in the cochlea. *Nature Neuroscience, 12*(4), 444–453.

Moser, T., & Beutner, D. (2000). Kinetics of exocytosis and endocytosis at the cochlear inner hair cell afferent synapse of the mouse. *Proceedings of the National Academy of Sciences of the United States of America, 97*(2), 883–888.

Nouvian, R., Beutner, D., Parsons, T. D., & Moser, T. (2006). Structure and function of the hair cell ribbon synapse. *Journal of Membrane Biology, 209*(2–3), 153–165.

Obholzer, N., Wolfson, S., Trapani, J. G., Mo, W., Nechiporuk, A., Busch-Nentwich, E., Seiler, C., Sidi, S., Söllner, C., Duncan, R. N., Boehland, A., & Nicolson, T. (2008). Vesicular glutamate transporter 3 is required for synaptic transmission in zebrafish hair cells. *Journal of Neuroscience, 28*(9), 2110–2118.

Osman, A. A., Schrader, A. D., Hawkes, A. J., Akil, O., Bergeron, A., Lustig, L. R., & Simmons, D. D. (2008). Muscle-like nicotinic receptor accessory molecules in sensory hair cells of the inner ear. *Molecular and Cellular Neurosciences, 38*(2), 153–169.

Pangrsic, T., Lasarow, L., Reuter, K., Takago, H., Schwander, M., Riedel, D., Frank, T., Tarantino, L. M., Bailey, J. S., Strenzke, N., Brose, N., Müller, U., Reisinger, E., & Moser, T. (2010). Hearing requires otoferlin-dependent efficient replenishment of synaptic vesicles in hair cells. *Nature Neuroscience, 13*(7), 869–876.

Parsons, T. D., Lenzi, D., Almers, W., & Roberts, W. M. (1994). Calcium-triggered exocytosis and endocytosis in an isolated presynaptic cell: capacitance measurements in saccular hair cells. *Neuron, 13*(4), 875–883.

Platzer, J., Engel, J., Schrott-Fischer, A., Stephan, K., Bova, S., Chen, H., Zheng, H., & Striessnig, J. (2000). Congenital deafness and sinoatrial node dysfunction in mice lacking class D L-type Ca^{2+} channels. *Cell, 102*(1), 89–97.

Ramakrishnan, N. A., Drescher, M. J., & Drescher, D. G. (2009). Direct interaction of otoferlin with syntaxin 1A, SNAP-25, and the L-type voltage-gated calcium channel Cav1.3. *Journal of Biological Chemistry, 284*(3), 1364–1372.

Roberts, W. M. (1993). Spatial calcium buffering in saccular hair cells. *Nature, 363*(6424), 74–76.

Roberts, W. M., Jacobs, R. A., & Hudspeth, A. J. (1990). Colocalization of ion channels involved in frequency selectivity and synaptic transmission at presynaptic active zones of hair cells. *Journal of Neuroscience, 10*(11), 3664–3684.

Roux, I., Safieddine, S., Nouvian, R., Grati, M., Simmler, M.-C., Bahloul, A., Perfettini, I., Le Gall, M., Rostaing, P., Hamard, G., Triller, A., Avan, P., Moser, T., & Petit, C. (2006). Otoferlin, defective in a human deafness form, is essential for exocytosis at the auditory ribbon synapse. *Cell, 127*(2), 277–289.

Ruel, J., Bobbin, R. P., Vidal, D., Pujol, R., & Puel, J. L. (2000). The selective AMPA receptor antagonist GYKI 53784 blocks action potential generation and excitotoxicity in the guinea pig cochlea. *Neuropharmacology, 39*(11), 1959–1973.

Ruel, J., Wang, J., Rebillard, G., Eybalin, M., Lloyd, R., Pujol, R., & Puel, J.-L. (2007). Physiology, pharmacology and plasticity at the inner hair cell synaptic complex. *Hearing Research, 227*(1–2), 19–27.

Ruel, J., Emery, S., Nouvian, R., Bersot, T., Amilhon, B., Van Rybroek, J. M., Rebillard, G., Lenoir, M., Eybalin, M., Delprat, B., Sivakumaran, T. A., Giros, B., El Mestikawy, S., Moser, T., Smith, R. J. H., Lesperance, M. M., & Puel, J.-L. (2008). Impairment of SLC17A8 encoding vesicular glutamate transporter-3, VGLUT3, underlies nonsyndromic deafness DFNA25 and inner hair cell dysfunction in null mice. *American Journal of Human Genetics, 83*(2), 278–292.

Safieddine, S., & Wenthold, R. J. (1999). SNARE complex at the ribbon synapses of cochlear hair cells: Analysis of synaptic vesicle- and synaptic membrane-associated proteins. *European Journal of Neuroscience, 11*(3), 803–812.

Schmitz, F., Königstorfer, A., & Südhof, T. C. (2000). RIBEYE, a component of synaptic ribbons: A protein's journey through evolution provides insight into synaptic ribbon function. *Neuron, 28*(3), 857–872.

Schnee, M. E., Lawton, D. M., Furness, D. N., Benke, T. A., & Ricci, A. J. (2005). Auditory hair cell–afferent fiber synapses are specialized to operate at their best frequencies. *Neuron, 47*(2), 243–254.

Schug, N., Braig, C., Zimmermann, U., Engel, J., Winter, H., Ruth, P., Blin, N., Pfister, M., Kalbacher, H., & Knipper, M. (2006). Differential expression of otoferlin in brain, vestibular system, immature and mature cochlea of the rat. *European Journal of Neuroscience, 24*(12), 3372–3380.

Seal, R. P., Akil, O., Yi, E., Weber, C. M., Grant, L., Yoo, J., Clause, A., Kandler, K., Noebels, J. L., Glowatzki, E., Lustig, L. R., & Edwards, R. H. (2008). Sensorineural deafness and seizures in mice lacking vesicular glutamate transporter 3. *Neuron, 57*(2), 263–275.

Sidi, S., Busch-Nentwich, E., Friedrich, R., Schoenberger, U., & Nicolson, T. (2004). Gemini encodes a zebrafish L-type calcium channel that localizes at sensory hair cell ribbon synapses. *Journal of Neuroscience, 24*(17), 4213–4223.

Siegel, J. H. (1992). Spontaneous synaptic potentials from afferent terminals in the guinea pig cochlea. *Hearing Research, 59*(1), 85–92.

Sjostrand, F. S. (1953). The ultrastructure of the inner segments of the retinal rods of the guinea pig eye as revealed by electron microscopy. *Journal of Cellular Physiology, 42*(1), 45–70.

Sobkowicz, H. M., Rose, J. E., Scott, G. E., & Slapnick, S. M. (1982). Ribbon synapses in the developing intact and cultured organ of Corti in the mouse. *Journal of Neuroscience, 2*(7), 942–957.

Spassova, M., Eisen, M. D., Saunders, J. C., & Parsons, T. D. (2001). Chick cochlear hair cell exocytosis mediated by dihydropyridine-sensitive calcium channels. *Journal of Physiology, 535*(Pt 3), 689–696.

Sridhar, T. S., Brown, M. C., & Sewell, W. F. (1997). Unique postsynaptic signaling at the hair cell efferent synapse permits calcium to evoke changes on two time scales. *Journal of Neuroscience, 17*(1), 428–437.

Starr, P. A., & Sewell, W. F. (1991). Neurotransmitter release from hair cells and its blockade by glutamate-receptor antagonists. *Hearing Research, 52*(1), 23–41.

Taberner, A. M., & Liberman, M. C. (2005). Response properties of single auditory nerve fibers in the mouse. *Journal of Neurophysiology, 93*(1), 557–569.

Trapani, J. G., Obholzer, N., Mo, W., Brockerhoff, S. E., & Nicolson, T. (2009). Synaptojanin1 is required for temporal fidelity of synaptic transmission in hair cells. *PloS Genetics, 5*(5), e1000480.

Varga, R., Kelley, P. M., Keats, B. J., Starr, A., Leal, S. M., Cohn, E., & Kimberling, W. J. (2003). Non-syndromic recessive auditory neuropathy is the result of mutations in the otoferlin (OTOF) gene. *Journal of Medical Genetics, 40*(1), 45–50.

Weisz, C., Glowatzki, E., & Fuchs, P. (2009). The postsynaptic function of type II cochlear afferents. *Nature, 461*(7267), 1126–1129.

Westerman, L. A., & Smith, R. L. (1984). Rapid and short-term adaptation in auditory nerve responses. *Hearing Research, 15*(3), 249–260.

Yasunaga, S., Grati, M., Chardenoux, S., Smith, T. N., Friedman, T. B., Lalwani, A. K., Wilcox, E. R., & Petit, C. (2000). OTOF encodes multiple long and short isoforms: Genetic evidence that the long ones underlie recessive deafness DFNB9. *American Journal of Human Genetics, 67*(3), 591–600.

Zenisek, D., Horst, N. K., Merrifield, C., Sterling, P., & Matthews, G. (2004). Visualizing synaptic ribbons in the living cell. *Journal of Neuroscience, 24*(44), 9752–9759.

Zhang, Z., Bhalla, A., Dean, C., Chapman, E. R., & Jackson, M. B. (2009). Synaptotagmin IV: A multifunctional regulator of peptidergic nerve terminals. *Nature Neuroscience, 12*(2), 163–171.

Chapter 4
The Endbulbs of Held

Paul B. Manis, Ruili Xie, Yong Wang, Glen S. Marrs, and George A. Spirou

1 Introduction

A remarkable nerve terminal is present at the endings of the auditory nerve fibers (ANFs) in the anterior ventral cochlear nucleus (AVCN), called the "endbulb of Held." The endbulbs are complex synaptic endings that provide a coordinated release from multiple presynaptic sites of neurotransmitter onto their target postsynaptic cells, the globular and spherical bushy cells of the cochlear nucleus. These synapses play a key role in bringing a precisely timed representation of sound into the central auditory system. Traditionally, the endbulbs, owing to their large size and the presence of multiple presynaptic release zones, were thought to provide a "secure" synapse between auditory nerve fibers and the target neurons, the globular and spherical bushy cells. However, this view has been strongly challenged by several recent observations. While the endbulbs are indeed a particularly strong synapse, they are subject to dynamic regulation of transmitter release probability and receptor sensitivity, and their ability to initiate action potentials in the postsynaptic cell is not immune to postsynaptic inhibition. Integration by convergence of endbulb synapses onto target cells is an important part of central auditory processing. In particular, cells postsynaptic to the endbulbs can fire more precisely to specific temporal features of acoustic stimuli than their individual auditory nerve fiber inputs. The endbulb synapses are found widely in mammals including humans (Adams 1986), as well as in birds (Carr and Boudreau 1991; Koppl 1994) and reptiles (Browner and Marbey 1988; Szpir et al. 1990), but their presence in amphibians is less clear (Lewis et al. 1980; Feng and Lin 1996).

P.B. Manis (✉)
Department of Otolaryngology/Head and Neck Surgery,
UNC Chapel Hill, G127 Physician's Office Building., CB#7070,
Chapel Hill, NC 27599–7070, USA
e-mail: pmanis@med.unc.edu

L.O. Trussell et al. (eds.), *Synaptic Mechanisms in the Auditory System*,
Springer Handbook of Auditory Research 41, DOI 10.1007/978-1-4419-9517-9_4,
© Springer Science+Business Media, LLC 2012

This chapter reviews the information transfer across the endbulb, as viewed from the standpoint of the spike trains of auditory nerve fibers and their target bushy cells, in Sect. 2. In Sect. 3, the anatomical organization of endbulbs is outlined, along with their variation. In Sects 4 and 5, the fundamental features of synaptic transmission and the dynamics of transmission between the endbulb and the target cell are reviewed. Sect. 6 discusses modulation of synaptic transmission at the endbulb. In the summary (Sect. 7), several outstanding problems surrounding endbulb transmission are presented.

2   Single-Unit Studies

The earliest studies of endbulb function depended on the single-unit technique, in which the extracellular electrical voltage around a cell is recorded with a metal electrode. Multiple features in the extracellular voltage waveform can be related to underlying current generators, including the action potential invasion of the presynaptic terminal, the postsynaptic currents, the postsynaptic action potential, and the dendritic currents (Fig. 4.1). The concept that endbulb synapses contacting spherical and globular bushy cells can faithfully transmit the spike timing of their incoming AN afferents is largely derived from the observations that postsynaptic spikes are preceded by "prepotentials" that reflect currents produced around the large endbulb of Held as it is invaded by the nerve action potential (Pfeiffer 1966; Goldberg and Brownell 1973; Bourk 1976). Evidence that prepotentials arise from the endbulb or a similar presynaptic structure includes their absence with antidromic stimulation, as measured in the medial nucleus of the trapezoid body (MNTB) (Guinan and Li 1990), and their occasional presence in the absence of a putative postsynaptic spike. Recent detailed analyses of the relation between prepotentials and firing in both the AVCN and MNTB provide strong support of the original hypotheses (Englitz et al. 2009; Lorteije et al. 2009) and further clarify the observations on the likelihood that a presynaptic spike fails to elicit a postsynaptic action potential at this synapse.

The shape of the extracellular waveform associated with the endbulbs and postsynaptic cells has several stereotypical features that are common across species (Fig. 4.1). Assuming that the recording electrode is close to the cell body, the shape of the waveform can be interpreted based on the locations of current sources and sinks at different times during the invasion of the presynaptic action potential and the activation of the postsynaptic receptors and ion channels. The initial positive presynaptic waveform that defines the "prepotential," or the "P" component, arises from the current sources that are generated when the action potential nears the terminal; because the density of sodium channels in the terminal itself appears to be low (as shown at the calyx of Held in the MNTB (Leão et al. 2004)), the terminal serves as a current source back to the last myelin heminode, where the highest density of sodium channels is located.

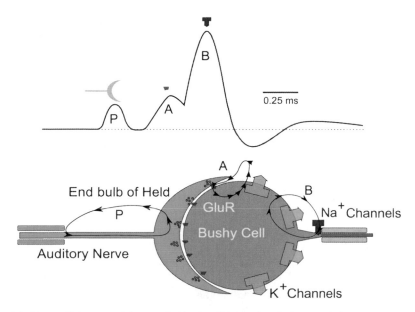

Fig. 4.1 Extracellular potentials recorded in the vicinity of endbulbs and their target bushy cells. The top drawing shows a typical waveform with three components. The lower cartoon shows the anatomical arrangement of the synapse and the bushy cell, along with the three principal currents across the cell membrane that generated as an action potential invades the terminal, releases transmitter and activates the postsynaptic receptors, and initiates an action potential. The "P" component is attributed to the action potential invasion of the presynaptic terminal, the "A" component is attributed to the inward currents associated with the opening of postsynaptic ionotropic glutamate receptors (labeled GluR), and the "B" component is associated with the opening of voltage-gated sodium channels in the axon hillock and first heminode of Ranvier. See text for further details

The next positivity in the waveform, the "A" component, represents a current source generated by the opening of the postsynaptic receptors (labeled "GluR" in Fig. 4.1). Because the postsynaptic density is covered by the structure of the endbulb itself, the principal current sinks should occur underneath the endbulb, while the remainder of the membrane of the postsynaptic cell serves as a current source. The synaptic current sinks would be apparent in the extracellular field only at the lateral margins of the endbulb, and at that point they would be opposed by capacitative current sources arising through the membrane of the postsynaptic cell. Thus, the net extracellular potential will (usually, depending on the location of the recording electrode) be dominated by the current sources and will appear as an extracellular positivity.

The synaptic current is followed by a biphasic waveform, or the "B" component, reflecting the initiation of the postsynaptic action potential, which may be at the first heminode or even at the first full node, followed by invasion of the cell body and activation of somatic sodium channels (producing a large negative wave). The action potential waveform may also appear to merge with the synaptic potential, depending on the latency between the excitatory postsynaptic potential (EPSP) and spike generation. The components of these waveforms will vary depending on the position of the electrode relative to the cell and the endbulbs, but the general structure should always be present, even if it is sometimes difficult to separate some components from the recording noise. The signs of events may also change, depending on the location of the recording electrode, and examples of waveforms with different signs are present in the literature (Pfeiffer 1966; Bourk 1976). A recent in vitro study combining extracellular and whole-cell recording most clearly provides the strongest and unequivocal experimental separation of the different components of the waveform (Typlt et al. 2010) and is consistent with the interpretation in Fig. 4.1.

If the endbulb synapses are truly secure, then it follows that their postsynaptic target neurons should fire with "primary-like" response patterns. This supposition is partially true, in that neurons with prepotentials tend to have primary-like (or primary-like with notch) poststimulus time histograms in many species (Bourk 1976; Blackburn and Sachs 1989; Winter and Palmer 1990). However, even within the AVCN, the morphology of the prepotential–action potential complex varies among cells, or even within recordings from a single cell (Bourk 1976), and this variation might provide clues to the number and organization of synaptic inputs. Bourk (1976) defined five classes of prepotentials based on their amplitudes. Unfortunately, this classification has not subsequently been used, and usually prepotentials are reported as either present or not detected. Prepotentials have infrequently been associated with single units that have unusual response properties, so cells receiving large endings, which may not necessarily be endbulbs of Held, can also exhibit a prepotential. These observations imply either that these cells are not bushy cells or that bushy cells have a greater variety of response patterns than is commonly assumed. Because such observations are rare, it will be difficult to clarify their significance. However, this is an important caveat when interpreting the cellular origin of extracellular recordings.

A second approach used to study endbulb synaptic function has been to use brain slices of the cochlear nucleus where individual, identified, postsynaptic cells can be recorded intracellularly (Oertel 1983). In these preparations, EPSPs produced by shocks to the ANFs in cochlear nucleus brain slices are usually suprathreshold, although weaker, subthreshold EPSPs or excitatory postsynaptic currents (EPSCs) can be also be observed. The origin of the smaller events is not entirely clear because stimulation of the nerve stump, or of the nerve fibers within the ventral cochlear nucleus (VCN) proper, can also evoke firing in other types of axons or cells that provide excitatory inputs to the bushy cells. However, the frequent observation that there are small endings from auditory nerve fibers onto bushy cells (see Sect. 3) could explain the presence of weaker inputs. Evoked synaptic currents are large compared to spontaneous release events; even the smaller currents are about 300–500 pA compared to ~100 pA for the miniature events. Furthermore, the glutamate receptors are rapidly desensitizing (Sect. 3) so that they produce only a brief current injection into the postsynaptic cell.

As discussed later, bushy cells in brain slices show good entrainment of postsynaptic spiking to afferent stimulation for frequencies at least as high as 300 Hz (Zhang and Trussell 1994a, b; Isaacson and Walmsley 1995b, 1996). However, several observations clearly show that bushy neurons in vivo do not function as true relays. First, Carney (1990) examined the sensitivity of cells in the cat AVCN to acoustic stimuli that contained rapid phase shifts placed over a narrow frequency range. Sensitivity of a cell's response to the frequency position of the phase shift relative to the characteristic frequency (CF) can be interpreted as evidence for convergence across auditory nerve fibers representing different frequency channels. A subset of cells in the posterior region of the AVCN showed sensitivity to this phase shift, and slightly more than half of these cells had prepotentials. The location and presence of the prepotential suggest that they were globular bushy cells receiving endbulb synapses. Second, the responses of AVCN cells with prepotentials can show non-monotonic rate-level functions and single-tone response suppression, while neither of these features is present in ANF responses (Winter and Palmer 1990). Third, the output of the globular bushy cells, as measured from their axons in the trapezoid body, can show phase locking to low-frequency stimuli that is greater than that of individual auditory nerve fibers (Joris et al. 1994), which most likely requires integration of multiple inputs (Rothman et al. 1993; Rothman and Young 1996). Fourth, bushy cell responses can be modified by blocking inhibition (Caspary et al. 1994; Kopp-Scheinpflug et al. 2002). Fifth, as reported in gerbils, prepotentials can occur in the absence of the "B" component of the extracellular potential, suggesting failures of the synapse to generate a spike (Englitz et al. 2009; Typlt et al. 2010). Finally, intracellular recordings from globular bushy cells in vivo reveal subthreshold EPSPs during tonal stimulation as well as in silence, and these can be also be seen distinct from action potentials in axonal recordings (Smith and Rhode 1987; Rhode 2008). All of these observations suggest that, despite the large synaptic currents generated by endbulb synapses, significant postsynaptic integration can and does occur. Integration depends on the anatomical structure of endbulbs, afferent convergence, the time course of synaptic conductances, and the short-term dynamics of the synapses. These are discussed in the next sections.

3   Neuroanatomy of Auditory Nerve Innervation of Bushy Cells

3.1   Discovery of Endbulbs and Their Innervation of Bushy Cells

Large nerve terminals were first documented in association with the innervating ANFs within the AVCN in the late nineteenth century, establishing this neural territory as a rich brain region for systematic anatomic investigation of the nervous system even into modern times. Initial observations by Held (1891) and Kolliker (1896) revealed that auditory nerve fibers bifurcated into ascending (anterior) and descending (posterior) branches on entry into the VCN and that the anterior branches terminated in large endings (Held 1891). Further observation revealed that these large endings emerged only from the ascending branch, were of varied and generally increasing size along the ascending branch, and were uniquely apposed to post-synaptic cell bodies (Ramon y Cajal 1896). These terminals were described as

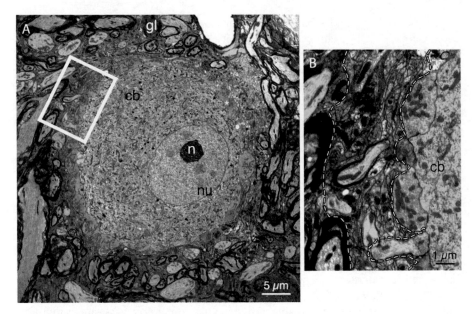

Fig. 4.2 Structural characteristics of GBCs and adjacent neuropil. (**a**) GBCs exhibit a round cell body (cb) with eccentrically located nucleus (nu), smooth and nonindented nuclear membrane, prominent nucleolus (n), and absence of prominent stacks of rough endoplasmic reticulum. In this section nearly the entire somatic surface is contacted by auditory nerve fiber terminals (type LS, colored blue), excitatory inputs likely of noncochlear origin (type SS, colored blue with asterisk), and inhibitory terminals (type PL, colored red) that form a layer around the cell body. (**b**) Surrounding the nerve terminal layer is a layer of complex cellular processes (between *dashed lines*) that is penetrated by axons leading to nerve terminals (*blue*) or high-order dendritic branches of adjacent cells. Outside of this layer, as viewed in panel (**a**), the cell is surrounded by axons of varying diameter that form bundles or are more dispersed, tertiary dendrites primarily originating from neighboring cells (*green*), glial cell bodies (gl), and capillaries (at *top* of image) (Reprinted with permission from Spirou et al. 2005)

bulbs, clubs, or chalices, several of which converged onto some cells to form nests or baskets around the cell body (Lorente de Nó 1981; Ramon y Cajal 1995). Subsequently, auditory nerve fibers were classified as yielding only large chalices, both large and small chalices, or neither (Lorente de Nó 1981). These compiled observations relied heavily on Golgi-stained material from a variety of mammalian species including dogs, cats, mice, and rabbits, usually from immature brains. A more detailed view of the endbulb can be obtained with electron microscopy, in which the endings can be seen to consist of multiple active sites on the soma, interspersed with putative inhibitory synapses (see Fig. 4.2).

Systematic description of anatomically classified cell types and regions in the mammalian VCN made possible the association of cell morphology with both the structural features of auditory nerve terminals and their physiological classifications. Cells with short, brush-like dendrites [called brush cells by Lorente de Nó (1981) and more recently bushy cells by Brawer et al. (1974)] were divided into two populations and linked with their appearance and terminology as spherical and globular cells in Nissl stain (Harrison and Warr 1962; Osen 1969a). Later studies showed that spherical and globular cells could be distinguished by their unique distributions of rough endoplasmic reticulum when viewed using electron microscopy (Cant and Morest 1979; Tolbert and Morest 1982). Pre- and postsynaptic cellular features are regionally matched, whereby spherical bushy cells and large endbulbs cluster rostrally, and globular bushy cells and their associated smaller terminals, termed modified endbulbs (Harrison and Irving 1965), cluster caudally in the AVCN (Brawer and Morest 1975). Regional association with primary-like and primary-notch physiological types (Kiang et al. 1965) further solidified the notion that large and small endbulbs should underlie measurable differences in information processing by spherical and globular bushy cells.

3.2   Variations of Innervation with Characteristic Frequency and Spontaneous Discharge Rate

Given the regional positions of relatively distinct spherical and globular bushy cell populations that appeared to cross the entire auditory-nerve fiber array, an early question was whether peripheral inputs are focally mapped and patterned, providing a precise spatial-functional relationship between apical-basal location of hair cells (and associated characteristic frequency) and synaptic position within VCN cell populations. Initially, focal cochlear ablations (Sando 1965; Osen 1970; Moskowitz and Liu 1972; Noda and Pirsig 1974) and, later, single cell tracing with horseradish peroxidase (HRP) techniques related VCN targets to cochlear position and showed a clear spatial relationship between auditory nerve terminal arbors and CF (Liberman 1982; Fekete et al. 1984).

The demonstration that auditory nerve fibers of a given CF can also be physiologically segregated on the basis of their spontaneous rate (SR) of firing in the absence of sound suggested that even within VCN isofrequency laminae, the pattern of innervation could vary with additional physiological parameters. In cats, three categories of SR are apparent, with low (<0.5 Hz), medium (0.5–18 Hz), and high (>18 Hz) discharge rates (Kiang 1965; Liberman 1978). The SR is correlated with the threshold at which a fiber begins to respond to sound (Liberman 1978) and co-varies with a multitude of important physiologic parameters including dynamic range (Schalk and Sachs 1980), maximum discharge rate (Liberman 1978), adaptation rate (Rhode and Smith 1985), and recovery from noise masking (Costalupes et al. 1984). Therefore, SR distinctions likely underlie fundamental processing differences among auditory nerve fibers that could extend to processing in the CN through specialized innervation patterns, targets, and terminal features.

The application of intracellular HRP-labeling techniques in combination with physiological characterization in cats revealed structural correlations with auditory nerve fiber CF and SR. Ryugo and colleagues performed a series of analyses of branching patterns, terminal size, and terminal number across these functional categories.

Each fiber gave rise to nerve terminals ranging in size from boutons to large endbulbs. Branching in the auditory-nerve root region was similar across CF, except that low-CF fibers (<0.5 kHz) formed two major ascending axons after branching in the nerve root (4/5 axons), and each axon delivered a large terminal endbulb (Ryugo and Rouiller 1988). In an analysis oriented toward correlation of ending morphology with postsynaptic cell type (Liberman 1991), in concurrence with Ryugo and colleagues (Rouiller et al. 1986; Sento and Ryugo 1989), showed that endbulbs were largest at CFs between 1 and 4 kHz, and the largest endbulbs innervated the largest spherical cells. Large terminal and collateral endbulbs onto spherical bushy cells were analyzed separately. Large endbulbs were reportedly larger in the high SR population (Ryugo et al. 1996), but contrasting results have been reported (Sento and Ryugo 1989; Liberman 1991), due in part to the difficulty of acquiring large sample sizes in these technically challenging experiments.

Another key observation was that within SR categories, the length and complexity of auditory nerve branches did not vary across CF. However, within the VCN, low to medium SR fibers branched more, had longer collaterals, and generated more nerve terminals (Fekete et al. 1984; Ryugo and Rouiller 1988). All SR groups equally innervated spherical bushy cells, but the high SR group preferentially innervated globular bushy cells (Liberman 1991).

Convergence of different auditory nerve fiber inputs onto individual bushy cells influences the detection of temporal coincidence, temporal firing precision, the dynamic range, and the structure of the frequency response area. Because spherical bushy cells receive somatic contacts from 1 to 2 large endbulbs in cat, the pattern of convergence among auditory nerve SR categories with their different acoustic response sensitivities has emerged as a fundamental question. Endbulb structure was shown to be finer and more complex in the low and medium SR populations. Using this information (Sento and Ryugo 1989; Ryugo and Sento 1991), convergence onto individual spherical bushy cells appears to be conserved among auditory-nerve fibers of the same SR group. The distribution of SRs in the spherical bushy cell population is roughly similar to that of auditory nerve fibers, except at low CF (<2–3 kHz) where fewer low SR fibers are encountered (Bourk 1976; Blackburn and Sachs 1989; Spirou et al. 1990). These data were interpreted as indicating that single endbulb innervation of spherical bushy cells occurred preferentially via high SR auditory-nerve fibers (Spirou et al. 1990).

The somewhat smaller modified endbulbs (Fig. 4.2) dominate the population of complex endings in contact with globular bushy cell somata (Liberman 1991), although some large collateral endbulbs may contribute to this population (Rouiller et al. 1986). The number of modified endbulbs per auditory nerve fiber does not vary with CF or SR category [~2.85 (Rouiller et al. 1986)], so initial estimates of convergence based on numbers of auditory-nerve fibers and globular bushy cells in cat yielded values of about 20 per cell (Spirou et al. 1990; Liberman 1991). Direct counting of inputs using combined light and electron microscopic analysis revealed an average value of 22.9 auditory-nerve fiber inputs to a single cell (Spirou et al. 2005), but a wide range of convergence was seen (9–69 ANFs). Because morphological features of modified endbulbs have not been associated with SR and the number of converging inputs is relatively large, convergence of SR categories onto globular bushy cells is difficult to estimate. Light microscopic estimates of inputs onto globular cell bodies identified using a Nissl stain yielded an 8:1 high:low SR innervation ratio (Liberman 1991). The estimated SR distribution for globular bushy cells, however, does not reflect this weighting of high SR inputs and, indeed, a bias toward low SR occurs at low frequencies [<2 kHz (Blackburn and Sachs 1989); <6 kHz (Spirou et al. 1990)]. These differences may also reflect a contribution of spontaneously active inhibitory inputs that suppress spontaneous and driven rates.

3.3 Large and Small Spherical Bushy Cells

In her naming of the spherical cell type, Osen (1969b) distinguished qualitatively a region at the anterior pole of the cat VCN where cells appeared larger, on average, than the remaining, more caudally located cells. Although this distinction is based on the profile of cells in the coronal plane, careful measurements did reveal a 30% greater cross-sectional area of cell bodies in the rostral VCN region of terminal endbulbs, in comparison to more caudal regions of the AVCN (Cant 1991). In cats, both spherical and globular cell regions exhibited a tonotopic or frequency-dependent decrease in soma size. Because the anterior AVCN is selectively innervated by lower CF fibers, a distinction in size exists between large cells anterior and nearby posterior small cells. Therefore, this observation is in general agreement with the association of the largest endbulbs with low CFs and the largest terminals with the largest postsynaptic cells (Sento and Ryugo 1989; Liberman 1991).

3.4 The Rodent Cochlear Nucleus

The carefully elaborated structure-function correlations for endbulbs and bushy cells in cats remain the template for comparison with other species, even though rodents, especially rats and mice, are increasingly used for studies of the lower auditory CNS. The mouse VCN exhibits the same cell types and subdivisions as cat in Nissl- and Golgi-stained sections (Martin 1981; Webster and Trune 1982). Spherical and globular bushy cells can be distinguished, and the globular bushy cells are most dense in the auditory-nerve root region. However, there are no detailed maps of the general borders and an analysis of the potential intermixing of cell types has not been performed. Applying the regional distributions of cells in cat to the mouse VCN, Webster and Trune (1982) reported a more subtle distinction in cell size between spherical and globular bushy cells than occurs in cat, perhaps related to the limited representation of low frequencies in the mouse audible spectrum. The rat VCN, which contains spherical and globular cells cytologically similar to cats and mice, was parceled into zones that correspond generally to locations of low and high CF spherical cells (zones III and I) and of globular cells within and posterior to the nerve root [zone II (Harrison and Irving 1965)]. A detailed correlation of auditory nerve terminal parameters with cellular region in the VCN has not been performed in rodents, due to the technical challenge of marking single auditory-nerve fibers of known physiological properties. However, the modified endbulbs in rodents with normal hearing (Tsuji and Liberman 1997; Limb and Ryugo 2000; Nicol and Walmsley 2002) are generally similar to those in cats.

3.5   *Ultrastructure and Activity*

Synaptic features of high SR endbulbs appear designed to promote high efficacy neurotransmission, in that synapses are more numerous and are associated with greater numbers of synaptic vesicles when compared to low to medium SR fibers (Ryugo et al. 1996). A larger endbulb size for high SR fibers would transform a similar percent mitochondrial volume into a nearly 50% increase in absolute mito-chondrial volume relative to low SR fibers. Mitochondria function to generate ATP and buffer Ca^{2+} in nerve terminals. They can affect short-term plasticity in large nerve terminals via Ca^{2+} buffering (Billups and Forsythe 2002), but their role in ATP generation likely modulates synaptic vesicle recruitment from reserve pools only at high activity rates (Verstreken et al. 2005). Likely, then, mitochondrial ATP genera-tion plays important roles at endbulb terminals. Indeed, mitochondria in endbulbs and the related calyces of Held take up calibrated positions near the presynaptic membrane adjacent to active zones (Cant and Morest 1979; Tolbert and Morest 1982; Rowland et al. 2000; Perkins et al. 2010). High SR endbulbs have smaller synapses but double the total synaptic contact area. Given the convergence of audi-tory-nerve fibers of the same SR category onto single spherical bushy cells, synapse size appears to correlate with overall postsynaptic activity. The increase in synapse size in deafness models, and reversibility of this parameter via electrical stimulation of the auditory nerve, argues for use or activity-dependent regulation of size (Ryugo et al. 1997; O'Neil et al. 2010).

The dendrites of both spherical and globular bushy cells extend away from the cell body but pass near somata of neighboring cells. This orientation places them in proximity to large and modified endbulb somatic inputs. Numerous studies have documented multiple synaptic sites within single endbulb swellings that contact not only somata but dendrites of neighboring cells (Cant and Morest 1979; Ostapoff and Morest 1991; Ryugo and Sento 1991; Spirou et al. 2005; Gomez-Nieto and Rubio 2009). High SR auditory nerve fibers have a greater tendency to make such contacts in the spherical bushy cell regions (Ryugo et al. 1996). The contribution of these synaptic inputs to bushy cell physiology is not understood and currently is not fac-tored into models of bushy cell activity.

4 Synaptic Transmission at the Endbulb

4.1 Neurotransmitters Released

Although early studies suggested aspartate as a possible neurotransmitter at auditory nerve endings (Wenthold and Gulley 1977), such a role in synaptic transmission has been downplayed due to an inconsistent association between aspartate and presynaptic vesicles at excitatory terminals. It is now widely accepted that glutamate is the neurotransmitter that mediates synaptic transmission at the endbulb synapses (Jackson et al. 1985; Martin 1985; Raman and Trussell 1992; Hackney et al. 1996; Wang et al. 1998). Once released, glutamate binds to and activates the AMPA and NMDA types of glutamate receptors to drive excitatory responses in postsynaptic cells.

The fast component of synaptic transmission at the endbulb relies principally on the AMPA receptors. AMPA receptors are a tetramer of independent subunits, named GluR1–GluR4 (Hollmann and Heinemann 1994; Shepherd and Huganir 2007). AMPARs at mature endbulb synapses are principally composed of GluR3 and GluR4 subunits (Wang et al. 1998; Ravindranathan et al. 2000), although the postsynaptic cells also express mRNA for GluR2 (Hunter et al. 1993). Although GluR2 subunits are functionally present in immature endbulbs, their contribution declines dramatically during development (Sugden et al. 2002). GluR4 subunits are associated with rapid gating of AMPARs, while the presence of GluR2 subunits in the complex leads to slower gating (Geiger et al. 1995). The abundance of GluR4 and the scarcity of GluR2 subunits at endbulb synapses thus contribute to fast EPSC kinetics. In addition, GluR subunits in mature endbulbs exist mostly in the flop isoform (rather than the flip isoform found earlier in development), and this further contributes to the rapid desensitization of AMPARs (Geiger et al. 1995). The rapid desensitization of the receptors is critical in minimizing the time window for temporal summation between independent inputs and for communicating rapid changes in afferent spike times to the postsynaptic cell (Lawrence and Trussell 2000; Gardner et al. 2001; Sugden et al. 2002). The low abundance of GluR2 subunits in endbulb AMPARs also results in a high channel permeability to calcium and a voltage-dependent block by intracellular polyamines (Bowie and Mayer 1995; Bassani et al. 2009) and is reflected by the presence of inwardly rectifying current–voltage relationships (Lawrence and Trussell 2000; Ravindranathan et al. 2000; Wang and Manis 2005).

The other principal ionotropic receptor type at endbulb synapses is the NMDAR. NMDARs at the endbulb synapses are composed of obligate but nonconducting NR1 subunits in conjunction with NR2A and/or NR2B subunits (Joelson and Schwartz 1998; Tang and Carr 2004). A low level of NMDAR conductance remains in mature endbulbs, which participates in an activity-dependent manner that can promote firing probability and improve temporal precision (Pliss et al. 2009).

4.2   Development and Maturation

Along with the morphological changes of endbulbs during development (Limb and Ryugo 2000; Ryugo et al. 2006), synaptic transmission also undergoes substantial developmental changes that result in increased synaptic efficacy, improved fidelity, and temporal precision (Bellingham et al. 1998; Brenowitz and Trussell 2001a). As demonstrated in chicks and rodents, ANFs initially contact small somatic and dendritic appendages near the base of the primary bushy cell dendrite (Jhaveri and Morest 1982; Mattox et al. 1982; Neises et al. 1982). Initial growth of the terminal entails increased contact area with the cell body through expansion over its surface and resorption of many appendages that may pull the terminal to the cell body. During this developmental process, there are four main changes in synaptic transmission. First, the postsynaptic EPSCs significantly increase in amplitude and decay more quickly (Wu and Oertel 1987; Bellingham et al. 1998; Lu et al. 2007). Second, spontaneous mEPSCs increase in frequency and quantal size (Bellingham et al. 1998; Lu et al. 2007). The increase in frequency likely reflects an increase in the number of functional synaptic release sites. Third, synaptic depression at high frequencies decreases to maintain larger steady-state EPSCs. Fourth, the input resistance and time constant of the postsynaptic cells decreases (Brenowitz and Trussell 2001a), likely in part due to an increase in the expression of potassium channels (Caminos et al. 2005; Bortone et al. 2006).

One of the most significant developmental changes associated with physiological maturation is the reciprocal up-regulation of fast AMPAR-mediated transmission and down-regulation of slower NMDAR-mediated transmission (Bellingham et al. 1998; Lu and Trussell 2007). Accompanying these changes are alterations in receptor subunit composition. In chick nucleus magnocellularis (NM), the ratio of NMDAR-versus AMPAR-mediated EPSC amplitude at endbulb synapses decreases from about 1 to almost 0 between E12 and E19. In part, this is driven by a 30-fold increase in amplitude of AMPAR-mediated EPSCs. Consequently, the amplitudes of EPSPs in mature endbulbs are well above the threshold to trigger action potentials, which establishes a secure synaptic transmission where presynaptic inputs can nearly always trigger postsynaptic spikes, even at high frequencies where there is strong synaptic depression (Brenowitz and Trussell 2001b). The EPSC kinetics also become faster during development as AMPARs mature, and EPSC decay time constants are approximately 200 µs (at the temperatures typically used in slice experiments, 31–35°C) in mature endbulbs. As discussed in the previous section, during development the endbulbs expand in size and increase in structural complexity (Limb and Ryugo 2000). Associated with these maturational changes is an increase in the number of release sites and the density of presynaptic vesicles (Ryugo et al. 2006). All of these changes suggest that there is a larger releasable vesicle pool in mature endbulbs, which likely contributes to larger EPSCs, as well as an increased frequency of spontaneous mEPSCs (Lu et al. 2007). These developmental changes appear to be similar at the endbulbs in mice (Wu and Oertel 1987; Lu et al. 2007), rats (Bellingham et al. 1998), and chickens (Brenowitz and Trussell 2001a) and have been observed in the calyx of Held (see Borst and Rusu, Chap. 5).

NMDARs play important roles in early development and provide a substrate for coincidence detection of activity (depolarization) with synaptic inputs (Perez-Otano and Ehlers 2004). In the avian NM, NMDAR-mediated EPSCs decrease by tenfold at E19 from its peak at E14. A similar decrease over the first several postnatal weeks is seen in rats (Bellingham et al. 1998) and mice (Pliss et al. 2009). At many other synapses, the NR2B subunits are gradually replaced by NR2A subunits during development (Perez-Otano and Ehlers 2004), which results in faster kinetics of NMDAR mediated EPSCs. However, the existence of such a switch is unclear at endbulb synapses, because the number of NMDARs decreases dramatically during the same age period (Joelson and Schwartz 1998). Ifenprodil, a selective antagonist of NR2B containing receptors, blocks the same portion of NMDAR mediated EPSCs at different developmental stages (Lu and Trussell 2007), which suggests that the NMDAR subunit composition does not change substantially as receptor numbers decrease.

Synaptic activity is essential for the proper development and maturation of endbulbs. Deprivation of auditory nerve inputs in deaf *shaker-2* mice show markedly reduced endbulb branching, which indicates auditory inputs help to drive synaptic growth (Limb and Ryugo 2000). In avians, endbulb synapses associated with weak inputs follow a developmental path that includes a delayed decrease of NMDAR-mediated transmission; these weak synapses appear to be selectively eliminated (Lu and Trussell 2007). By the time the synaptic maturation has reached a steady state, only one to three endbulbs converge onto each postsynaptic cell. This conclusion is, however, somewhat complicated by observations that there is a tonotopic pattern of convergence in avians (Fukui and Ohmori 2004).

4.3 Role of Receptor Desensitization

Glutamate receptors at the endbulbs, especially AMPARs, can desensitize to accumulated glutamate in the synaptic (Trussell et al. 1993; Otis and Trussell 1996). Changes in receptor desensitization are regulated during development in concert with the changes in receptor composition. The speed of desensitization gradually increases with age as GluR subunits switch from flip to flop form (Mosbacher et al. 1994; Lawrence and Trussell 2000), and the influence of desensitization decreases (Brenowitz and Trussell 2001a). Functionally, receptor desensitization was shown to accelerate the decay of evoked EPSCs, by limiting receptor availability and thus reducing the response to the low level of glutamate that remained at the cleft due to delayed clearance (Otis and Trussell 1996). The role of receptor desensitization may also be influenced by maturational changes in receptor composition or by glutamate uptake mechanisms (Turecek and Trussell 2000). In one group of studies (Isaacson and Walmsley 1996; Oleskevich et al. 2000; Lawrence et al. 2003; Yang and Xu-Friedman 2008), receptor desensitization was shown to contribute to synaptic depression at the endbulbs. Specifically, receptor desensitization is responsible for the early fast phase of synaptic depression as demonstrated by Yang and

Xu-Friedman (2008) in P15–P21 mice. However, a significant contribution of receptor desensitization was not detected in other studies (Bellingham and Walmsley 1999; Wang and Manis 2008). These contradicting results are likely due to differences in the stimulus patterns and underlying assumptions about the actions of the pharmacologic agents used. Chanda and Xu-Friedman (2010) used a combination of cyclothiazide and γ-D-glutamylglycine (γ-DGG), two drugs that manipulate receptor desensitization through different mechanisms, to provide evidence that receptor desensitization at endbulb synapses can be seen in mice from P5 to P40. Desensitization appears to contribute to synaptic depression at all ages tested, reduces spike probability in bushy cells, and increases spike latency and jitter. These results suggest that, under appropriate conditions, receptor desensitization may dynamically affect sound processing in mature endbulbs.

4.4 Kinetics

In order to precisely encode temporal information, endbulbs are specialized both anatomically and biophysically to become one of the fastest synapses in the central nervous system (Trussell 1999). Spontaneous mEPSCs at mature endbulbs can decay with a time constant of less than 200 μs at physiological temperature in mice (Wang and Manis 2005; Lu et al. 2007) and chick (Brenowitz and Trussell 2001a). The decay time constant of evoked EPSCs is a little slower than that of mEPSCs (Brenowitz and Trussell 2001a), probably because of a slight asynchrony of release at multiple synaptic sites. Nonetheless, the fast kinetics of EPSCs at endbulbs are essential to allow phase-locking to acoustic signals up to several kHz.

The brief time course of endbulb EPSPs is influenced by multiple factors, including the speed of glutamate release, diffusion, and reuptake; kinetics of glutamate receptors upon ligand binding; and the intrinsic biophysical properties of postsynaptic cells. In addition, a delayed clearance of glutamate at the cleft slows EPSC decays in chick NM endbulbs, which is relieved by AMPAR desensitization (Otis and Trussell 1996; Otis et al. 1996). The deactivation kinetics of AMPARs is fast enough not to limit the speed of EPSCs even in response to brief quantal release of glutamate (Otis and Trussell 1996). Finally, the fast kinetics of EPSPs is also determined by the intrinsic properties of bushy cells and chick NM cells, which have low input resistances and short membrane time constants (Oertel 1983; Wang and Manis 2006). All of these factors combine to minimize the effect of membrane capacitance on the EPSP time course, and thereby to minimize temporal summation in the postsynaptic cells (Rothman and Manis 2003a). The intrinsic properties are achieved by expressing a specific combination of voltage-sensitive conductances, including both low voltage-activated potassium channels (Manis and Marx 1991; Reyes et al. 1994; Rathouz and Trussell 1998; Rothman and Manis 2003a; Cao et al. 2007; Manis 2008) and hyperpolarization-activated cation channels (Schwarz and Puil 1997; Oertel et al. 2008) that are active near the resting potential and shape the voltage response to endbulb synaptic conductances.

4.5   Estimates of Release Probability

In order to produce large EPSCs for reliable transmission, the vesicle release probability at endbulb synapses is higher than at most other CNS synapses. Estimates of release probability are 0.6 in rats (2.0 mM calcium, P10–12) (Oleskevich et al. 2000), 0.5 in CBA mice (2.0 mM calcium, P11–16) (Oleskevich and Walmsley 2002), 0.65 in normal-hearing young DBA/2j mice (2.0 mM calcium, P17–25) (Wang and Manis 2005), and 0.69 in chick NM endbulbs (3.0 mM calcium, E18) (Brenowitz and Trussell 2001a). For comparison, the release probably is only about 0.1–0.2 at hippocampal synapses (Murthy et al. 1997). Measurements made in near-physiological Ca^{2+} have shown that endbulb synaptic release probability at rest is still about 0.5–0.6 (Oleskevich et al. 2000; Wang and Manis 2005; Chanda and Xu-Friedman 2010). Interestingly, this release probability is high compared to a resting release probability of ~0.2 in near physiological $(Ca)^{2+}$ (Koike-Tani et al. 2008) for the calyx synapse in the MNTB. As a major determinant of synaptic strength, the release probability at endbulb synapse is regulated during development and by the level of physiological activity. Release probability decreases with age during normal development (Brenowitz and Trussell 2001a) and declines during age-related hearing loss (Wang and Manis 2005). However, in congenitally deaf *dn/dn* mice that never hear, the release probability of endbulb synapse is 0.8, which is exceptionally high compared to 0.5 in age-matched CBA mice (Oleskevich and Walmsley 2002).

4.6   Physiological Estimates of Convergence

A number of attempts have been made to estimate the number of endbulbs converging onto single postsynaptic cells using physiological methods in rodents and birds. Assuming that different auditory nerve fibers have different firing thresholds, electrical stimulation with graded intensities or durations should successively recruit inputs to the postsynaptic cells, which would be visible as step-wise increases in EPSC or EPSP amplitudes. The number of discrete steps should correspond to the number of inputs that converge on the postsynaptic cell. In mature chick endbulbs, an average of 2.2 inputs converge onto each NM cell, down from four inputs at E14 (Jackson and Parks 1982). In mice, ~60% of bushy cells appear to receive three or fewer inputs, while the other 40% receive more than four (Cao and Oertel 2010), and interestingly this difference was correlated with other aspects of the intrinsic physiology. It is possible that this represents a partition between spherical and globular bushy cells, although in the mouse the spherical bushy cell population, defined as those cells that project to the lateral or medial superior olivary nuclei, is small and confined to the most rostral pole of the nucleus (Webster and Trune 1982). These variations could also reflect the diversity of convergence ratios onto globular bushy cells, as observed in cat (Spirou et al. 2005). However, a study in rats using this method showed that in 90% of bushy cells only a single input is detected (Isaacson and Walmsley 1995a). Two difficulties with using graded stimuli are that multiple fibers may be recruited at even the smallest experimentally feasible steps and some fibers may be cut between the stimulating and recording site. In each case, the method may fail to fully count inputs. Lu and Trussell (2007) studied spontaneous EPSCs in chick NM by applying the potassium channel blocker 4-aminopyridine to induce spontaneous (and uncorrelated) firing in the auditory nerve fibers. This method yielded an average of four inputs per NM cell at E12 and 1.8 inputs at E16. This method requires each input to have a statistically distinguishable conductance and so could also undercount inputs. In post-hatching chicks, there seems to be a gradient of convergence, such that the lowest-frequency cells receive more small inputs, based on both stimulus–response functions and anatomical labeling of the afferent fibers (Fukui and Ohmori 2004). However, no estimates of convergence for the low-frequency cells were provided.

5   Synaptic Dynamics at Endbulbs

Although information in the nervous system is carried along axons by action potentials, the representation of that information depends on the strength of synaptic conductances that are generated in postsynaptic cells by different patterns of presynaptic activity. Consequently, the temporal dynamics of synapses play a critical role in information processing and must be considered in the context of the patterns of activity that drive them (Abbott and Regehr 2004; Grande and Spain 2005). EPSPs and EPSCs at the calyx and endbulb of Held terminals show a significant reduction in amplitude during repeated high-frequency activation. This initially seemed curious, as it suggests that synaptic transmission should fade over time in the presence of strong driving stimuli. For example, the firing rates of auditory nerve fibers to acoustic stimuli can exceed 400 Hz for extended periods of time, which produces strong synaptic depression in vitro, yet these high firing rates are clearly used to encode information about the acoustic environment. Auditory nerve fibers also have a range of relatively high spontaneous rates in silence (Liberman and Oliver 1984; Taberner and Liberman 2005), from 0.1 to more than 100 Hz. Thus, the dynamics of communication between the auditory nerve and the target cochlear nucleus neurons has become an area of considerable interest. The principal approach is to use brain slice preparations to evaluate synaptic transmission at endbulb synapses during high-frequency stimulation with different stimulus patterns. There are two important caveats to the studies that will be discussed herein. First, most experiments use an artificial cerebrospinal fluid with elevated divalent concentrations relative to the "natural" cerebrospinal fluid. This increases release probability from rest but also changes the dynamics of synaptic depression and facilitation in complex ways. Second, the contribution of neuromodulatory and tonic receptor activation is altered in the slice, and this also can influence the release dynamics. Thus some caution is warranted when extrapolating these results to the in vivo situation. Studies of release dynamics have used three different stimuli: paired pulses, regular trains, and Poisson trains that mimic auditory nerve firing statistics.

5.1 Paired-Pulse Facilitation and Depression

Paired-pulse facilitation and depression are common forms of short-term synaptic plasticity (Zucker and Regehr 2002). The paired-pulse ratio measures the amplitude of the second EPSP relative to the first evoked EPSP when the postsynaptic neuron is activated by two presynaptic stimuli delivered in short succession. Although facilitation is generally thought to be due to residual Ca^{2+} accumulation in the presynaptic terminal after the first stimulus and is a function of Ca^{2+} buffer kinetics (Atluri and Regehr 1996), its time course can also be affected by partial saturation of terminal Ca^{2+} buffer (Rozov et al. 2001). Paired-pulse depression can result from one or a combination of events that would reduce the postsynaptic response. These events include postsynaptic receptor desensitization (as discussed earlier), presynaptic depletion of releasable vesicle pool, decreases in presynaptic calcium channel open probability due to calcium-dependent inactivation, intraterminal signaling cascades activated by Ca^{2+} or modulated by Na^+/K^+ exchange, or retrograde signaling from the postsynaptic terminal. Not surprisingly, the paired-pulse ratio varies with experimental conditions that alter the release probability. Mammalian endbulb synapses show strong paired-pulse depression for interpulse intervals between 3 and 200 ms (Bellingham and Walmsley 1999; Oleskevich et al. 2000; Wang and Manis 2005; Chanda and Xu-Friedman 2010). The lack of paired-pulse facilitation at shorter intervals indicates that the Ca^{2+} buffering capability in the endbulb terminal is high and may reflect underlying facilitation that is masked by receptor desensitization. The endbulb paired-pulse ratio has been shown to correlate with the presynaptic release probability when release probabilities are altered by changing external Ca^{2+} in the bath (Oleskevich et al. 2000). However, the paired-pulse ratio is variable from cell to cell (or synapse to synapse) under identical experimental conditions, indicating that resting release probability also probably varies from cell to cell. When the release probability is high, the correlation between paired-pulse amplitudes and release probability breaks down, and postsynaptic mechanisms can make a significant contribution to the paired-pulse depression, likely through receptor saturation and desensitization (discussed later). Since AMPA receptors recover rapidly from desensitization (~10 ms) (Trussell and Fischbach 1989; Chanda and Xu-Friedman 2010), paired-pulse depression at longer intervals is probably not exclusively due to desensitization. Another factor that can contribute to the endbulb paired-pulse depression is a transient depression of release, which has been shown to depend on the level of presynaptic Ca^{2+} entry (Bellingham and Walmsley 1999). This reduction of presynaptic Ca^{2+} entry for the second pulse may be due to Ca^{2+}-dependent Ca^{2+} channel inactivation at the endbulb synaptic terminal (Wang et al. 2010), as suggested for the calyx of Held in the MNTB (Forsythe et al. 1998; Xu and Wu 2005).

5.2   High-Frequency Trains

In vitro, trains of high-frequency ($\geq$100 Hz) stimulation of endbulb synapses result in pronounced rate-dependent EPSC depression (Oleskevich and Walmsley 2002; Wang and Manis 2008; Yang and Xu-Friedman 2009; Wang et al. 2010). Altering the initial release probability by changing external Ca^{2+} changes the rate of synaptic depression as well as the final steady-state depression level. While receptor desensitization may contribute to the initial phase of the fast depression during high-frequency stimulus trains, steady-state synaptic depression with and without the desensitization blocker cyclothiazide reaches the same level for short trains of 100–200 ms (Chanda and Xu-Friedman 2010). Therefore, the depletion of presynaptic vesicle pools and the rate of pool replenishment, rather than postsynaptic receptor desensitization, probably determine the final level of synaptic depression during high-frequency trains. Synaptic facilitation is only observed at a higher stimulation rate (300 Hz) and when receptor desensitization and saturation are blocked by a low-affinity competitive AMPA receptor blocker γ-DGG (Wang and Manis 2008; Chanda and Xu-Friedman 2010). Therefore, under normal operating conditions, receptor desensitization outweighs presynaptic facilitation, so that high-frequency trains do not result in synaptic facilitation in the endbulb synapse. The endbulb synapse also exhibits a second, additional, slow depression when stimulation trains are prolonged beyond 200 ms (Wang and Manis 2008). Similar additional slow depression has been observed in the calyx of Held in the MNTB (Wong et al. 2003) and is attributed to the effect of presynaptic terminal calcium channel inhibition mediated by presynaptic glutamate autoreceptors (mGluRs) involving G-protein activation (Billups et al. 2005; Takago et al. 2005). However, these modulatory pathways have not yet been demonstrated in the endbulb of Held synapses.

Compared to endbulb synapses from younger animals (Oleskevich and Walmsley 2002; Yang and Xu-Friedman 2008), the amount of synaptic depression in more mature endbulbs is smaller (~40% versus 80–90% for short trains at 100 Hz) (Wang and Manis 2008). This could be explained by a greater role for postsynaptic receptor desensitization at immature synapses (Yang and Xu-Friedman 2008), a difference in initial release probability between ages (Bellingham et al. 1998), a difference in intrinsic Ca^{2+} buffering capability and in Ca^{2+} sensitivity for vesicle release as in the MNTB (Felmy and Schneggenburger 2004; Fedchyshyn and Wang 2005), or a difference in Ca^{2+}-dependent recovery from depression (Wang and Manis 2008; Yang and Xu-Friedman 2008).

Even though the endbulb synapse is generally considered a strong synapse, it often fails to drive action potentials during high rates of activity in vitro (Wang and Manis 2006; Yang and Xu-Friedman 2009; Chanda and Xu-Friedman 2010) and in vivo (Kopp-Scheinpflug et al. 2002, 2003; Englitz et al. 2009). Failure to elicit action potentials can even occur for lower-frequency stimuli (25 Hz) when tested with minimal stimulation that putatively activates a single input (Wang et al. 2010). Thus, under most conditions, increasing depression leads to lower probability of bushy neuron spiking (Wang and Manis 2008; Yang and Xu-Friedman 2008). However, a lower probability of firing can result in improved information transfer, because depression can also indirectly enhance the postsynaptic neuron spike time precision by suppressing highly active inputs that do not carry phase-locked information (Yang and Xu-Friedman 2009).

5.3   Poisson Trains

Tone-evoked, nonphase-locked, and spontaneous spike trains recorded in vivo auditory nerve fibers in post-hearing animals have a Poisson-like interspike interval distribution (Kiang 1965). The spontaneous activity is generated by stochastic spontaneous neurotransmitter release at the ribbon synapse between the inner hair cell and the auditory nerve fiber (Sewell 1984) and is influenced by the mechano-electric transduction channels in the stereocilia of the hair cell (Farris et al. 2006). AN spike trains, when not strongly phase-locked, have irregular interspike intervals that can be described as a Poisson process with a dead time due to the refractory period (Young and Barta 1986). When the spike trains have a mean rate of 100 Hz, half of the intervals are shorter than 10 Hz and can drive the synapse at very high rates for short periods, leading to periods of strong synaptic depression even when the mean rate would not be expected to produce much depression. Synaptic depression and recovery from depression operate on different time scales and may have multiple components. Consequently, the pattern of activity creates a complex set of fluctuations in synaptic strength that depends on the history of activity, and it may engage components that include slow vesicle replenishment. Because the transmitter release process is not linear in the time domain, it is not clear whether the response to regular trains is predictive of responses under different stimulus patterns. When the endbulb synapse is stimulated with an in vivo–like Poisson train at average rates of 100 and 200 Hz (for 500 ms), synaptic depression at the end of Poisson trains appears to match the depression levels at the end of regular trains of the same rates (Wang and Manis 2008). However, one prominent, and not unexpected, difference between Poisson and regular trains is the larger fluctuation of individual EPSC amplitudes in Poisson trains than in regular trains. Similar observations have been made in the calyx synapse in the MNTB (Hermann et al. 2007). The fluctuations arise from the irregular interspike interval in the Poisson train, which in turn determines the amount and rate of Ca^{2+}-dependent synaptic recovery during the train. Spontaneous firing rates of auditory nerve fibers in mammals vary from less than 1 spike/s to more than 100 spikes/s. Sustained presynaptic spontaneous activity appears to be poised to depress the endbulb synapses in ways similar to what has been observed in the MNTB calyx (Hermann et al. 2007). Spontaneous activity at endbulbs in vivo does not always drive the postsynaptic cell to spike due to the presence of inhibitory inputs (Kopp-Scheinpflug et al. 2002; Englitz et al. 2009). In the MNTB, rate-dependent depression of release from calyceal terminals during spontaneous activity may not be an important factor because release probability is low in vivo (Lorteije et al. 2009). However, the high resting release probability of endbulb synapses, compared to MNTB calyces under similar conditions, suggests that spontaneous activity should still lead to depression at the endbulb, even under in vivo conditions. Indeed, even relatively low rates of activity produce depression at the endbulbs (Wang et al. 2010), suggesting that the endbulbs of medium and high SR fibers may operate in a chronically depressed state. However, it is still unclear to what extent spontaneous activity modulates the reliability and precision of endbulb synaptic transmission.

5.4   Recovery from Depression

Auditory nerve fibers can have sustained sound-driven discharge rates up to 400 Hz (Sachs and Abbas 1974; Liberman 1978; Winter et al. 1990; Ohlemiller et al. 1991; Taberner and Liberman 2005). Moreover, the spontaneous firing rates of single ANFs can exceed 100 spikes/s. In vitro and in vivo recordings have revealed that the response entrainment (measured as the ratio of spikes generated per presynaptic stimulus) of bushy cells to ANF stimulation falls below 50% when trains of shock stimuli are delivered at 200 Hz and higher (Kopp-Scheinpflug et al. 2002; Wang and Manis 2006; Englitz et al. 2009). Thus, recovery from severe synaptic depression during high rates of ANF activity can significantly shape the functional relationship between the ANF and the bushy neuron.

Although high ANF firing rates cause greater depression at the endbulb synapse, an activity-dependent increase in terminal Ca^{2+} accumulation appears to facilitate recovery from this depression (Wang and Manis 2008; Yang and Xu-Friedman 2008). This phenomenon was first described in the MNTB calyx of Held (Wang and Kaczmarek 1998). Yang and Xu-Friedman (2008) showed that the Ca^{2+}-dependent recovery is critical to counterbalance the effect of predominately depletion-based synaptic depression at the endbulb synapse. A simple vesicle depletion-based model can describe the steady-state endbulb synaptic depression observed experimentally if a Ca^{2+}-facilitated replenishment of releasable vesicle pool is included (Yang and Xu-Friedman 2008). It appears that the fast recovery mechanism requires intracellular Ca^{2+} accumulation to reach a certain "threshold" that is achieved only at higher firing rates, and the fast recovery time constant seems to be "fixed" under different intracellular Ca^{2+} conditions, for example, 200 versus 300 Hz (Wang and Manis 2008). This is consistent with the observation that a Ca^{2+}/calmodulin signaling mechanism is likely involved in the fast recovery in both the endbulb (Wang and Manis 2008) and the MNTB calyx terminals (Sakaba and Neher 2001).

5.5   Recovery with Poisson-Like Stimulus Trains

Poisson spike trains in vivo may not have same recovery dynamics as seen in regularly spaced spike train in vitro, because during a Poisson spike train, periods of brief bursts of activity would have interspike intervals that are much shorter than the mean interval, along with periods of slower firing that would result in intervals longer than the mean interval. If the high-frequency bursts of spikes drive a nonlinear process that decays slowly relative to the mean interspike interval, then the net accumulation of Ca^{2+} during a Poisson spike train and a regular train may be different. Thus the rate of recovery from depression may depend on moment-to-moment firing patterns of a spike train. Endbulb synapses do show a more rapid recovery following a Poisson train at 100 Hz compared to a regular train at the same frequency (Wang and Manis 2008). This suggests that the Poisson train results in a higher average terminal $(Ca^{2+})_i$ than the regular train. Similarly, less synaptic depression is evident with 200-Hz Poisson trains, perhaps because rapid recovery also occurs during the lulls in activity.

5.6 Presynaptic Modulation

Presynaptic modulation of release can have significant consequences for the functional synaptic relationship between neurons. In the MNTB calyx of Held, presynaptic modulation may involve activation of presynaptic autoreceptors such as mGluR receptors (von Gersdorff et al. 1997), AMPA receptors (Takago et al. 2005), glycine receptors (Turecek and Trussell 2001), and $GABA_B$ receptors (Caspary et al. 1984; Brenowitz et al. 1998). The targets of many of these autoreceptors are very likely the voltage-gated Ca^{2+} channels at the endbulb terminal. However, modulation at the endbulb synapse has not been extensively studied. Endbulb presynaptic release probability is modulated in an activity-dependent fashion in bushy neurons of the nucleus magnocellularis in chickens (Brenowitz et al. 1998; Brenowitz and Trussell 2001a). Activation of presynaptic $GABA_B$ receptors greatly reduces the presynaptic release probability, which in turn reduces the postsynaptic EPCS amplitude. Paradoxically, the response of bushy neurons in nucleus magnocellularis to high-frequency train stimulation is enhanced in the presence of $GABA_B$ receptor agonist baclofen (Brenowitz and Trussell 2001a, b). This is attributed to reduced transmitter depletion, as well as to activation of $GABA_B$ receptors on the relief of postsynaptic AMPA receptor desensitization.

Repetitive stimulation of the auditory nerve can cause a $GABA_B$-dependent inhibition of release from glycinergic terminals on bushy neurons (Lim et al. 2000), suggesting that transmitter spillover might also modulate other terminals, such as the endbulbs. Activation of $GABA_B$ receptors near VCN neurons in vivo depresses responses to sound (Caspary et al. 1984) throughout the duration of a stimulus. Even when presynaptic $GABA_B$ receptors are activated, some rate-dependent synaptic depression may still occur.

Are there other factors that are released by the endbulb synapses that might influence the activity or operation of their postsynaptic targets? ATP is released along with glutamate at presynaptic terminals (Abbracchio et al. 2009) and can trigger either ionotropic (P2X) or metabotropic (P2Y) receptors on postsynaptic cells. Purinergic receptors of the P2X2 and P2Y1 classes are present in bushy cells of the gerbil AVCN, and their pharmacologic activation modulates the intrinsic excitability of the bushy cells and induces intracellular increases in Ca^{2+} (Milenkovic et al. 2009). It is not yet clear what role ATP release from the endbulb has in activating these receptors. However, this observation raises the interesting issue of whether there are other kinds of interactions between the endbulb and its target cells that are independent of the activation of postsynaptic ionotropic receptors and what role such interactions might play in auditory information processing.

6 Trophic Importance of Endbulb Synapses

The large size of the endbulb and its perisomatic location suggest that activity in the terminal might have an unusually important influence on the metabolic state of the postsynaptic cell. Removal of the auditory nerve before the onset of hearing leads to considerable cell death in both avian and mammalian cochlear nuclei (Parks 1979; Webster 1983; Hashisaki and Rubel 1989), especially of those cells postsynaptic to endbulb synapses. In a paradoxical manner, the removal of eighth nerve activity (in the form of evoked transmitter release) in avians led to an increase in postsynaptic calcium (Zirpel and Rubel 1996; Zirpel et al. 1998), which then engaged cell-death signaling pathways. This could be prevented by stimulation of the auditory nerve in brain slices (Zirpel and Rubel 1996), but not by antidromic activation of the post-synaptic cells, suggesting that an anterograde mechanism was involved. Activation of metabotropic glutamate receptors via glutamate release from the endbulb depressed calcium influx through voltage-gated calcium channels (Lachica et al. 1995; Lu and Rubel 2005), and activation of these receptors prevented release of calcium from intracellular stores (Kato and Rubel 1999). These observations, while carried out in young chicks, raise a number of interesting questions regarding the role of G-protein coupled signaling systems at this synapse. So far, no comparable studies appear to have been carried out in mammals; in the VCN the only physio-logical study of metabotropic glutamate receptors examined cells posterior to the nerve root in gerbils, and therefore probably did not examine modulation of endbulb synapses (Sanes et al. 1998).

7 Summary and Issues

Projections of auditory-nerve fibers onto spherical and globular bushy cells via end-bulb and modified endbulb terminals are among the best characterized, in terms of the fundamentals of their structure and function, in the CNS. Nonetheless, critical questions remain regarding explicit targeting and distinctions of particular SR-related auditory-nerve terminals onto these two types of bushy cells and how such spatio-topic mapping collectively supports auditory processing. In particular, the role of distinct terminal convergence ratios for spherical and globular bushy cells remains to be explored. Anatomical estimates of convergence vary widely, and physiological estimates (primarily in rodents) often are much smaller than expected from structural analyses. Insight into the function of these neural contacts has benefited from direct structure-function associations for single auditory-nerve fibers and a good understanding of the biophysical properties of bushy cells. A next experimental phase that incorporates simultaneous monitoring of all converging auditory-nerve fibers with a thorough catalog of their synaptic parameters via large-volume ultrastructural reconstruction could provide a larger perspective on auditory processing through these relatively secure neural connections, in dynamic acoustic environments.

Convergence, with respect to auditory nerve fibers of different SRs, onto individual cells, is still a major unsolved problem. It has been suggested that synaptic depression might allow neurons receiving a mixture of high- and low-rate inputs (such as from high- and low-spontaneous-rate ANFs) to "listen" to the low-rate inputs, because the high-rate inputs would depress more than the low-rate inputs, and thus the balance of synaptic conductances could favor the low-rate inputs (Abbott and Regehr 2004). Such a mechanism could be envisioned in the convergence of high- and low-spontaneous-rate auditory nerve fiber endbulbs onto single postsynaptic cells, where it might extend the dynamic range of sound intensity representation in a single cell. In addition, because synaptic depression itself can be regulated by presynaptic modulation (see the earlier discussion), the relative balance of such selective listening could be dynamically regulated, either in an activity-dependent manner or from descending inputs. Interestingly, converging endbulbs (termed "sibling inputs") onto a single target cell are more similar in paired-pulse depression and tetanic depression than expected from the overall distribution of these measures at single sites (Yang and Xu-Friedman 2009). However, it is unclear whether this is a function of presynaptic history of activity (e.g., dominated by spontaneous rate or a related variable) or shared retrograde signaling from the postsynaptic cell to the endbulb. In addition, structural evidence so far suggests that endbulb convergence tends to be from fibers of similar spontaneous rate (Ryugo and Sento 1991).

Models that have investigated convergence and temporal coding in the VCN so far have all used synaptic conductances that are linear and time-invariant (e.g., show no synaptic release dynamics) (Carney 1992; Li and Young 1993; Rothman et al. 1993; Kuhlmann et al. 2002; Rothman and Manis 2003b). Because the synaptic transfer function is nonlinear and shows strong time dependent effects, the contribution of time- and rate-dependent synaptic depression on information transfer from the AN to the VCN needs to be evaluated more critically. Experimental conditions used in brain-slice experiments may not match those in vivo however (Rothman et al. 1993; Rothman and Manis 2003b; Lorteije et al. 2009), and so experimental data that can be used to develop acceptable kinetic models of release and recovery are limited. It may be that in vivo release at endbulb synapses is best represented by experiments in $1–1.5$ mM Ca^{2+}, where there is less depression than under conditions used in most available data sets. Nonetheless, depression is still present and needs to be incorporated.

The possibility of multiquantal release at single sites at the mammalian endbulb synapse has also been suggested recently (Chanda and Xu-Friedman 2010). Studies at the hair-cell auditory nerve synapse have convincingly shown the presence of multiquantal release at the first synapse in the auditory pathway (Glowatzki and Fuchs 2002; Li et al. 2009). Analysis of the mEPSC amplitudes against decay time constant, however (Wang and Manis 2005), suggest that multiquantal events do not occur frequently during spontaneous release at endbulb synapses in spite of the large mEPSCs. Endbulbs also do not have the ribbon structure typical of hair cell synapses. A proper analysis will require recording from the endbulb terminal, which has so far only been accomplished in birds (Sivaramakrishnan and Laurent 1995). In that study, however, neither agatoxin-IVA nor conotoxin GVIA had a large effect on the mEPSC amplitude distribution, unlike calcium channel block at hair cell synapses (Li et al. 2009), which seems to lessen the possibility of multiquantal release at endbulbs.

Finally, the endbulb appears to be the potential target of a number of neuromodulatory systems whose importance is not yet known. For example, endocannabinoid receptors have been suggested by immunocytochemistry to be present in neurons in the VCN (Zheng et al. 2007; Baek et al. 2008), although the cellular and subcellular localization of these receptors has not been rigorously distinguished. Bushy cells contain both NOS and scGMP (Burette et al. 2001), while auditory nerve fibers contain NOS, suggesting that activation of these signaling systems may also play a role in early auditory processing. The presence of NT3 in presumptive ANFs and TrkC in bushy cells (Feng et al. 2009) in adult mice and of NT3 and BDNF in gerbils (Tierney et al. 2001) also indicates a potential role in regulating transmission at these terminals. It is likely that these ligands and receptors all play important roles in the adaptation of endbulb synapses to changing acoustic environments and peripheral hearing loss (Suneja et al. 2005), and are awaiting exploration of their functional roles.

Acknowledgments This work was supported by NIDCD grant R01DC004551 to PBM (PBM, RX); a grant from the Deafness Research Foundation to RX; NIDCD grant R03DC008190 to YW; grants P20 RR015774 to the Sensory Neuroscience Research Center at WVU and R01 DC007695 to GAS; and F32 DC010546 to GSM.

References

Abbott, L. F., & Regehr, W. G. (2004). Synaptic computation. *Nature, 431*(7010), 796–803. doi: 10.1038/nature03010.

Abbracchio, M. P., Burnstock, G., Verkhratsky, A., & Zimmermann, H. (2009). Purinergic signalling in the nervous system: An overview. *Trends in Neurosciences, 32*(1), 19–29. doi: 10.1016/j.tins.2008.10.001.

Adams, J. C. (1986). Neuronal morphology in the human cochlear nucleus. *Archives of Otolaryngology Head Neck Surgery, 112*(12), 1253–1261.

Atluri, P. P., & Regehr, W. G. (1996). Determinants of the time course of facilitation at the granule cell to Purkinje cell synapse. *Journal of Neuroscience, 16*(18), 5661–5671.

Baek, J. H., Zheng, Y., Darlington, C. L., & Smith, P. F. (2008). Cannabinoid CB2 receptor expression in the rat brainstem cochlear and vestibular nuclei. *Acta Otolaryngologica, 128*(9), 961–967.

Bassani, S., Valnegri, P., Beretta, F., & Passafaro, M. (2009). The GLUR2 subunit of AMPA receptors: Synaptic role. *Neuroscience, 158*(1), 55–61.

Bellingham, M. C., Lim, R., & Walmsley, B. (1998). Developmental changes in EPSC quantal size and quantal content at a central glutamatergic synapse in rat. *Journal of Physiology, 511*(Pt. 3), 861–869.

Bellingham, M. C., & Walmsley, B. (1999). A novel presynaptic inhibitory mechanism underlies paired pulse depression at a fast central synapse. *Neuron, 23*(1), 159–170.

Billups, B., & Forsythe, I. D. (2002). Presynaptic mitochondrial calcium sequestration influences transmission at mammalian central synapses. *Journal of Neuroscience, 22*(14), 5840–5847.

Billups, B., Graham, B. P., Wong, A. Y., & Forsythe, I. D. (2005). Unmasking group III metabotropic glutamate autoreceptor function at excitatory synapses in the rat CNS. *Journal of Physiology, 565*(Pt. 3), 885–896. doi: 10.1113/jphysiol.2005.086736.

Blackburn, C. C., & Sachs, M. B. (1989). Classification of unit types in the anteroventral cochlear nucleus: PST histograms and regularity analysis. *Journal of Neurophysiology, 62*(6), 1303–1329.

Bortone, D. S., Mitchell, K., & Manis, P. B. (2006). Developmental time course of potassium channel expression in the rat cochlear nucleus. *Hearing Research, 211*(1–2), 114–125. doi: 10.1016/j.heares.2005.10.012.

Bourk, T. R. (1976). Electrical responses of neural units in the anteroventral cochlear nucleus of the cat. Ph.D. dissertation, Massachussetts Institute of Technology.

Bowie, D., & Mayer, M. L. (1995). Inward rectification of both AMPA and kainate subtype glutamate receptors generated by polyamine-mediated ion channel block. *Neuron, 15*(2), 453–462.

Brawer, J. R., & Morest, D. K. (1975). Relations between auditory nerve endings and cell types in the cat's anteroventral cochlear nucleus seen with the Golgi method and Nomarski optics. *Journal of Comparative Neurology, 160*(4), 491–506. doi: 10.1002/cne.901600406.

Brawer, J. R., Morest, D. K., & Kane, E. C. (1974). The neuronal architecture of the cochlear nucleus of the cat. *Journal of Comparative Neurology, 155*(3), 251–300. doi: 10.1002/cne.901550302.

Brenowitz, S., David, J., & Trussell, L. (1998). Enhancement of synaptic efficacy by presynaptic GABA(B) receptors. *Neuron, 20*(1), 135–141.

Brenowitz, S., & Trussell, L. O. (2001a). Maturation of synaptic transmission at end-bulb synapses of the cochlear nucleus. *Journal of Neuroscience, 21*(23), 9487–9498.

Brenowitz, S., & Trussell, L. O. (2001b). Minimizing synaptic depression by control of release probability. *Journal of Neuroscience, 21*(6), 1857–1867.

Browner, R. H., & Marbey, D. (1988). The nucleus magnocellularis in the red-eared turtle, Chrysemys scripta elegans: Eighth nerve endings and neuronal types. *Hearing Research, 33*(3), 257–271.

Burette, A., Petrusz, P., Schmidt, H. H., & Weinberg, R. J. (2001). Immunohistochemical localization of nitric oxide synthase and soluble guanylyl cyclase in the ventral cochlear nucleus of the rat. *Journal of Comparative Neurology, 431*(1), 1–10. doi: 10.1002/1096-9861(20010226).

Caminos, E., Vale, C., Lujan, R., Martinez-Galan, J. R., & Juiz, J. M. (2005). Developmental regulation and adult maintenance of potassium channel proteins (Kv 1.1 and Kv 1.2) in the cochlear nucleus of the rat. *Brain Research, 1056*(2), 118–131. doi: 10.1016/j.brainres.2005.07.031.

Cant, N. B. (1991). Projections to the lateral and medial superior olivary nuclei from the spherical and globular bushy cells of the anteroventral cochlear nucleus. In Altschuler, R. A., Bobbin, R. P., Clopton, B. M. & Hoffman, D. W. (Eds.), *Neurobiology of hearing: The central auditory system* (pp. 99–119). New York: Raven Press.

Cant, N. B., & Morest, D. K. (1979). The bushy cells in the anteroventral cochlear nucleus of the cat. A study with the electron microscope. *Neuroscience, 4*(12), 1925–1945.

Cao, X. J., & Oertel, D. (2010). Auditory nerve fibers excite targets through synapses that vary in convergence, strength and short-term plasticity. *Journal of Neurophysiology.104*(5), 2308–2320. doi: 10.1152/jn.00451.2010.

Cao, X. J., Shatadal, S., & Oertel, D. (2007). Voltage-sensitive conductances of bushy cells of the mammalian ventral cochlear nucleus. *Journal of Neurophysiology, 97*(6), 3961–3975. doi: 10.1152/jn.00052.2007.

Carney, L. H. (1990). Sensitivities of cells in anteroventral cochlear nucleus of cat to spatiotemporal discharge patterns across primary afferents. *Journal of Neurophysiology, 64*(2), 437–456.

Carney, L. H. (1992). Modelling the sensitivity of cells in the anteroventral cochlear nucleus to spatiotemporal discharge patterns. *Philosophical Transactions of the Royal Society of London. Series B, Biological Sciences, 336*(1278), 403–406. doi: 10.1098/rstb.1992.0075.

Carr, C. E., & Boudreau, R. E. (1991). Central projections of auditory nerve fibers in the barn owl. *Journal of Comparative Neurology, 314*(2), 306–318. doi: 10.1002/cne.903140208.

Caspary, D. M., Backoff, P. M., Finlayson, P. G., & Palombi, P. S. (1994). Inhibitory inputs modulate discharge rate within frequency receptive fields of anteroventral cochlear nucleus neurons. *Journal of Neurophysiology, 72*(5), 2124–2133.

Caspary, D. M., Rybak, L. P., & Faingold, C. L. (1984). Baclofen reduces tone-evoked activity of cochlear nucleus neurons. *Hearing Research, 13*(2), 113–122.

Chanda, S., & Xu-Friedman, M. A. (2010). A low-affinity antagonist reveals saturation and desensitization in mature synapses in the auditory brainstem. *Journal of Neurophysiology, 103*, 1915–1926. doi: 10.1152/jn.00751.2009.

Costalupes, J. A., Young, E. D., & Gibson, D. J. (1984). Effects of continuous noise backgrounds on rate response of auditory nerve fibers in cat. *Journal of Neurophysiology, 51*(6), 1326–1344.

Englitz, B., Tolnai, S., Typlt, M., Jost, J., & Rubsamen, R. (2009). Reliability of synaptic transmission at the synapses of Held in vivo under acoustic stimulation. *PLoS One, 4*(10), e7014. doi: 10.1371/journal.pone.0007014.

Farris, H. E., Wells, G. B., & Ricci, A. J. (2006). Steady-state adaptation of mechanotransduction modulates the resting potential of auditory hair cells, providing an assay for endolymph [Ca2+]. *Journal of Neuroscience, 26*(48), 12526–12536. doi: 10.1523/JNEUROSCI. 3569-06.2006.

Fedchyshyn, M. J., & Wang, L. Y. (2005). Developmental transformation of the release modality at the calyx of Held synapse. *Journal of Neuroscience, 25*(16), 4131–4140. doi: 10.1523/ JNEUROSCI.0350-05.2005.

Fekete, D. M., Rouiller, E. M., Liberman, M. C., & Ryugo, D. K. (1984). The central projections of intracellularly labeled auditory nerve fibers in cats. *Journal of Comparative Neurology, 229*(3), 432–450. doi: 10.1002/cne.902290311.

Felmy, F., & Schneggenburger, R. (2004). Developmental expression of the Ca^{2+}-binding proteins calretinin and parvalbumin at the calyx of Held of rats and mice. *European Journal of Neuroscience, 20*(6), 1473–1482. doi: 10.1111/j.1460-9568.2004.03604.x.

Feng, A. S., & Lin, W. Y. (1996). Neuronal architecture of the dorsal nucleus (cochlear nucleus) of the frog, Rana pipiens pipiens. *Journal of Comparative Neurology, 366*(2), 320–334. doi: 10.1002/(SICI)1096-9861(19960304)366:2.

Feng, J., Bendiske, J., & Morest, D. K. (2009). Postnatal development of NT3 and TrkC in mouse ventral cochlear nucleus. *Journal of Neuroscience Research, 88*(1), 86–94.

Forsythe, I. D., Tsujimoto, T., Barnes-Davies, M., Cuttle, M. F., & Takahashi, T. (1998). Inactivation of presynaptic calcium current contributes to synaptic depression at a fast central synapse. *Neuron, 20*(4), 797–807.

Fukui, I., & Ohmori, H. (2004). Tonotopic gradients of membrane and synaptic properties for neurons of the chicken nucleus magnocellularis. *Journal of Neuroscience, 24*(34), 7514.

Gardner, S. M., Trussell, L. O., & Oertel, D. (2001). Correlation of AMPA receptor subunit composition with synaptic input in the mammalian cochlear nuclei. *Journal of Neuroscience, 21*(18), 7428–7437.

Geiger, J. R., Melcher, T., Koh, D. S., Sakmann, B., Seeburg, P. H., & Jonas, P. (1995). Relative abundance of subunit mRNAs determines gating and Ca2+ permeability of AMPA receptors in principal neurons and interneurons in rat CNS. *Neuron, 15*(1), 193–204.

Glowatzki, E., & Fuchs, P. A. (2002). Transmitter release at the hair cell ribbon synapse. *Nature Neuroscience, 5*(2), 147–154. doi: 10.1038/nn796.

Goldberg, J. M., & Brownell, W. E. (1973). Discharge characteristics of neurons in anteroventral and dorsal cochlear nuclei of cat. *Brain Research, 64*, 35–54.

Gomez-Nieto, R., & Rubio, M. E. (2009). A bushy cell network in the rat ventral cochlear nucleus. *Journal of Comparative Neurology, 516*(4), 241–263. doi: 10.1002/cne.22139.

Grande, L. A., & Spain, W. J. (2005). Synaptic depression as a timing device. *Physiology (Bethesda), 20*, 201–210. doi: 10.1152/physiol.00006.2005.

Guinan, J. J. Jr., & Li, R. Y. (1990). Signal processing in brainstem auditory neurons which receive giant endings (calyces of Held) in the medial nucleus of the trapezoid body of the cat. *Hearing Research, 49*(1–3), 321–334.

Hackney, C. M., Osen, K. K., Ottersen, O. P., Storm-Mathisen, J., & Manjaly, G. (1996). Immunocytochemical evidence that glutamate is a neurotransmitter in the cochlear nerve: A quantitative study in the guinea-pig anteroventral cochlear nucleus. *European Journal of Neuroscience, 8*(1), 79–91.

Harrison, J. M., & Irving, R. (1965). The anterior ventral cochlear nucleus. *Journal of Comparative Neurology, 124*, 15–41.

Harrison, J. M., & Warr, W. B. (1962). A study of the cochlear nuclei and ascending auditory pathways of the medulla. *Journal of Comparative Neurology, 119*, 341–379.

Hashisaki, G., & Rubel, E. (1989). Age-related effects of unilateral cochlea removal on anteroventral cochlear nucleus in developing gerbils. *Journal of Comparative Neurology, 283*, 465–473.

Held, H. (1891). Die centralen Bahnen des Nervus acusticus bei der Katze. *Archives of Anatomy Physiology Anatomy Abtil, 15*, 271–291.

Hermann, J., Pecka, M., von Gersdorff, H., Grothe, B., & Klug, A. (2007). Synaptic transmission at the calyx of Held under in vivo like activity levels. *Journal of Neurophysiology, 98*(2), 807–820. doi: 10.1152/jn.00355.2007.

Hollmann, M., & Heinemann, S. (1994). Cloned glutamate receptors. *Annual Review of Neuroscience, 17*(1), 31–108.

Hunter, C., Petralia, R. S., Vu, T., & Wenthold, R. J. (1993). Expression of AMPA-selective glutamate receptor subunits in morphologically defined neurons of the mammalian cochlear nucleus. *Journal of Neuroscience, 13*(5), 1932–1946.

Isaacson, J. S., & Walmsley, B. (1995a). Counting quanta: Direct measurements of transmitter release at a central synapse. *Neuron, 15*(4), 875–884.

Isaacson, J. S., & Walmsley, B. (1995b). Receptors underlying excitatory synaptic transmission in slices of the rat anteroventral cochlear nucleus. *Journal of Neurophysiology, 73*(3), 964–973.

Isaacson, J. S., & Walmsley, B. (1996). Amplitude and time course of spontaneous and evoked excitatory postsynaptic currents in bushy cells of the anteroventral cochlear nucleus. *Journal of Neurophysiology, 76*(3), 1566–1571.

Jackson, H., Nemeth, E. F., & Parks, T. N. (1985). Non-N-methyl-D-aspartate receptors mediating synaptic transmission in the avian cochlear nucleus: Effects of kynurenic acid, dipicolinic acid and streptomycin. *Neuroscience, 16*(1), 171–179.

Jackson, H., & Parks, T. N. (1982). Functional synapse elimination in the developing avian cochlear nucleus with simultaneous reduction in cochlear nerve axon branching. *Journal of Neuroscience, 2*(12), 1736–1743.

Jhaveri, S., & Morest, D. K. (1982). Sequential alterations of neuronal architecture in nucleus magnocellularis of the developing chicken: An electron microscope study. *Neuroscience, 7*(4), 855–870.

Joelson, D., & Schwartz, I. R. (1998). Development of N-methyl-D-aspartate receptor subunit immunoreactivity in the neonatal gerbil cochlear nucleus. *Microscopy Research and Technique, 41*(3), 246–262. doi: 10.1002/(SICI)1097-0029(19980501)41:3.

Joris, P. X., Carney, L. H., Smith, P. H., & Yin, T. C. (1994). Enhancement of neural synchronization in the anteroventral cochlear nucleus. I. Responses to tones at the characteristic frequency. *Journal of Neurophysiology, 71*(3), 1022–1036.

Kato, B., & Rubel, E. (1999). Glutamate regulates IP3-type and CICR stores in the avian cochlear nucleus. *Journal of Neurophysiology, 81*(4), 1587.

Kiang, N. Y.-S. (1965). *Discharge patterns of single fibers in the cat's auditory nerve*. Cambridge: MIT Press.

Kiang, N. Y.-S., Pfeiffer, R. R., Warr, W. B., & Backus, A. S. (1965). Stimulus coding in the cochlear nucleus. *Annals of Otology Rhinology and Laryngology, 74*, 463–485.

Koike-Tani, M., Kanda, T., Saitoh, N., Yamashita, T., & Takahashi, T. (2008). Involvement of AMPA receptor desensitization in short-term synaptic depression at the calyx of Held in developing rats. *Journal of Physiology, 586*(9), 2263–2275. doi: 10.1113/jphysiol.2007.142547.

Kolliker, A. (1896). *Handbuch der Gewebelehre des Menschem Bd. 2.* Leipzig: Wilhelm Engelman.

Kopp-Scheinpflug, C., Dehmel, S., Dorrscheidt, G. J., & Rubsamen, R. (2002). Interaction of excitation and inhibition in anteroventral cochlear nucleus neurons that receive large endbulb synaptic endings. *Journal of Neuroscience, 22*(24), 11004–1018.

Kopp-Scheinpflug, C., Lippe, W. R., Dorrscheidt, G. J., & Rubsamen, R. (2003). The medial nucleus of the trapezoid body in the gerbil is more than a relay: Comparison of pre- and postsynaptic activity. *JARO: Journal of the Association for Research in Otolaryngology, 4*(1), 1–23. doi: 10.1007/s10162-002-2010-5.

Koppl, C. (1994). Auditory nerve terminals in the cochlear nucleus magnocellularis: Differences between low and high frequencies. *Journal of Comparative Neurology, 339*(3), 438–446. doi: 10.1002/cne.903390310.

Kuhlmann, L., Burkitt, A. N., Paolini, A., & Clark, G. M. (2002). Summation of spatiotemporal input patterns in leaky integrate-and-fire neurons: Application to neurons in the cochlear nucleus receiving converging auditory nerve fiber input. *Journal of Computational Neuroscience, 12*(1), 55–73.

Lachica, E., Rubsamen, R., Zirpel, L., & Rubel, E. (1995). Glutamatergic inhibition of voltage-operated calcium channels in the avian cochlear nucleus. *Journal of Neuroscience, 15*(3), 1724.

Lawrence, J. J., Brenowitz, S., & Trussell, L. O. (2003). The mechanism of action of aniracetam at synaptic alpha-amino-3-hydroxy-5-methyl-4-isoxazolepropionic acid (AMPA) receptors: Indirect and direct effects on desensitization. *Molecular Pharmacology, 64*(2), 269–278. doi: 10.1124/mol.64.2.269.

Lawrence, J. J., & Trussell, L. O. (2000). Long-term specification of AMPA receptor properties after synapse formation. *Journal of Neuroscience, 20*(13), 4864–4870.

Leão, R. N., Oleskevich, S., Sun, H., Bautista, M., Fyffe, R. E., & Walmsley, B. (2004). Differences in glycinergic mIPSCs in the auditory brain stem of normal and congenitally deaf neonatal mice. *Journal of Neurophysiology, 91*(2), 1006–1012. doi: 10.1152/jn.00771.2003.

Lewis, E. R., Leverenz, E. L., & Koyama, H. (1980). Mapping functionally identified auditory afferents from their peripheral origins to their central terminations. *Brain Research, 197*(1), 223–229.

Li, G. L., Keen, E., Andor-Ardo, D., Hudspeth, A. J., & von Gersdorff, H. (2009). The unitary event underlying multiquantal EPSCs at a hair cell's ribbon synapse. *Journal of Neuroscience, 29*(23), 7558–7568. doi: 10.1523/JNEUROSCI.0514-09.2009.

Li, J., & Young, E. D. (1993). Discharge-rate dependence of refractory behavior of cat auditory-nerve fibers. *Hearing Research, 69*(1–2), 151–162.

Liberman, M. C. (1978). Auditory-nerve response from cats raised in a low-noise chamber. *Journal of the Acoustic Society of America, 63*(2), 442–455.

Liberman, M. C. (1982). The cochlear frequency map for the cat: Labeling auditory-nerve fibers of known characteristic frequency. *Journal of the Acoustic Society of America, 72*(5), 1441–1449.

Liberman, M. C. (1991). Central projections of auditory-nerve fibers of differing spontaneous rate. I. Anteroventral cochlear nucleus. *Journal of Comparative Neurology, 313*(2), 240–258. doi: 10.1002/cne.903130205.

Liberman, M. C., & Oliver, M. E. (1984). Morphometry of intracellularly labeled neurons of the auditory nerve: Correlations with functional properties. *Journal of Comparative Neurology, 223*(2), 163–176. doi: 10.1002/cne.902230203.

Lim, R., Alvarez, F. J., & Walmsley, B. (2000). GABA mediates presynaptic inhibition at glycinergic synapses in a rat auditory brainstem nucleus. *Journal of Physiology, 525*(Pt. 2), 447–459.

Limb, C. J., & Ryugo, D. K. (2000). Development of primary axosomatic endings in the anteroventral cochlear nucleus of mice. *JARO: Journal of the Association for Research in Otolaryngology, 1*(2), 103–119.

Lorente de Nó, R. (1981). *The primary acoustic nuclei.* New York: Raven Press.

Lorteije, J. A., Rusu, S. I., Kushmerick, C., & Borst, J. G. (2009). Reliability and precision of the mouse calyx of Held synapse. *Journal of Neuroscience, 29*(44), 13770–13784. doi: 10.1523/JNEUROSCI.3285-09.2009.

Lu, T., & Trussell, L. O. (2007). Development and elimination of endbulb synapses in the chick cochlear nucleus. *Journal of Neuroscience, 27*(4), 808–817. doi: 10.1523/JNEUROSCI.4871-06.2007.

Lu, Y., Harris, J. A., & Rubel, E. W. (2007). Development of spontaneous miniature EPSCs in mouse AVCN neurons during a critical period of afferent-dependent neuron survival. *Journal of Neurophysiology, 97*(1), 635–646. doi: 10.1152/jn.00915.2006.

Lu, Y., & Rubel, E. W. (2005). Activation of metabotropic glutamate receptors inhibits high-voltage-gated calcium channel currents of chicken nucleus magnocellularis neurons. *Journal of Neurophysiology, 93*(3), 1418.

Manis, P. B. (2008). Biophysical specializations of neurons that encode timing. In Bausbaum, A.I., Akimichi, K., Shepherd, G. M., & Westheimer, G. (Eds.), *The senses: A comphrehensive reference* (Vol. 3, pp. 565–586). San Diego: Academic Press.

Manis, P. B., & Marx, S. O. (1991). Outward currents in isolated ventral cochlear nucleus neurons. *Journal of Neuroscience, 11*(9), 2865–2880.

Martin, M. R. (1981). Morphology of the cochlear nucleus of the normal and reeler mutant mouse. *Journal of Comparative Neurology, 197*(1), 141–152. doi: 10.1002/cne.901970111.

Martin, M. R. (1985). Evidence for an excitatory amino acid as the transmitter of the auditory nerve in the in vitro mouse cochlear nucleus. *Hearing Research, 20*(3), 215–220.

Mattox, D. E., Neises, G. R., & Gulley, R. L. (1982). A freeze-fracture study of the maturation of synapses in the anteroventral cochlear nucleus of the developing rat. *Anatomical Record, 204*(3), 281–287. doi: 10.1002/ar.1092040313.

Milenkovic, I., Rinke, I., Witte, M., Dietz, B., & Rubsamen, R. (2009). P2 receptor-mediated signaling in spherical bushy cells of the mammalian cochlear nucleus. *Journal of Neurophysiology, 102*(3), 1821–1833. doi: 10.1152/jn.00186.2009.

Mosbacher, J., Schoepfer, R., Monyer, H., Burnashev, N., Seeburg, P. H., & Ruppersberg, J. P. (1994). A molecular determinant for submillisecond desensitization in glutamate receptors. *Science, 266*(5187), 1059–1062.

Moskowitz, N., & Liu, J. C. (1972). Central projections of the spiral ganglion of the squirrel monkey. *Journal of Comparative Neurology, 144*(3), 335–344. doi: 10.1002/cne.901440305.

Murthy, V. N., Sejnowski, T. J., & Stevens, C. F. (1997). Heterogeneous release properties of visualized individual hippocampal synapses. *Neuron, 18*(4), 599–612.

Neises, G. R., Mattox, D. E., & Gulley, R. L. (1982). The maturation of the end bulb of Held in the rat anteroventral cochlear nucleus. *Anatatomical Record, 204*(3), 271–279. doi: 10.1002/ar.1092040312.

Nicol, M. J., & Walmsley, B. (2002). Ultrastructural basis of synaptic transmission between endbulbs of Held and bushy cells in the rat cochlear nucleus. *Journal of Physiology, 539*(Pt. 3), 713–723.

Noda, Y., & Pirsig, W. (1974). Anatomical projection of the cochlea to the cochlear nuclei of the guinea pig. *Archives of Otorhinolaryngology, 208*(2), 107–120.

O'Neil, J. N., Limb, C. J., Baker, C. A., & Ryugo, D. K. (2010). Bilateral effects of unilateral cochlear implantation in congenitally deaf cats. *Journal of Comparative Neurology, 518*(12), 2382–2404. doi: 10.1002/cne.22339.

Oertel, D. (1983). Synaptic responses and electrical properties of cells in brain slices of the mouse anteroventral cochlear nucleus. *Journal of Neuroscience, 3*(10), 2043–2053.

Oertel, D., Shatadal, S., & Cao, X. J. (2008). In the ventral cochlear nucleus Kv1. 1 and subunits of HCN1 are colocalized at surfaces of neurons that have low-voltage-activated and hyperpolarization-activated conductances. *Neuroscience, 154*(1), 77–86.

Ohlemiller, K. K., Echteler, S. M., & Siegel, J. H. (1991). Factors that influence rate-versus-intensity relations in single cochlear nerve fibers of the gerbil. *Journal of the Acoustic Society of America, 90*(1), 274–287.

Oleskevich, S., Clements, J., & Walmsley, B. (2000). Release probability modulates short-term plasticity at a rat giant terminal. *Journal of Physiology, 524*(Pt. 2), 513–523.

Oleskevich, S., & Walmsley, B. (2002). Synaptic transmission in the auditory brainstem of normal and congenitally deaf mice. *Journal of Physiology, 540*(Pt. 2), 447–455.

Osen, K. K. (1969a). Cytoarchitecture of the cochlear nuclei in the cat. *Journal of Comparative Neurology, 136*(4), 453–484. doi: 10.1002/cne.901360407.

Osen, K. K. (1969b). The intrinsic organization of the cochlear nuclei. *Acta Otolaryngologica, 67*(2), 352–359.

Osen, K. K. (1970). Course and termination of the primary afferents in the cochlear nuclei of the cat. An experimental anatomical study. *Archives Italiennes de Biologie, 108*(1), 21–51.

Ostapoff, E. M., & Morest, D. K. (1991). Synaptic organization of globular bushy cells in the ventral cochlear nucleus of the cat: A quantitative study. *Journal of Comparative Neurology, 314*(3), 598–613. doi: 10.1002/cne.903140314.

Otis, T. S., & Trussell, L. O. (1996). Inhibition of transmitter release shortens the duration of the excitatory synaptic current at a calyceal synapse. *Journal of Neurophysiology, 76*(5), 3584–3588.

Otis, T. S., Wu, Y. C., & Trussell, L. O. (1996). Delayed clearance of transmitter and the role of glutamate transporters at synapses with multiple release sites. *Journal of Neuroscience, 16*(5), 1634–1644.

Parks, T. N. (1979). Afferent influences on the development of the brain stem auditory nuclei of the chicken: Otocyst ablation. *Journal of Comparative Neurology, 183*, 665–677.

Perez-Otano, I., & Ehlers, M. (2004). Learning from NMDA receptor trafficking: Clues to the development and maturation of glutamatergic synapses. *Neurosignals, 13*, 175–189.

Perkins, G. A., Tjong, J., Brown, J. M., Poquiz, P. H., Scott, R. T., Kolson, D. R. (2010). The micro-architecture of mitochondria at active zones: Electron tomography reveals novel anchoring scaffolds and cristae structured for high-rate metabolism. *Journal of Neuroscience, 30*(3), 1015–1026. doi: 10.1523/JNEUROSCI.1517-09.2010.

Pfeiffer, R. R. (1966). Anteroventral cochlear nucleus: Wave forms of extracellularly recorded spike potentials. *Science, 154*(749), 667–668.

Pliss, L., Yang, H., & Xu-Friedman, M. A. (2009). Context-dependent effects of NMDA receptors on precise timing information at the endbulb of Held in the cochlear nucleus. *Journal of Neurophysiology, 102*(5), 2627–2637. doi: 10.1152/jn.00111.2009.

Raman, I. M., & Trussell, L. O. (1992). The kinetics of the response to glutamate and kainate in neurons of the avian cochlear nucleus. *Neuron, 9*(1), 173–186.

Ramon y Cajal, S. (1896). *Beitrag zum Studium der Medulla oblongata, des Kleinhirns und des Ursprungs der Gehirnnerven*: Verlag von Johann Ambrosius Barth.

Ramon y Cajal, S. (1995). Histology of the Nervous System of Man and Vertebrates (Swanson, N., & Swanson, L., trans. Vol. 1). New York: Oxford University Press.

Rathouz, M., & Trussell, L. (1998). Characterization of outward currents in neurons of the avian nucleus magnocellularis. *Journal of Neurophysiology, 80*(6), 2824.

Ravindranathan, A., Donevan, S. D., Sugden, S. G., Greig, A., Rao, M. S., & Parks, T. N. (2000). Contrasting molecular composition and channel properties of AMPA receptors on chick auditory and brainstem motor neurons. *Journal of Physiology, 523*(Pt. 3), 667–684.

Reyes, A. D., Rubel, E. W., & Spain, W. J. (1994). Membrane properties underlying the firing of neurons in the avian cochlear nucleus. *Journal of Neuroscience, 14*(9), 5352.

Rhode, W. S. (2008). Response patterns to sound associated with labeled globular/bushy cells in cat. *Neuroscience, 154*(1), 87–98.

Rhode, W. S., & Smith, P. H. (1985). Characteristics of tone-pip response patterns in relationship to spontaneous rate in cat auditory nerve fibers. *Hearing Research, 18*(2), 159–168.

Rothman, J. S., & Manis, P. B. (2003a). Kinetic analyses of three distinct potassium conductances in ventral cochlear nucleus neurons. *Journal of Neurophysiology, 89*(6), 3083–3096. doi: 10.1152/jn.00126.2002.

Rothman, J. S., & Manis, P. B. (2003b). The roles potassium currents play in regulating the electrical activity of ventral cochlear nucleus neurons. *Journal of Neurophysiology, 89*(6), 3097–3113. doi: 10.1152/jn.00127.2002.

Rothman, J. S., & Young, E. D. (1996). Enhancement of neural synchronization in computational models of ventral cochlear nucleus bushy cells. *Auditory Neuroscience, 2*, 47–62.

Rothman, J. S., Young, E. D., & Manis, P. B. (1993). Convergence of auditory nerve fibers onto bushy cells in the ventral cochlear nucleus: Implications of a computational model. *Journal of Neurophysiology, 70*(6), 2562.

Rouiller, E. M., Cronin-Schreiber, R., Fekete, D. M., & Ryugo, D. K. (1986). The central projections of intracellularly labeled auditory nerve fibers in cats: An analysis of terminal morphology. *Journal of Comparative Neurology, 249*(2), 261–278. doi: 10.1002/cne.902490210.

Rowland, K. C., Irby, N. K., & Spirou, G. A. (2000). Specialized synapse-associated structures within the calyx of Held. *Journal of Neuroscience, 20*(24), 9135–9144.

Rozov, A., Burnashev, N., Sakmann, B., & Neher, E. (2001). Transmitter release modulation by intracellular Ca2+ buffers in facilitating and depressing nerve terminals of pyramidal cells in layer 2/3 of the rat neocortex indicates a target cell-specific difference in presynaptic calcium dynamics. *Journal of Physiology, 531*(Pt, 3), 807–826.

Ryugo, D. K., Montey, K. L., Wright, A. L., Bennett, M. L., & Pongstaporn, T. (2006). Postnatal development of a large auditory nerve terminal: The endbulb of Held in cats. *Hearing Research, 216–217*, 100–115. doi: 10.1016/j.heares.2006.01.007.

Ryugo, D. K., Pongstaporn, T., Huchton, D. M., & Niparko, J. K. (1997). Ultrastructural analysis of primary endings in deaf white cats: Morphologic alterations in endbulbs of Held. *Journal of Comparative Neurology, 385*(2), 230–244. doi: 10.1002/(SICI)1096-9861(19970825)385:2.

Ryugo, D. K., & Rouiller, E. M. (1988). Central projections of intracellularly labeled auditory nerve fibers in cats: Morphometric correlations with physiological properties. *Journal of Comparative Neurology, 271*(1), 130–142. doi: 10.1002/cne.902710113.

Ryugo, D. K., & Sento, S. (1991). Synaptic connections of the auditory nerve in cats: Relationship between endbulbs of Held and spherical bushy cells. *Journal of Comparative Neurology, 305*(1), 35–48. doi: 10.1002/cne.903050105.

Ryugo, D. K., Wu, M. M., & Pongstaporn, T. (1996). Activity-related features of synapse morphology: A study of endbulbs of Held. *Journal of Comparative Neurology, 365*(1), 141–158. doi: 10.1002/(SICI)1096-9861(19960129)365:1.

Sachs, M. B., & Abbas, P. J. (1974). Rate versus level functions for auditory-nerve fibers in cats: Tone-burst stimuli. *Journal of the Acoustic Society of America, 56*(6), 1835–1847.

Sakaba, T., & Neher, E. (2001). Calmodulin mediates rapid recruitment of fast-releasing synaptic vesicles at a calyx-type synapse. *Neuron, 32*(6), 1119–1131.

Sando, I. (1965). The anatomical interrelationships of the cochlear nerve fibers. *Acta Otolaryngologica, 59*, 417–436.

Sanes, D. H., McGee, J. A., & Walsh, E. J. (1998). Metabotropic glutamate receptor activation modulates sound level processing in the cochlear nucleus. *Journal of Neurophysiology, 80*(1), 209.

Schalk, T. B., & Sachs, M. B. (1980). Nonlinearities in auditory-nerve fiber responses to bandlimited noise. *Journal of the Acoustic Society of America, 67*(3), 903–913.

Schwarz, D. W. F., & Puil, E. (1997). Firing properties of spherical bushy cells in the anteroventral cochlear nucleus of the gerbil. *Hearing Research, 114*(1–2), 127–138.

Sento, S., & Ryugo, D. K. (1989). Endbulbs of Held and spherical bushy cells in cats: Morphological correlates with physiological properties. *Journal of Comparative Neurology, 280*(4), 553–562. doi: 10.1002/cne.902800406.

Sewell, W. F. (1984). The relation between the endocochlear potential and spontaneous activity in auditory nerve fibres of the cat. *Journal of Physiology, 347*, 685–696.

Shepherd, J. D., & Huganir, R. L. (2007). The cell biology of synaptic plasticity: AMPA receptor trafficking. *Annual Review of Cell and Developmental Biology, 23*, 613–643.

Sivaramakrishnan, S., & Laurent, G. (1995). Pharmacological characterization of presynaptic calcium currents underlying glutamatergic transmission in the avian auditory brainstem. *Journal of Neuroscience, 15*(10), 6576–6585.

Smith, P. H., & Rhode, W. S. (1987). Characterization of HRP-labeled globular bushy cells in the cat anteroventral cochlear nucleus. *Journal of Comparative Neurology, 266*, 360–375.

Spirou, G. A., Brownell, W. E., & Zidanic, M. (1990). Recordings from cat trapezoid body and HRP labeling of globular bushy cell axons. *Journal of Neurophysiology, 63*(5), 1169–1190.

Spirou, G. A., Rager, J., & Manis, P. B. (2005). Convergence of auditory-nerve fiber projections onto globular bushy cells. *Neuroscience, 136*(3), 843–863. doi: 10.1016/j.neuroscience.2005.08.068.

Sugden, S. G., Zirpel, L., Dietrich, C. J., & Parks, T. N. (2002). Development of the specialized AMPA receptors of auditory neurons. *Journal of Neurobiology, 52*(3), 189–202. doi: 10.1002/neu.10078.

Suneja, S. K., Yan, L., & Potashner, S. J. (2005). Regulation of NT-3 and BDNF levels in guinea pig auditory brain stem nuclei after unilateral cochlear ablation. *Journal of Neuroscience Research, 80*(3), 381–390.

Szpir, M. R., Sento, S., & Ryugo, D. K. (1990). Central projections of cochlear nerve fibers in the alligator lizard. *Journal of Comparative Neurology, 295*(4), 530–547. doi: 10.1002/cne.902950403.

Taberner, A. M., & Liberman, M. C. (2005). Response properties of single auditory nerve fibers in the mouse. *Journal of Neurophysiology, 93*(1), 557–569. doi: 10.1152/jn.00574.2004.

Takago, H., Nakamura, Y., & Takahashi, T. (2005). G protein-dependent presynaptic inhibition mediated by AMPA receptors at the calyx of Held. *Proceedings of the National Academy of Sciences of the United States of America, 102*(20), 7368–7373. doi: 10.1073/pnas.0408514102.

Tang, Y. Z., & Carr, C. E. (2004). Development of NMDA R1 expression in chicken auditory brainstem. *Hearing Research, 191*(1–2), 79–89. doi: 10.1016/j.heares.2004.01.007.

Tierney, T. S., Doubell, T. P., Xia, G., & Moore, D. R. (2001). Development of brain-derived neurotrophic factor and neurotrophin-3 immunoreactivity in the lower auditory brainstem of the postnatal gerbil. *European Journal of Neuroscience, 14*(5), 785–793.

Tolbert, L. P., & Morest, D. K. (1982). The neuronal architecture of the anteroventral cochlear nucleus of the cat in the region of the cochlear nerve root: Electron microscopy. *Neuroscience, 7*(12), 3053–3067.

Trussell, L. O. (1999). Synaptic mechanisms for coding timing in auditory neurons. *Annual Review of Physiology, 61*, 477–496. doi: 10.1146/annurev.physiol.61.1.477.

Trussell, L. O., & Fischbach, G. D. (1989). Glutamate receptor desensitization and its role in synaptic transmission. *Neuron, 3*(2), 209–218.

Trussell, L. O., Zhang, S., & Raman, I. M. (1993). Desensitization of AMPA receptors upon multiquantal neurotransmitter release. *Neuron, 10*(6), 1185–1196.

Tsuji, J., & Liberman, M. C. (1997). Intracellular labeling of auditory nerve fibers in guinea pig: central and peripheral projections. *Journal of Comparative Neurology, 381*(2), 188–202. doi: 10.1002/(SICI)1096-9861(19970505)381:2.

Turecek, R., & Trussell, L. O. (2000). Control of synaptic depression by glutamate transporters. *Journal of Neuroscience, 20*(5), 2054.

Turecek, R., & Trussell, L. O. (2001). Presynaptic glycine receptors enhance transmitter release at a mammalian central synapse. *Nature, 411*(6837), 587–590. doi: 10.1038/35079084.

Typlt, M., Haustein, M. D., Dietz, B., Steinert, J. R., Witte, M., & Englitz, B. (2010). Presynaptic and postsynaptic origin of multicomponent extracellular waveforms at the endbulb of Held–spherical bushy cell synapse. *European Journal of Neuroscience, 31*(9), 1574–1581. doi: 10.1111/j.1460-9568.2010.07188.x.

Verstreken, P., Ly, C. V., Venken, K. J., Koh, T. W., Zhou, Y., & Bellen, H. J. (2005). Synaptic mitochondria are critical for mobilization of reserve pool vesicles at Drosophila neuromuscular junctions. *Neuron, 47*(3), 365–378. doi: 10.1016/j.neuron.2005.06.018.

von Gersdorff, H., Schneggenburger, R., Weis, S., & Neher, E. (1997). Presynaptic depression at a calyx synapse: The small contribution of metabotropic glutamate receptors. *Journal of Neuroscience, 17*(21), 8137–8146.

Wang, L. Y., & Kaczmarek, L. K. (1998). High-frequency firing helps replenish the readily releasable pool of synaptic vesicles. *Nature, 394*(6691), 384–388. doi: 10.1038/28645.

Wang, Y., & Manis, P. B. (2005). Synaptic transmission at the cochlear nucleus endbulb synapse during age-related hearing loss in mice. *Journal of Neurophysiology, 94*(3), 1814–1824. doi: 10.1152/jn.00374.2005.

Wang, Y., & Manis, P. B. (2006). Temporal coding by cochlear nucleus bushy cells in DBA/2J mice with early onset hearing loss. *Journal of the Association for Research in Otolaryngology, 7*(4), 412–424. doi: 10.1007/s10162-006-0052-9.

Wang, Y., & Manis, P. B. (2008). Short-term synaptic depression and recovery at the mature mammalian endbulb of Held synapse in mice. *Journal of Neurophysiology, 100*(3), 1255–1264. doi: 10.1152/jn.90715.2008.

Wang, Y., Ren, C., & Manis, P. B. (2010). Endbulb synaptic depression within the range of presynaptic spontaneous firing and its impact on the firing reliability of cochlear nucleus bushy neurons. *Hearing Research, 270*(1–2), 101–109. doi: 10.1016/j.heares.2010.09.003.

Wang, Y. X., Wenthold, R. J., Ottersen, O. P., & Petralia, R. S. (1998). Endbulb synapses in the anteroventral cochlear nucleus express a specific subset of AMPA-type glutamate receptor subunits. *Journal of Neuroscience, 18*(3), 1148–1160.

Webster, D. B. (1983). Late onset of auditory deprivation does not affect brainstem auditory neuron soma size. *Hearing Research, 12*(1), 145–147.

Webster, D. B., & Trune, D. R. (1982). Cochlear nuclear complex of mice. *American Journal of Anatomy, 163*(2), 103–130. doi: 10.1002/aja.1001630202.

Wenthold, R. J., & Gulley, R. L. (1977). Aspartic acid and glutamic acid levels in the cochlear nucleus after auditory nerve lesion. *Brain Research, 138*(1), 111–123.

Winter, I. M., & Palmer, A. R. (1990). Responses of single units in the anteroventral cochlear nucleus of the guinea pig. *Hearing Research, 44*(2–3), 161–178.

Winter, I. M., Robertson, D., & Yates, G. K. (1990). Diversity of characteristic frequency rate-intensity functions in guinea pig auditory nerve fibres. *Hearing Research, 45*(3), 191–202.

Wong, A. Y., Graham, B. P., Billups, B., & Forsythe, I. D. (2003). Distinguishing between presynaptic and postsynaptic mechanisms of short-term depression during action potential trains. *Journal of Neuroscience, 23*(12), 4868–4877.

Wu, S. H., & Oertel, D. (1987). Maturation of synapses and electrical properties of cells in the cochlear nuclei. *Hearing Research, 30*(1), 99–110.

Xu, J., & Wu, L. G. (2005). The decrease in the presynaptic calcium current is a major cause of short-term depression at a calyx-type synapse. *Neuron, 46*(4), 633–645. doi: 10.1016/j.neuron.2005.03.024.

Yang, H., & Xu-Friedman, M. A. (2008). Relative roles of different mechanisms of depression at the mouse endbulb of Held. *Journal of Neurophysiology, 99*(5), 2510–2521. doi: 10.1152/jn.01293.2007.

Yang, H., & Xu-Friedman, M. A. (2009). Impact of synaptic depression on spike timing at the endbulb of Held. *Journal of Neurophysiology, 102*(3), 1699–1710. doi: 10.1152/jn.00072.2009.

Young, E. D., & Barta, P. E. (1986). Rate responses of auditory nerve fibers to tones in noise near masked threshold. *Journal of the Acoustic Society of America, 79*(2), 426–442.

Zhang, S., & Trussell, L. O. (1994a). A characterization of excitatory postsynaptic potentials in the avian nucleus magnocellularis. *Journal of Neurophysiology, 72*(2), 705–718.

Zhang, S., & Trussell, L. O. (1994b). Voltage clamp analysis of excitatory synaptic transmission in the avian nucleus magnocellularis. *Journal of Physiology, 480*(Pt. 1), 123–136.

Zheng, Y., Baek, J. H., Smith, P. F., & Darlington, C. L. (2007). Cannabinoid receptor down-regulation in the ventral cochlear nucleus in a salicylate model of tinnitus. *Hearing Research, 228*(1–2), 105–111. doi: 10.1016/j.heares.2007.01.028.

Zirpel, L., Lippe, W. R., & Rubel, E. W. (1998). Activity-dependent regulation of [Ca2+] i in avian cochlear nucleus neurons: Roles of protein kinases A and C and relation to cell death. *Journal of Neurophysiology, 79*(5), 2288.

Zirpel, L., & Rubel, E. W. (1996). Eighth nerve activity regulates intracellular calcium concentration of avian cochlear nucleus neurons via a metabotropic glutamate receptor. *Journal of Neurophysiology, 76*(6), 4127.

Zucker, R. S., & Regehr, W. G. (2002). Short-term synaptic plasticity. *Annual Reviews of Physiology, 64*, 355–405. doi: 10.1146/annurev.physiol.64.092501.114547.

Chapter 5
The Calyx of Held Synapse

J.G.G. Borst and S.I. Rusu

1 Introduction

Specialized synapses are employed in the central auditory system to preserve timing information, often transmitting at the high firing frequencies that typically accompany sound transduction in the periphery. A unique example of such a specialized synapse is the calyx of Held synapse. It is a giant axosomatic synapse within the superior olivary complex (SOC) that functions as a fast sign-inverting relay, providing inhibition to most auditory nuclei of the auditory brainstem. Because of its size – it could well be the largest terminal in the mammalian brain – it has become a popular model system to study mechanisms of transmitter release. Figure 5.1 illustrates the most important property that allows it to act as a relay: it harbors hundreds of synaptic active zones.

Apart from its significance for sound localization, the properties of the calyx of Held make it popular with physiologists studying synaptic transmission. Indeed, the unique accessibility of the presynaptic terminal has allowed a detailed dissection of the biophysical mechanisms underlying glutamate release, including its dependence on firing frequency. Its popularity thus stems from a number of technical advantages, including the feasibility of presynaptic patch-clamp recordings (Forsythe 1994), monitoring transmitter release by combining presynaptic recording with recordings from its postsynaptic partner (Borst et al. 1995), or by the use of imaging methods (Kay et al. 1999) or capacitance measurements (Sun and Wu 2001). Even patching the presynaptic release face of the terminal has been achieved (He et al. 2009). Furthermore, using viruses it has become possible to use the calyx as an exogenous expression system (Wimmer et al. 2004).

J.G.G. Borst(✉)
Department of Neuroscience, Erasmus MC, University Medical Center Rotterdam,
Dr. Molewaterplein 50, GE Rotterdam 3015, The Netherlands
e-mail: g.borst@erasmusmc.nl

L.O. Trussell et al. (eds.), *Synaptic Mechanisms in the Auditory System*,
Springer Handbook of Auditory Research 41, DOI 10.1007/978-1-4419-9517-9_5,
© Springer Science+Business Media, LLC 2012

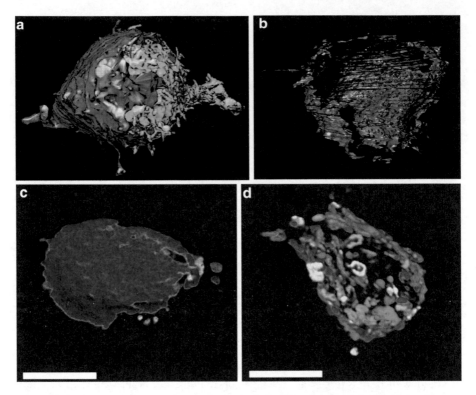

Fig. 5.1 The calyx of Held has many active zones. (**a**) Surface view of a three-dimensional reconstruction of a calyx of Held (yellow) and a principal neuron (blue) from a P9 rat. (**b**) En face view on the synaptic surface of the same calyx (yellow) as in A, showing the distribution of synaptic contacts highlighted in red. (**c**) Three-dimensional rendering of a confocal stack from a P11 rat calyx containing synaptic vesicles labeled with synaptophysin–EGFP. Each vesicle cluster is represented by a different color. Despite the large number of active zones, most vesicles are in a single large cluster. (**d**) In contrast, in a P22 calyx, a clustering algorithm identifies many different clusters. Scale bars in C and D are 5 μm (**a** and **b** are reproduced with permission from (Sätzler et al. 2002), **c** and **d** from (Wimmer et al. 2006))

These technical advantages have made it a popular model system for brain slice studies. However, the calyx has distinct advantages for in vivo recordings as well, again because of its large size. With a single recording it is possible to sample both the presynaptic spike and the postsynaptic potential in vivo (Guinan and Li 1990). Moreover, with in vivo juxtacellular recordings, an estimate of both release and postsynaptic excitability can be obtained (Lorteije et al. 2009).

The unique size and shape of the calyx is also advantageous for developmental studies. Calyces can be identified unambiguously in Golgi preparations or by immunocytochemistry. Around the turn of the previous century, the calyx was a popular subject for neuroanatomists using the newly developed Golgi stain (reviewed in Morest 1968b). An example by Ramon y Cajal is shown in Fig. 5.2. Because calyces

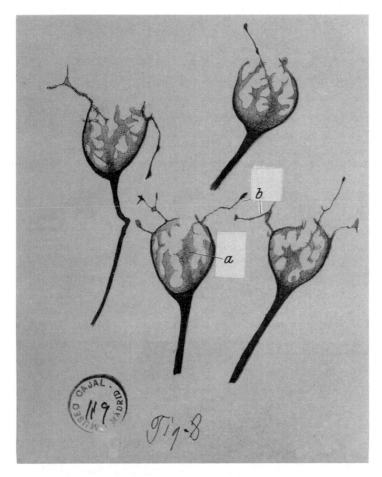

Fig. 5.2 Morphology of the calyx of Held. Drawing of Golgi staining of calyces of a kitten (Ramón y Cajal 1899, 1904). (**a**) Calyx; (**b**) Calyceal collaterals (Reproduction of the original drawing by permission of the Cajal Legacy, Instituto Cajal (CSIC), Madrid, Spain)

contact a relatively homogeneous population of postsynaptic cells, which can be readily identified by their distinct eccentric nucleus immediately after birth, the ontogeny of their form and function can be traced systematically.

This chapter presents an overview of the current state of knowledge about this synapse. It starts with a description of the anatomy of the adult calyx of Held and its role in auditory signaling. It then presents what is known about the physiology of the immature calyx of Held, because, for technical reasons, most of the brain slice studies have been performed on rats or mice before hearing onset. In the last section, the maturation of the calyx into a relay synapse is discussed.

2   Form and Function of the Adult Calyx of Held Synapse

2.1   Comparative Anatomy of the Medial Nucleus of the Trapezoid Body

Calyces are found in the medial nucleus of the trapezoid body (MNTB), the most medial nucleus of the SOC. In most mammals, the number of cells in the MNTB ranges between 2,000 and 6,000 (Irving and Harrison 1967; Kulesza et al. 2002). The MNTB is relatively poorly developed in primates. Among primate species, a progressive reduction of the MNTB has occurred during phylogeny (Hilbig et al. 2009). There is some discussion about whether in humans the MNTB contains only a few dispersed cells (Moore 1987) or is still clearly present as a distinct, small nucleus (Richter et al. 1983; Kulesza 2008). The large majority of neurons in the medial nucleus of the trapezoid body are principal neurons (Morest 1968a; Casey and Feldman 1982). Their morphology and main afferent and efferent projections are highly conserved between species (Lenn and Reese 1966; Morest 1968b; Kuwabara et al. 1991; Kuwabara and Zook 1991).

2.2   Anatomy of the Mature Calyx of Held

2.2.1   Afferent Connections

Each principal cell in the MNTB is contacted by a single giant axosomatic terminal that forms the calyx of Held (Morest 1968b). These large-diameter axons originate from the contralateral cochlear nucleus (Harrison and Warr 1962), in particular from the globular bushy cells (GBCs) of the anteroventral cochlear nucleus (AVCN) and, to a lesser extent, the posteroventral cochlear nucleus (PVCN; Manis et al., Chap. 4). Most axons form only a single calyx, but rarely they can form two or even three (Held 1893; Kuwabara et al. 1991; Rodriguez-Contreras et al. 2006). Interestingly, although the GBCs contact other nuclei, mostly of the ipsilateral and contralateral SOC, they form a calyx-type synapse only on the principal neurons of the MNTB, indicating that a target-specific signal is involved in its formation (Thompson and Schofield 2000).

2.2.2   Macroscopic Anatomy of the Calyx

A calyx encloses the cell body of a principal cell, covering about half of its surface (Casey and Feldman 1988). Already from its first description by Hans von Held in 1893, it was clear that the morphology of the calyx is complex. Each axonal ending divides into several branches, which can further divide into smaller second- and third-order branches, giving the appearance of a floral calyx (Morest 1968b). Second- and third-order branches typically consist of ellipsoid swellings, often linked by narrow necks; these swellings vary in size from bouton-like to much larger and could constitute distinct biochemical compartments (Rowland et al. 2000; Wimmer et al. 2006; Spirou et al. 2008). In addition, the calyx forms collateral arborizations that can end in conventional boutons and contact other principal neurons (Fig. 5.2; Lenn and Reese 1966; Rodriguez-Contreras et al. 2008).

2.2.3   Microscopic Anatomy of the Adult Calyx

The adult calyx contains several anatomical specializations that are related to its unusually large size or are thought to be important for its function as a relay synapse. As mentioned earlier, the most important specialization of the calyx is its large number of active zones (Fig. 5.1). Whereas most terminals in the brain typically have only a single active zone, estimates of the total number of active zones in the calyx of Held range from 300 to 700 (Sätzler et al. 2002; Taschenberger et al. 2002; Wimmer et al. 2006; Dondzillo et al. 2010). The active zones are similar to those of conventional small nerve terminals, although they are relatively small, and each active zone contains about three docked vesicles (Rowland et al. 2000; Taschenberger et al. 2002). There are about 100 vesicles per active zone (Wimmer et al. 2006), for a total of about 80,000 vesicles (Sätzler et al. 2002; de Lange et al. 2003; Wimmer et al. 2006). The synaptic vesicles are relatively large (~50 nm; Lenn and Reese 1966; Nakajima 1971; Casey and Feldman 1985; Taschenberger et al. 2002), which is in agreement with the relatively large quantal size found in the adult synapse (Lorteije et al. 2009). The synaptic vesicle clusters are predominantly located in the swellings within the higher-order branches of the terminal (Wimmer et al. 2006). Within these swellings, synaptic vesicles typically form donut-like assemblies around a central cluster of interconnected mitochondria (Wimmer et al. 2006). In addition, the calyx contains dense-core vesicles, whose content and function are unknown (Lenn and Reese 1966; Nakajima 1971).

The calyx contains many microtubules, especially at the base, as well as neurofilaments (Lenn and Reese 1966). Apart from providing stability, these cytoskeletal elements may restrict the distribution of the synaptic vesicle pool and localize mitochondria (Perkins et al. 2010). The calyx is attached to its principal cell via numerous puncta adherentia (Rowland et al. 2000; Sätzler et al. 2002). The calyx contains many mitochondria, especially in the central regions (Lenn and Reese 1966; Nakajima 1971; Kil et al. 1995); a subset of these are attached to puncta adherentia via a cytoskeletal network called the mitochondria-associated adherens complex. This network places the mitochondria in the vicinity of the active zone, which could allow them to help the calyx meet metabolic demands, buffer Ca^{2+}, or synthesize glutamate (Rowland et al. 2000; Perkins et al. 2010).

2.3   The Synaptic Cleft

Extended extracellular spaces are interposed at several places between the pre- and postsynaptic membrane (Nakajima 1971). These spaces are probably not a fixation artifact, as they have also been observed in the bushy cells of the AVCN that were fixed in the absence of chemicals (Tatsuoka and Reese 1989). These spaces may help to make the synapse more resistant to depletion of Ca^{2+} from the synaptic cleft (Borst and Sakmann 1999a). The perineuronal net that surrounds the calyx and the principal neuron may have a similar function (Härtig et al. 2001).

2.4 Glia of the MNTB

Glial processes are in intimate contact with the calyceal synapse (Nakajima 1971). These processes originate from two types of glial cells, astrocytes and NG2 glial cells (Müller et al. 2009). The NG2 glial cells do not express glutamate trans-porters, but instead have α-amino-3-hydroxyl-5-methyl-4-isoxazole-propionate–type (AMPA) glutamate receptors; they receive small, fast, glutamatergic inputs, most likely originating from the calyx of Held (Müller et al. 2009). The astrocytes express both glutamate receptors and glutamate transporters. Their transporters are localized in processes next to the region of the principal cell not contacted by the calyx and in processes next to the calyx itself but are not present in the synaptic cleft (Elezgarai et al. 2001; Renden et al. 2005; Ford et al. 2009). This distribution of transporters might prevent spillover of glutamate to adjacent synapses. Glutamate transporters are not present on either pre- or postsynaptic membranes (Palmer et al. 2003), and thus astrocytes are responsible for clearing all glutamate from the MNTB. In addition, MNTB astrocytes release D-serine, which can act as a co-agonist on N-methyl-D-aspartic acid (NMDA) receptors, suggesting astrocytes may facilitate some aspects of transmission besides transmitter clearance (Reyes-Haro et al. 2010).

2.5 Principal Neurons of the MNTB

2.5.1 Macroscopic Anatomy of Principal Neurons

Principal neurons are characterized by their globular or oval perikaryon of 15–20 μm containing an eccentric nucleus; they have relatively short, slender dendrites that branch profusely in the vicinity of the cell (Morest 1968a; Sommer et al. 1993; Smith et al. 1998). The distal dendrites are only sparsely innervated (Smith et al. 1998). Apart from the large calyceal input, the principal neurons also receive somatic inhibitory inputs, whose origin and function are unknown (Nakajima 1971; Moore and Moore 1987; Awatramani et al. 2004).

2.5.2 Effect of Principal Neurons on Target Nuclei

The principal neurons of the MNTB are glycinergic (Moore and Caspary 1983; Wenthold et al. 1987; Wu and Kelly 1995). They provide tonotopic inhibitory projections to several ipsilateral neighboring nuclei within the SOC, including the lateral superior olive (LSO), the superior paraolivary nucleus (SPON), the medial superior olive (MSO), and at least five other ipsilateral periolivary nuclei (Thompson and Schofield 2000). They also provide projections to extrinsic nuclei including the nucleus of the lateral lemniscus (Glendenning et al. 1981; Spangler et al. 1985; Kuwabara and Zook 1991) and efferent projections to the cochlea (Robertson 1985; Riemann and Reuss 1998). Not all projections described here necessarily originate from the principal neurons, as the nonprincipal neurons can also project outside of the MNTB (Banks and Smith 1992).

Its function has been relatively well studied in the two most prominent inputs, the SPON and the LSO, and is less well defined in the MSO. Most cells in the SOC have a low intracellular chloride concentration. The glycinergic inputs from the MNTB will therefore hyperpolarize and inhibit these cells. Cells in the rat SPON have low spontaneous rates but fire at stimulus offset. They are tonically inhibited by the glycinergic inputs from the principal neurons of the MNTB; this inhibition is lifted when the spontaneous firing of principal neurons is briefly suppressed following tones (Kulesza et al. 2003, 2007; Kadner et al. 2006). Because the neurons in the SPON form an important source of GABAergic inhibition to the inferior colliculus (Saldaña et al. 2009), this circuit can contribute to duration tuning or gap detection in the inferior colliculus, by inhibiting IC neurons at the offset of tones (Kadner and Berrebi 2008).

LSO cells receive excitatory input from, among others, the glutamatergic spherical bushy cells of the ipsilateral AVCN and inhibitory inputs indirectly from the contralateral GBCs via the ipsilateral MNTB (Manis et al., Chap. 4; Trussell, Chap. 7). Neurons in the LSO are thus specialized for detecting interaural level differences (ILDs) as ipsilateral sound-evoked excitatory responses can be inhibited by sound at the contralateral ear (Galambos et al. 1959; Boudreau and Tsuchitani 1968; Guinan et al. 1972b). ILDs are one of the primary acoustical cues for sound location. LSO cells can compute the position in azimuth for complex sounds over a wide range of sound intensities. Consequently, the inputs from both ears have to be compared in temporal register, despite the sign-inverting extra synapse for the contralateral input to the LSO (Joris and Yin 1995). A match of the arrival times of the contralateral and the ipsilateral inputs is accomplished by the large caliber of the axons of the globular bushy cells and the principal cells of the MNTB and the small synaptic delay at the calyx synapse. On average, the inhibition arrives only ~0.2 ms later than the excitation (Joris and Yin 1995; Tollin and Yin 2005). Because of these temporal features, especially low-frequency units in the LSO are sensitive to interaural time differences (ITDs) as well. The precision of the ITD function of the LSO cells is on the order of 1 ms (Joris and Yin 1995). For its function in the LSO, timing is thus important for the calyx synapse, especially for the low-frequency cells. Many cells in the LSO are not spontaneously active or have low spontaneous firing rates (Tsuchitani and Boudreau 1966; Guinan et al. 1972b), a surprising fact as these cells receive a strong input from SBCs, which typically have high spontaneous firing rates (Manis et al., Chap. 4). Tonic inhibition due to spontaneous activity from the MNTB may thus contribute to the low firing frequencies in the LSO (Boudreau and Tsuchitani 1968).

ITD is the primary cue for localizing low-frequency sounds in the horizontal plane. Cells in the MSO are innervated by spherical bushy cells from both cochlear nuclei (Manis et al., Chap. 4) and are believed to be coincidence detectors that are specialized in detecting the precise arrival time of the excitatory inputs from both ears (MacLeod and Carr, Chap. 6). Different cells in the MSO respond optimally to different delays, and it is proposed that the set of best delays provides a map of ITDs (Jeffress 1948). In addition to the excitatory inputs from both ears, MSO neurons receive glycinergic somatic inputs arising mainly from the ipsilateral MNTB (contralateral ear) and from the ipsilateral lateral nucleus of the trapezoid body (ipsilateral ear; Grothe et al. 2010). The possible function of the inhibitory input from the MNTB depends to a large extent on its timing relative to the excitation, which is currently unknown because of the technical difficulties associated with making recordings from MSO cells. If inhibition would precede excitation it could effectively delay the effect of the excitation. It could thus have a similar function as a delay line, providing an elegant explanation for the observation that the best delay is typically observed when contralateral arrival precedes ipsilateral (Grothe et al. 2010). However, this idea still lacks conclusive experimental evidence, and alternative functions for inhibition have been proposed (Joris and Yin 2007).

In summary, the many efferent connections of the principal neurons indicate that the most important function of the principal neurons is to provide a rapid inhibitory input to auditory brainstem nuclei. It functions as a sign-converting relay station between the cochlear nucleus and the LSO, which is used for detecting interaural level differences. However, the prominent projection to the monaural SPON indicates that the function of the calyx synapse extends beyond a binaural function. For both of these major projections, both a tonic and a phasic component are essential. The precision of the phasic component appears to be submillisecond, but it is far removed from the microsecond resolution that can be represented within the MSO (Zhou et al. 2005).

2.5.3 Sound Responses of Principal Neurons

Firing Patterns of the Principal Neurons in Response to Sound

The vast majority of cells in the MNTB are excited only by contralateral sounds, in agreement with their innervation (Goldberg and Brown 1968; Guinan et al. 1972b). The thick, heavily myelinated axons of the GBCs allow a minimum response latency of only 3–5 ms, close to the delay of GBCs but shorter than for many other SOC nuclei (Goldberg and Brown 1968; Sommer et al. 1993; Smith et al. 1998; Kopp-Scheinpflug et al. 2008).

The MNTB is tonotopically organized, with the gradient from low to high characteristic frequencies running from lateral to medial (Guinan et al. 1972b; Vater and Feng 1990; Sommer et al. 1993). The frequency response area of the principal neuron is similar to that of the GBCs. Most principal neurons show a V-shaped frequency response area (Guinan et al. 1972b). They have monotonic rate-level functions at the characteristic frequency (Kopp-Scheinpflug et al. 2008). There is a preponderance of high-frequency units, which is inherited from the GBCs and is illustrated by the more densely populated ventro-medial part of the MNTB (Guinan et al. 1972b). The principal neurons show primarylike, primarylike-with-notch, or phase-locked responses to tones (Guinan et al. 1972a, b; Smith et al. 1998), similar to the responses of the GBCs (Manis et al., Chap. 4). The spontaneous activity varies greatly between principal neurons, ranging from <1 to >100 Hz; on average it is about 30 Hz (Kopp-Scheinpflug et al. 2008). During tone presentations, the sustained rates can go up to around 300 Hz (Spirou et al. 1990; Kopp-Scheinpflug et al. 2008; Lorteije et al. 2009). At high spike rates, the synaptic delay of the calyx can increase considerably (Guinan and Li 1990; McLaughlin et al. 2008; Tolnai et al. 2009).

In the cat, low-frequency MNTB units can show very good phase locking, which is inherited from GBCs (Tollin and Yin 2005). The rat MNTB cells can show phase locking up to 2 kHz, with a timing jitter of ~0.15 ms; below 1 kHz, phase locking is slightly better than in the auditory nerve but is comparable to the bushy cells (Paolini et al. 2001).

The Complex Waveform of Principal Neurons

Principal neurons in the MNTB can be identified during in vivo recordings by their complex waveforms (Guinan et al. 1972b). These waveforms consist of a presynaptic component, generally called the prespike because it results from the action potential of the calyx of Held and a postsynaptic component (Guinan and Li 1990). During extracellular recordings, the shape of the complex waveform is variable between units (Guinan and Li 1990; Kopp-Scheinpflug et al. 2003b; McLaughlin et al. 2008), but during juxtacellular (loose-patch) recordings with glass electrodes the complex waveform has a characteristic shape, with the postsynaptic component consisting of a component resulting from the calyceal EPSP and a component resulting from the action potential; the shape of this waveform is similar between species (Guinan and Li 1990; Green and Sanes 2005; Lorteije et al. 2009). Under these conditions the complex waveform can be used as a measure for both presynaptic release and postsynaptic excitability (Lorteije et al. 2009). The similarities between the complex waveform recorded from the calyx synapse and the endbulb of Held synapse are discussed in Chap. 4.

Noncalyceal Excitatory Inputs of Principal Neurons

The principal cells also receive noncalyceal excitatory inputs (Vater and Feng 1990). Electrically evoked action potentials or spontaneous action potentials can sometimes lack a prespike, indicating that the noncalyceal excitatory inputs can be suprathreshold (Guinan and Li 1990; Wu and Kelly 1991; Banks and Smith 1992; Smith et al. 1998; Hamann et al. 2003). However, this is not true for acoustically evoked action potentials, indicating that the calyx plays an essential role in the generation of acoustically evoked action potentials in the principal neurons (Guinan and Li 1990).

Reliability and Precision of the Calyx Synapse

Despite its giant size, the calyx synapse does show failures of transmission. Postsynaptic failures were observed in about half of the cells in the mouse (Lorteije et al. 2009); they are rare in the gerbil (Englitz et al. 2009), and they are observed only under extreme conditions in the cat (Guinan and Li 1990; McLaughlin et al. 2008). There are two main causes for failures. The first is a reduction of postsynaptic excitability, which depends strongly on firing rate and probably reflects Na^+ channel inactivation. The second is a small EPSP size, an effect that does not depend on recent history of activity (i.e., synaptic depression). Rather, the small EPSP size results from stochastic fluctuations in the number of released vesicles, which was estimated to be about 20 on average (Lorteije et al. 2009).

3   Development of the Calyx Synapse

In the next section, the developmental changes that are needed to acquire the cellular properties that characterize its adult function are described. The properties of the young calyx synapse have been well studied in rodents, especially in rats around P9. This is mainly for technical reasons, as it becomes progressively harder to make presynaptic recordings at later ages, because of increasing myelinization. The development of the calyx synapse is discussed in three parts: its formation, the properties of the immature calyx synapse, and the adaptations that are needed for its adult function.

3.1   Early Development of the Calyx Synapse

3.1.1   Formation of the Calyx Synapse

In the rat, neurons of both the ventral cochlear nucleus and the MNTB are born around E15 (Altman and Bayer 1980). Soon after, fibers from the cochlear nucleus start to grow into the ventral brainstem. Calyx-type synapses form exclusively on the contralateral MNTB. The mechanisms that prevent innervation of the ipsilateral MNTB are not yet known, except that ephrin-B reverse signaling is required (Hsieh et al. 2010). At birth, the principal neurons of the contralateral MNTB are already innervated by fibers from the cochlear nucleus (Kandler and Friauf 1993). These fibers are very dynamic and form many collaterals in the first postnatal days (Rodriguez-Contreras et al. 2008). The calyx forms at around P2–5 in rodents (Kandler and Friauf 1993; Kil et al. 1995; Rodriguez-Contreras et al. 2006), well before the onset of hearing, which in rats is at around P7 for bone conduction and P11 for air-borne sound (Geal-Dor et al. 1993). The young calyx has a cup-like form; it has many processes, which look like filopodia and have been called calyceal collaterals (Morest 1968b; Kandler and Friauf 1993; Kil et al. 1995). These collaterals can be formed de novo by early calyces, and they can make functional contacts with other principal neurons (Rodriguez-Contreras et al. 2008).

Although principal neurons can receive multiple large inputs during development (Bergsman et al. 2004; Hoffpauir et al. 2006), as of yet, no conclusive evidence for the formation of calyces from two different axons onto the same principal cell has been reported. This suggests that competition between calyces is not part of the normal developmental process, in contrast to the situation at, for example, the neuromuscular junction (Sanes and Yamagata 2009).

3.1.2 Electrical Activity of the Immature Calyx Synapse

The early connections that precede the calyx are small glutamatergic synapses containing both AMPA and NMDA receptors (Hoffpauir et al. 2006). Despite their small size, they can already drive the principal neurons, presumably because of the high input resistance of the early postnatal principal neurons (Rodriguez-Contreras et al. 2008). In the first days, the size of the postsynaptic calyceal currents rapidly increases (Chuhma and Ohmori 1998; Hoffpauir et al. 2006).

Until the onset of hearing, the calyx synapse has a bursting firing pattern (Sonntag et al. 2009); this bursting pattern is caused by ATP-driven calcium action potentials in hair cells, leading to a characteristic bursting pattern in the MNTB with characteristic intervals of 10, 100–300 ms, and several seconds (Tritsch et al. 2010). The development is accompanied by decreases in the time course of the EPSP, the pre- and postsynaptic action potential duration, and the synaptic delay (Chuhma and Ohmori 1998; Taschenberger and von Gersdorff 2000; Sonntag et al. 2009).

3.2 Synaptic Transmission in the Immature Calyx Synapse

3.2.1 Voltage-Dependent Ion Channels of the Immature Calyx of Held

Action Potential

The calyx has a brief overshooting action potential (Forsythe 1994). Small changes in its shape lead to changes in the Ca^{2+} currents during its repolarization phase and to large changes in the amount of transmitter release (Borst and Sakmann 1999b; Yang and Wang 2006). The action potential is followed by a depolarizing afterpotential (Borst et al. 1995). Following high-frequency stimuli, the terminal can show an afterhyperpolarization, which is largely due to a transporter current of the Na/K-ATPase (Kim et al. 2007).

Na+ Channels

Because of its large size, the calyx presents a considerable capacitive load (Borst and Sakmann 1998a; Sun et al. 2004). To overcome this load, the last axonal hemi-node is unusually long and contains a high density of Na+ channels (Leão et al. 2005a). Modeling suggests that the exclusion of Na+ channels from the calyx terminal produces an action potential waveform with a shorter half-width (Leão et al. 2005a). Some of the Na+ channels have a very negative activation voltage. This sustained current contributes to the resting membrane potential and lowers the resting conductance (Huang and Trussell 2008).

Presynaptic K+ Channels

Presynaptically, both low-threshold K+ channels and high-threshold delayed rectifier-type K+ channels are expressed. The low-threshold currents are dominated by Kv1.2 homomers; in addition, the calyx has Kv1.1 and Kv1.3 channels; both Kv1.1 and Kv1.2 channels are excluded from the calyx itself but are concentrated at the transition zone between the axon and the terminal (Dodson et al. 2003), whereas the Kv1.3 channels are located in the calyx itself (Gazula et al. 2010). A possible role of these low-threshold K+ channels is to reduce excitability and prevent firing on the depolarizing afterpotential (Ishikawa et al. 2003). Deletion of Kv1.1 leads to a loss of temporal fidelity and the inability to follow high-frequency amplitude-modulated sound stimulation in vivo (Kopp-Scheinpflug, Fuchs et al. 2003).

The high-threshold K+ current is mostly composed of Kv3.1, which gates rapidly but inactivates slowly (Dodson et al. 2003). It is largely excluded from the release face (Elezgarai et al. 2003). In addition, immunocytochemical evidence for the presence of Kv3.4 has been reported (Ishikawa et al. 2003). The calyx of Held also expresses a large-conductance Ca^{2+}-activated potassium (BK) channel, which is present at low density and does not contribute to transmitter release (Ishikawa et al. 2003). Finally, the calyx expresses a fast-gating hyperpolarization activated mixed cationic current I_h that contributes little to transmitter release but can influence the resting membrane potential (Cuttle et al. 2001) and contribute to a post-tetanic afterhyperpolarization (Kim et al. 2007).

Ca2+ Channels

The immature calyx of Held has high-voltage-activated Ca^{2+} currents, which are mostly located within the terminal but to a lesser extent on the axon as well (Borst et al. 1995; Helmchen et al. 1997). The channels consist of a mixture of N-, R-, and P/Q-type Ca channels; the P/Q-type channels couple more efficiently to release than the other types, suggesting that they are more concentrated at active zones (Wu et al. 1999; Inchauspe et al. 2007). During the repolarization phase of the action potential, these channels open and admit a Ca^{2+} charge of about 0.8 pC (Borst and Sakmann 1996, 1998a). However, the action potential is so brief that not all Ca^{2+} channels are opened (Borst and Sakmann 1998a). An increase in the intraterminal Ca^{2+} concentration can speed up activation of the Ca^{2+} channels, and as a result the Ca^{2+} current during an action potential can facilitate on repetitive activation (Borst and Sakmann 1998b; Cuttle et al. 1998). This effect is mediated by neuronal Ca^{2+} sensor 1 (Tsujimoto et al. 2002) and calmodulin (Nakamura et al. 2008) and occurs exclusively in P/Q-type Ca^{2+} channels (Inchauspe et al. 2004; Ishikawa et al. 2005).

Although these are all high-threshold Ca^{2+} channels, some Ca^{2+} influx can already occur at potentials as negative as $-\!-60$ mV (Awatramani et al. 2005), and the Ca^{2+} channels are thus sensitive to subthreshold voltage changes produced by neuromodulators. Ca^{2+} channels, and thus transmitter release, can be inhibited by a variety of neurotransmitters acting via G protein-coupled receptors, including noradrenalin (Leão and von Gersdorff 2002); serotonin (Mizutani et al. 2006); and γ-aminobutyric acid (GABA), glutamate, and adenosine (Barnes-Davies and Forsythe 1995). GABA can block almost half of the Ca^{2+} current and at least 80% of transmitter release via its action on $GABA_B$ receptors (Isaacson 1998; Takahashi et al. 1998). These receptors slow the activation of the Ca^{2+} channels, and this effect is voltage-dependent (Isaacson 1998). The block is mediated by β/γ G-protein subunits (Kajikawa et al. 2001). The calyx of Held also has type II/III metabotropic glutamate receptors, which can reduce transmitter release by an inhibitory effect on presynaptic Ca^{2+} currents (Barnes-Davies and Forsythe 1995; Takahashi et al. 1996; Elezgarai et al. 1999, 2001). The different types of metabotropic receptors can be differentially activated depending on the stimulus pattern (von Gersdorff et al. 1997; Iwasaki and Takahashi 2001; Billups et al. 2005; Renden et al. 2005). In contrast to $GABA_B$ and adenosine receptors, the metabotropic receptors are thought to preferentially couple to the presynaptic P/Q-type Ca^{2+} channels (Inchauspe et al. 2007). Adenosine, which is released presynaptically during high-frequency signaling (Wong et al. 2006), can inhibit Ca channels via type 1 adenosine (A1) receptors (Kimura et al. 2003). Finally, serotonin can inhibit the presynaptic Ca^{2+} channels via $5HT_{1B}$ receptors (Mizutani et al. 2006). Adenosine, noradrenalin, and serotonin block the Ca^{2+} currents by about 10–20% and release by about 40% (Leão and von Gersdorff 2002; Kimura et al. 2003; Mizutani et al. 2006) and their effects are thus more modest than those of GABA or glutamate.

Ligand-Gated Ion Channels

The calyx of Held has both ionotropic $GABA_A$ and glycine receptors (Turecek and Trussell 2001, 2002). Due to the relatively high Cl^- concentration in the terminal, the chloride current through these receptors is depolarizing (Price and Trussell 2006); it can thus enhance synaptic transmission at the calyx synapse.

3.2.2 Ca-secretion Coupling in the Immature Calyx of Held

After entry, calcium ions can bind to calcium-binding proteins, stay free, be taken up by intracellular organelles, or be extruded from the cytoplasm. The intracellular

Ca^{2+} dynamics of the calyx have been well studied, both the Ca^{2+} requirements of fast transmitter release and its involvement in various forms of short-term plasticity (STP).

Calcium-Binding Proteins, Buffering, and Clearance

The immature calyx has a relatively small calcium buffer capacity and fast clearance (Helmchen et al. 1997). It expresses the slow calcium-binding protein parvalbumin (Lohmann and Friauf 1996; Felmy and Schneggenburger 2004), which speeds the decay of short-term facilitation of transmitter release (Müller et al. 2007). In addition, a fraction of terminals express calretinin (Felmy and Schneggenburger 2004). Ca^{2+} loads are cleared from the calyx of Held by a variety of plasma membrane transporters (Kim et al. 2005). In addition, mitochondria can help to clear Ca^{2+} from the cytoplasm when the Ca^{2+} load is larger or prolonged (Billups and Forsythe 2002; Kim et al. 2005).

Calcium Sensitivity of Transmitter Release

Calcium ions control transmitter release at the calyx of Held. Small changes in the extracellular calcium concentration or the calcium influx lead to large changes in the amount of transmitter release (Barnes-Davies and Forsythe 1995; Borst and Sakmann 1996). At an extracellular calcium concentration of 2 mM, the calyx releases about 150–210 quanta, whereas at an extracellular calcium concentration of 0.25 mM, most action potentials fail to release even a single vesicle (Borst and Sakmann 1996).

Although heterogeneities in the Ca^{2+} influx within the calyx of Held have been observed with calcium imaging (Rodriguez-Contreras et al. 2008), it is technically not possible to measure the Ca^{2+} concentrations at the release sites. The Ca^{2+} requirements of transmitter release at the immature calyx have therefore been studied by uniformly raising the Ca^{2+} concentration by flash photolysis of caged calcium. In these experiments, the amount of transmitter release was quantified with capacitance studies, which use an estimate of membrane surface as a measure for release or by measuring excitatory postsynaptic currents (EPSCs) with simultaneous postsynaptic recordings. It was observed that a small elevation of the intraterminal Ca^{2+} concentration above the resting value already leads to an increase in transmitter release and that a uniform elevation of the Ca^{2+} concentration to 10–25 µM leads to release rates that match the peak rate reached during an action potential (Bollmann et al. 2000; Schneggenburger and Neher 2000). Of course, during an action potential each vesicle will experience a different Ca^{2+} concentration, because each vesicle will be at a different distance from open Ca^{2+} channels (Meinrenken et al. 2002), but these values nevertheless constrain diffusional models of calcium-secretion coupling.

One other finding from the flash photolysis experiments was that the relation between the Ca^{2+} concentration in the terminal and the release rates was highly nonlinear.

At Ca^{2+} concentrations between about 2–5 μM, the release scales with the fourth or fifth power of the Ca^{2+} concentration (Bollmann et al. 2000; Schneggenburger and Neher 2000). The calcium sensor controlling phasic transmitter release thus acts as a rapidly equilibrating, supralinear amplifier of the local Ca^{2+} transient (Bollmann and Sakmann 2005). A molecular explanation for this strongly supralinear relation is that multiple calcium ions have to bind to an isoform of synaptotagmin to trigger release. Synaptotagmin is a vesicle protein that binds to both phospholipids and soluble NSF attachment protein receptors (SNARE) complexes in a Ca^{2+}-regulated manner (Südhof and Rothman 2009). Although other synaptotagmin isoforms are also expressed (Xiao et al. 2010), a knock-out study showed that synaptotagmin-2 is essential for fast neurotransmitter release at the calyx of Held (Sun et al. 2007). Of the different isoforms, synaptotagmin-2 has the fastest kinetics (Xu et al. 2007), which is in line with the rapid relay function of the calyx synapse. At a Ca^{2+} concentration <2 μM, the relation between the Ca^{2+} concentration and release rate becomes less steep. This has been interpreted to signify that the release rates triggered by an individual calcium sensor such as synaptotagmin increase as it binds more calcium ions (Lou et al. 2005), or, alternatively, that at low Ca^{2+} concentrations release is driven by a calcium sensor other than synaptotagmin-2 (Sun et al. 2007). A possible candidate would be Doc2 (Groffen et al. 2010).

Calcium Domains

When a Ca^{2+} channel opens, a steep gradient of increased intracellular Ca^{2+} concentration quickly develops; this elevated Ca^{2+} concentration is called a calcium domain or nanodomain. If multiple nearby channels open, the individual nanodomains will overlap, and together these domains constitute a microdomain of elevated Ca^{2+}. The possibility to control both the membrane potential and the composition of the cytoplasm of the calyx of Held in slice preparations has allowed testing whether most vesicles are released by a concerted action of multiple channels or whether the opening of single Ca^{2+} channels is sufficient. These two configurations differ in the way the release face is organized. In the nanodomain configuration, individual Ca^{2+} channels and vesicles have to be quite close, whereas in the microdomain configuration this relation can be less tight. Several experiments suggest that most released vesicles in the immature calyx face microdomains instead of nanodomains. As many as 60 Ca^{2+} channels open for each vesicle that is released, although of course many of these Ca^{2+} channels do not contribute to its release (Borst and Sakmann 1996). Their open probability is close to maximal during the repolarization phase of the action potential (Borst and Sakmann 1998a). A change in the open probability, without a change in the time course of opening or the driving force, leads to supralinear changes in the number of released vesicles (Borst and Sakmann 1999b). The different Ca^{2+} channel subtypes that contribute to release can be differentially blocked using specific toxins. The sum of the block of release exceeds 100%, indicating that the release of some vesicles is controlled by more than one subtype of

Ca^{2+} channels (Wu et al. 1999). Finally, the release is blocked relatively efficiently by the slow calcium buffer ethylene glycol-bis(2-aminoethylether)-N,N,N',N'-tetraacetic acid (EGTA), suggesting that the diffusional distance between Ca^{2+} channels and vesicles is relatively large (Borst and Sakmann 1996). Together, these experiments suggest that in the immature calyx of Held most vesicles are released by a concerted action of calcium ions entering through clustered Ca^{2+} channels (Meinrenken et al. 2002).

3.2.3 Vesicle Pools

The Readily Releasable Pool

Under standard brain slice recording conditions (including 2 mM calcium in the extracellular medium), the EPSCs produced by the immature calyx synapse undergo strong synaptic depression during repetitive stimulation (Borst et al. 1995). Although other processes can also contribute to short-term synaptic depression, as detailed herein, its most important cause is a decrease in the number of releasable vesicles (Schneggenburger et al. 1999; Wu and Borst 1999). The vesicles that can be immediately released by a very large stimulus have been operationally defined as the readily releasable pool (RRP). As clear as this definition is, the size of this pool, its morphological equivalent, the biochemical reactions that determine the transitions within the RRP and between the RRP and the reserve pool from which the RRP is replenished, and the causes for the observed heterogeneity with respect to release probability are still debated. To measure its size, a very large stimulus is given. Often this is a high-frequency afferent tetanus that triggers a train of action potentials in the calyx of Held, which, by definition, leads to rapid exhaustion of the RRP. The size of the RRP is calculated as the ratio of the sum of the peak amplitudes of the evoked EPSCs and the quantal size, after correcting for fast replenishment during the train. A size of ~700 is obtained with this method (Schneggenburger et al. 1999). A possible confounder with this measurement is saturation or desensitization of the postsynaptic glutamate receptors, but methods to prevent this, or to correct for it, have been described (Neher and Sakaba 2001). An even more challenging finding is that the size of the estimate of the RRP depends to some extent on the initial release probability. At lower release probability the synaptic currents do not "catch up" with the currents evoked at higher release probability, and as a result the estimates of the RRP are lower (Wölfel et al. 2007). Indeed, when very large stimuli are given in simultaneous pre- and postsynaptic voltage clamp experiments or by flash photolysis of caged calcium, higher estimates of ~2,300 were obtained (Schneggenburger and Neher 2000; Sakaba and Neher 2001c). Even larger values were obtained when release rates were estimated using capacitance methods (Sun and Wu 2001).

The size of the readily-releasable pool is similar to the number of docked vesicles at the calyx of Held, suggesting that most docked vesicles are immediately releasable (Sätzler et al. 2002; Taschenberger et al. 2002). In addition, there does not seem to be a clear preferred spatial relationship of the vesicles in the recycling pool with the active zone (de Lange et al. 2003). Even though at present it cannot be excluded that the RRP does not rigorously correspond to a physical vesicle pool (Pan and Zucker 2009), the possibility of distinguishing between the number of releasable vesicles and their release probability at the calyx of Held has been quite informative. One important finding is that the vesicles in the RRP are heterogeneous with respect to release probability. The first indication that this is the case is that following a large depleting stimulus, the vesicles that show the most rapid replenishment have a relatively low release probability (Wu and Borst 1999). More detailed studies showed that based on the release time course following large depolarizing steps in the presence of the slow calcium buffer EGTA, the RRP can be subdivided into a fast- and a slow-releasing pool (Sakaba and Neher 2001c). Two possible causes can contribute to this heterogeneity: it can be positional, due to a larger distance from open Ca^{2+} channels, or biochemical, due to intrinsic differences in their Ca^{2+} sensitivity. As a result of the strong Ca^{2+} gradients around open Ca^{2+} channels, a contribution of positional heterogeneity is inevitable (Meinrenken et al. 2002) and is supported by experimental evidence (Wadel et al. 2007). Evidence for biochemical heterogeneity is also strong. Flash photolysis has shown heterogeneity in the Ca^{2+} sensitivity of the different vesicles in the RRP (Wölfel et al. 2007), although the calcium sensitivity of the slow-releasing pool is at most twofold lower than of the other vesicles in the RRP (Wadel et al. 2007). A further complicating issue is that even the fast-releasing pool has been shown to be heterogeneous with respect to release probability; again, at least part of the heterogeneity is due to differences in Ca^{2+} sensitivity (Müller et al. 2010). Whatever the cause of the differences in release probability of vesicles in the RRP, the difference in the apparent Ca^{2+} sensitivity of the vesicles in the RRP suggests that there is an additional slow priming (or "superpriming", Schlüter et al. 2006) step that further increases release probability of the vesicles in the RRP. To what extent this is primarily a biochemical maturation step or if it is due to a closer proximity is important for interpreting how the RRP replenishes after release.

Full recovery from synaptic depression is quite slow, taking several seconds (von Gersdorff et al. 1997). If all vesicles would recover as slowly as that, the RRP would be rapidly exhausted during high-frequency signaling, which would obviously be incompatible with the function of the calyx synapse as an auditory relay. However, the synapse is more resistant to synaptic depression than predicted purely on the basis of the slow time course of the recovery phase. Some vesicles can recover much more quickly, but it seems that these vesicles become part of the slow-releasing pool. Vesicles that show rapid replenishment have relatively low release probability; as a result they are not effectively released by an action potential (Wu and Borst 1999). More systematic studies showed that the fast-releasing pool replenishes slowly, whereas the slow-releasing pool replenishes rapidly (Sakaba and Neher 2001c). The functional significance of the slow-release pool is not yet clear. The release of the vesicles in the slow-releasing pool depends relatively strongly on the buildup of residual calcium (Sakaba and Neher 2001c). It has been proposed that this residual calcium can facilitate the release of the reluctant vesicles during high-frequency trains, thus counteracting depression and allowing the terminal to make use of this pool of rapidly replenishing vesicles (Wu and Borst 1999), but to what extent these vesicles can contribute to phasic release or if they only contribute to delayed (asynchronous) release is still debated (Sakaba 2006; Müller et al. 2010).

The accessibility of the calyx of Held to patch-clamp recording has made it possible to dissect some of the biochemical mechanisms that govern the replenishment of the RRP or of its subpools. Recovery of the RRP depends on ATP and an intact cytoskeleton (Sakaba and Neher 2003b). An important regulator of replenishment is Ca^{2+}. The replenishment of the fast-releasing vesicles of the RRP is sped up by Ca^{2+} (Wang and Kaczmarek 1998; Sakaba and Neher 2001a). Its dependence on Ca^{2+} is approximately linear (Hosoi et al. 2007) and involves the calcium-binding protein calmodulin (Sakaba and Neher 2001a), VAMP-2 (Sakaba et al. 2005), and synaptotagmin-2 (Young and Neher 2009). This process can speed up recovery substantially (Hosoi et al. 2007), and, depending on how Ca^{2+} accumulates presynaptically during trains, make the steady-state amplitude of phasic synaptic currents much less dependent on stimulus frequency (Neher and Sakaba 2008), which is obviously a desirable feature for a relay synapse such as the calyx synapse.

Second messenger pathways may modulate the dynamics of vesicle cycling. The replenishment of the fast-releasing pool depends on cAMP (Sakaba and Neher 2001b), acting via the cAMP-dependent guanine nucleotide exchange factor (Sakaba and Neher 2003a; Kaneko and Takahashi 2004). By contrast, replenishment of this pool is inhibited by myosin light chain kinase (Srinivasan et al. 2008). Interestingly, this kinase may also increase the size of the RRP during post-tetanic potentiation (Lee et al. 2008), as detailed later. Phorbol esters enhance transmitter release at many synapses including the calyx. They can change the calcium sensitivity of vesicles in the RRP via an action on both protein kinase C and munc-13 (Hori et al. 1999; Lou et al. 2005; Lou et al. 2008a). It is not yet known whether phorbol esters act preferentially on the fast-releasing vesicles.

3.2.4 Endocytosis

Using voltage clamp, it is possible to measure capacitance changes in the calyx that are associated with vesicle cycling. Despite the presence of the axon, which slows the charging of its membrane (Borst and Sakmann 1998a), these capacitance measurements can be used to measure changes in calyceal surface area resulting from exo- and endocytosis (Sun and Wu 2001; Sun et al. 2004). These measurements have shown that the time it takes before reuptake of vesicles can vary from a few seconds (at room temperature), a process named rapid endocytosis, to tens of seconds or longer, referred to as slow endocytosis (Sun and Wu 2001). The biochemical mechanisms that underlie these differences in time course are still heavily debated (Wu et al. 2007). Part of the interest for the distinction between fast and slow endocytosis has arisen from the idea that the fast form may reflect a form of endocytosis called kiss-and-run, in which vesicles do not fully collapse and thus recycle without the need for a clathrin-coat; by contrast, the slow form may represent the more classical form of endocytosis, in which vesicles need dynamin and clathrin for reuptake (He and Wu 2007). However, to what extent dynamin and clathrin are required for both forms is still unclear (Yamashita et al. 2005; Lou et al. 2008; Xu et al. 2008; Wu and Wu 2009).

Following very large stimuli, another form of endocytosis called bulk endocytosis is observed, in which large endosome-like structures directly bud off from the plasma membrane (de Lange et al. 2003; Wu and Wu 2007). Bulk endocytosis can be viewed as an emergency brake mechanism that prevents the terminal from becoming too large and may be less important at physiological temperatures and in more mature animals.

Intracellular Ca^{2+} can speed up the time course of endocytosis, probably via calmodulin (Wu et al. 2009), which in its turn may act via calcineurin (Yamashita et al. 2010). Both the replenishment of the RRP and endocytosis are thus stimulated strongly by Ca^{2+}/calmodulin. Although endocytosis is a rate-limiting step in the vesicle cycle and RRP replenishment needs free docking sites, RRP replenishment generally does not depend directly on endocytosis (Hosoi et al. 2009; Wu and Wu 2009). Wherever it has been possible to do similar experiments at other terminals, such as the bouton terminals of the hippocampus, similar results have generally been obtained to those in the calyx synapse, suggesting that endocytosis is mechanistically conserved (Dittman and Ryan 2009).

3.2.5   Postsynaptic Glutamate Receptors

Principal neurons of the MNTB have both AMPA- and NMDA-type glutamate receptors (Forsythe and Barnes-Davies 1993). The AMPA receptors show fast gating due to the relative abundance of the *flop* splice variant of the GluA4 subunit (Geiger et al. 1995). Following release of a vesicle, glutamate is thought to briefly reach mM concentrations in the synaptic cleft (Renden et al. 2005). Nevertheless, a single vesicle does not saturate the postsynaptic receptors opposite a release site (Ishikawa et al. 2002). However, under standard brain slice conditions, the release probability is quite high, and multiquantal release from single active zones may occur (Sun and Wu 2001). Because so many vesicles are released by a single action potential, the EPSP is generally suprathreshold. Since the soma of principal neurons has a low Na^+ channel density, the postsynaptic action potential is most likely always triggered in the axon, after which it back-propagates to the soma (Leão et al. 2005a; Leão et al. 2008).

3.2.6   Voltage-Dependent K^+ Channels of the Principal Neurons

The principal neuron contains both low- and high-threshold K^+ channels. The main role of the low-threshold K^+ channels is to dampen excitability, whereas the main role of the high-voltage-activated K^+ channels is to ensure rapid action potential repolarization (Brew and Forsythe 1995). A characteristic feature of the principal neuron is that the calyceal EPSP triggers only a single action potential (Wu and

Kelly 1991). This is due to the presence of dendrotoxin-sensitive low-threshold K^+ channels (Brew and Forsythe 1995), consisting of Kv1.1/Kv1.2 heteromers that are located at the initial segment (Dodson et al. 2002). ERG (*Ether-à-go-go*–related gene, Kv11) channels have a similar role as these Kv1 channels in the mouse MNTB (Hardman and Forsythe 2009). In addition, the Kv1 channels contribute to the threshold and amplitude of the action potentials in the principal neurons (Dodson et al. 2002; Klug and Trussell 2006). The axon initial segment also contains high-threshold Kv2 channels; they have slow gating, making them important during high-frequency signaling (Johnston et al. 2008). The Kv3 channels, with their high activation threshold and rapid gating, contribute to the brief duration of the postsynaptic action potentials, which is necessary to allow high firing rates (Brew and Forsythe 1995; Wang et al. 1998). Interestingly, the Kv3.1 channel is regulated in a complex way by activity. Medial high-frequency cells show higher expression of the high-voltage-activated K^+ channel Kv3.1 (Li et al. 2001), resulting in briefer postsynaptic action potentials medially (Brew and Forsythe 2005). This gradient depends on hearing (von Hehn et al. 2004; Strumbos et al. 2010b) and on the presence of the Fragile X mental retardation protein (Strumbos et al. 2010a). Intense electrical activity in the MNTB leads to dephosphorylation of this channel, which in its turn increases its activity, facilitating high-frequency spiking (Song et al. 2005). On the other hand, intense activity can also stimulate nitric oxide production in the principal neuron, whereas nitric oxide inhibits the postsynaptic Kv3 channels (Steinert et al. 2008). Finally, the principal neurons also have a high density of sodium-activated K^+ channels, which contribute to temporal fidelity (Yang et al. 2007). These channels are activated by the fragile X mental retardation protein and are especially prominent in the lateral low-frequency part of the MNTB (Brown et al. 2010).

3.2.7   Short-Term Plasticity at the Immature Calyx Synapse

In the immature calyx synapse, the strength of synaptic transmission depends strongly on recent history. This synapse displays several forms of short-term plasticity, and because of the accessibility of the calyx terminal to patch-clamp recordings, the underlying mechanisms are relatively well understood. The two main types of plasticity are short-term facilitation, which is an increase in strength, and short-term depression, which is a decrease in strength. In addition, following long trains, post-tetanic potentiation (PTP) has been observed.

Short-Term Synaptic Facilitation

Short-term facilitation is a short-lasting increase in synaptic transmission that can be observed during repetitive stimulation (Neher and Sakaba 2008). It is best observed when the release probability is low, because under these conditions synaptic depression is avoided (Borst et al. 1995). Short-term facilitation is a presynaptic form of STP, which results from an increase in the release probability of the vesicles in the RRP. If a second action potential is elicited after a brief interval, release will be about twice as large on average. By varying the interval between both pulses, its decay time course can be studied. In intact terminals it decays more or less exponentially with a time constant around 30 ms at room temperature; this time course matches the time course of the cytoplasmic Ca^{2+} clearance following an action potential (Müller et al. 2007). This correspondence suggests that synaptic facilitation is due to "residual" calcium, that is, the Ca^{2+} that has to be cleared from the presynaptic cytoplasm following an action potential (Katz and Miledi 1968). Small elevations of the presynaptic Ca^{2+} concentration can increase transmitter release (Awatramani et al. 2005). Residual calcium facilitates transmitter release in several different ways. First, it can increase Ca^{2+} influx by facilitating the presynaptic Ca^{2+} channels (Borst and Sakmann 1998b; Cuttle et al. 1998). The increased Ca^{2+} influx leads to increased release (Inchauspe et al. 2004, 2007). Second, the residual calcium will summate with the Ca^{2+} entering during the next action potential, leading to a somewhat higher Ca^{2+} concentration at the release sensor (Meinrenken et al. 2002). Although the relative importance of these different mechanisms is still debated, it seems that both contribute to facilitation, but they are insufficient to explain facilitation entirely (Müller et al. 2008). Therefore, additional mechanisms have been invoked. One possibility is that some of the Ca^{2+} binds to endogenous calcium buffers, and as a result these buffers can no longer compete with synaptotagmin for Ca^{2+} that enters during the next action potential (Felmy et al. 2003). However, the properties of these buffers that are needed for this scheme to work (Matveev et al. 2004) do not match known endogenous buffers at the calyx (Habets and Borst 2006; Müller et al. 2008). Other possibilities include the presence of a high-affinity calcium sensor separate from synaptotagmin-2 (Sun et al. 2007) or an action of Ca^{2+} on vesicles, for example, on a priming or a replenishment step, analogous to its action on the replenishment of the fast-releasing vesicles in the RRP.

Short-Term Synaptic Depression

Under standard brain slice conditions, the synaptic strength of EPSPs generated by the immature calyx progressively declines during ongoing presynaptic stimulation until a steady state is reached (Borst et al. 1995). Several mechanisms contribute to short-term depression. The main cause appears to be a depletion of the RRP, as discussed earlier. Other presynaptic mechanisms include a change in Ca^{2+} influx due to Ca^{2+} channel inactivation (Forsythe et al. 1998), which is especially important at low stimulation frequencies (Xu and Wu 2005), activation of presynaptic glutamate receptors (von Gersdorff et al. 1997), or depletion of Ca^{2+} from the synaptic cleft (Borst and Sakmann 1999a). At high stimulation frequencies the postsynaptic responsiveness decreases, due to glutamate receptor desensitization and saturation, presenting a sizable postsynaptic contribution to the synaptic depression in immature synapses (Neher and Sakaba 2001; Taschenberger et al. 2002; Wong et al. 2003).

Post-tetanic Potentiation

Following a long train of presynaptic action potentials, synaptic transmission may be enhanced for up to several minutes. This process is called post-tetanic potentiation and is accompanied by a long-lasting increase in presynaptic Ca^{2+} (Habets and Borst 2005; Korogod et al. 2005). The Ca^{2+} transient originates from mitochondria and builds up gradually during the tetanus (Lee et al. 2008). The potentiation of release is due to both an increase in the size of the RRP, especially in the fast-releasing vesicles, and an increase in the release probability of vesicles in the RRP (Habets and Borst 2005; Lee et al. 2010). The mechanisms underlying the increase in release probability are probably similar to the mechanisms underlying short-term synaptic facilitation. PTP is accompanied by an increase in action potential–evoked presynaptic Ca^{2+} influx (Habets and Borst 2006). Calmodulin is needed for PTP (Lee et al. 2010), but the involvement of protein kinase C is still debated (Korogod et al. 2007; Lee et al. 2008). The increase in the RRP may outlast the increase in the release probability and could be especially important at physiological temperatures (Habets and Borst 2007). The increase in RRP depends on activation of myosin light chain kinase (Lee et al. 2008).

3.3   Maturation of the Calyx Synapse

For technical reasons, most of the cell physiological work has been performed on calyces before the onset of hearing. Although in vivo the calyx fires at brief bursts of about 100 Hz before hearing onset (Tritsch et al. 2010), this is far removed from its adult role. Not surprisingly, the calyx synapse undergoes many changes that enable it to take up its role as an auditory relay. Many of these changes happen around the onset of hearing, marking this period as one of major reorganization.

3.3.1   Morphological Maturation of the Calyx Synapse

At about P8 in rodents, parts of the calyx become thinner, by P14 it becomes fenestrated, assuming its mature, floral-like structure (Kandler and Friauf 1993; Kil et al. 1995). Before hearing onset, the calyx can have many calyceal collaterals, but eventually most of these collaterals disappear (Morest 1968b; Kandler and Friauf 1993; Rodriguez-Contreras et al. 2008; Ford et al. 2009).

The surface area of the calyx peaks around the onset of hearing (Wimmer et al. 2006). At this time, the size of individual active zones has decreased, but their density remains unchanged (Taschenberger et al. 2002; Dondzillo et al. 2010). As with the decline in active zone size, synaptic vesicles are grouped into successively smaller clusters with maturation (Fig. 5.1; Wimmer et al. 2006), and this is accompanied by an increase in the number of docked vesicles (Taschenberger et al. 2002; Yang et al. 2010).

3.3.2 Physiological Maturation of the Calyx Synapse

To take up its relay function, the calyx has to transmit faster and become more resistant to fatigue. The increase in speed is to a large extent attained by changes in voltage-dependent ion channels, while the resistance to fatigue involves, among other things, changes in release probability.

Developmental Changes in Ion Channels

Both pre- and postsynaptic action potentials become faster during development, contributing to a strong decrease in synaptic delay (Taschenberger and von Gersdorff 2000). Both changes in Na^+ and in K^+ channels contribute to these changes. Presynaptic Na^+ channels recover more rapidly from inactivation (Leão et al. 2005a). The density of both Kv1 and Kv3 presynaptic K^+ currents becomes larger between P7 and P14, with little changes afterward (Nakamura and Takahashi 2007). Especially prominent is the increase in the rapidly activating Kv3 channels, which is paralleled by a strong increase in Kv3.1 labeling (Elezgarai et al. 2003); this change contributes to the decrease in the width of the presynaptic action potentials.

The presynaptic Ca^{2+} channels also undergo several changes during development. During early development, between P4 and P10, the size of the presynaptic Ca^{2+} currents increases (Chuhma and Ohmori 1998). Around hearing onset, the N- and R-channels disappear from the calyx, and the transmission becomes entirely dependent on P/Q-type Ca^{2+} channels (Iwasaki et al. 2000). The postsynaptic glutamate receptors also show several developmental changes. The gating kinetics of AMPAR EPSCs become gradually faster between P5 and P14 (Taschenberger and von Gersdorff 2000). These changes probably result from increased expression of the *flop* splice variant of GluA4 subunits after P8 (Caicedo and Eybalin 1999; Joshi et al. 2004; Koike-Tani et al. 2005). The increase in expression is mostly at postsynaptic densities; as a result there is increased compartmentalization of AMPA receptors in the adult principal cells (Hermida et al. 2006, 2010). The miniature EPSCs, which reflect the size of the transmitter quantum, show little change in amplitude until hearing onset (Chuhma and Ohmori 1998; Iwasaki and Takahashi 2001), but their amplitude becomes considerably larger afterward (Wang et al. 2008; Lorteije et al. 2009).

By contrast, the NMDA component of the EPSC is strongly reduced after hearing onset (Taschenberger and von Gersdorff 2000), although even in the adult synapse they remain present (Caicedo and Eybalin 1999). One function of the NMDA receptors in the adult synapse is to stimulate the production of nitric oxide in the principal neurons (Steinert et al. 2008, 2010). The NMDA-EPSCs also become faster (Taschenberger and von Gersdorff 2000), and a switch from NR2B to NR2A subunits contributes to the faster time course of NMDA EPSCs in adult animals (Steinert et al. 2010). The developmental decline in the NMDAR currents and their faster kinetics contribute to increased fidelity of transmission (Taschenberger and von Gersdorff 2000; Futai et al. 2001; Joshi and Wang 2002). Interestingly, this developmental down-regulation can be postponed by cochlear ablation before hearing onset (Futai et al. 2001).

Developmental Changes in Calcium-Secretion Coupling

A consequence of the developmental decrease in the width of the presynaptic action potential is that the influx of Ca^{2+} becomes much briefer and smaller in amplitude after hearing onset (Fedchyshyn and Wang 2005). Because of the strong dependence of release on Ca^{2+} influx, one would expect a strong reduction in the size of evoked EPSCs but little or no change is observed (Taschenberger and von Gersdorff 2000). The vesicles in the RRP indeed have a lower release probability (Iwasaki and Takahashi 2001; Taschenberger et al. 2002), to which a small decrease in the Ca^{2+} affinity of the phasic sensor of transmitter release after hearing onset contributes (Wang et al. 2008; Kochubey et al. 2009). However, the lower release probability is compensated for by an increase in the size of the RRP and a more efficient coupling between Ca^{2+} influx and transmitter release (Taschenberger and von Gersdorff 2000; Iwasaki and Takahashi 2001; Taschenberger et al. 2002). Based on a decreased sensitivity of transmitter release to the slow calcium buffer EGTA and a reduced slope in the relation between Ca^{2+} current and release rates, the most likely explanation for the more efficient coupling is a decrease in the average distance between Ca^{2+} channels and vesicles (Fedchyshyn and Wang 2005). A developmental reorganization of the localization of the filamentous protein septin 5 within the calyx contributes to this change in the coupling between Ca^{2+} channels and vesicles (Yang et al. 2010).

The rate of Ca^{2+} clearance from the terminal becomes much faster with development, occurring even before hearing onset, and this leads to a reduction in delayed (asynchronous) release (Chuhma and Ohmori 1998, 2001). After hearing onset, a further, small reduction in asynchronous release occurs (Taschenberger and von Gersdorff 2000; Taschenberger et al. 2005). Apart from changes in Ca^{2+} transporters, changes in the expression of presynaptic calcium-binding proteins can also contribute to the developmental changes in presynaptic Ca^{2+} handling. The slow calcium-binding protein parvalbumin increases strongly around hearing onset (Lohmann and Friauf 1996 ; Felmy and Schneggenburger 2004). Somewhat later, the concentration of the calcium-binding protein calretinin also increases, although its expression is more variable than that of parvalbumin (Felmy and Schneggenburger 2004). The increase in calcium-binding proteins may increase temporal precision during high-frequency trains (Fedchyshyn and Wang 2007).

After hearing onset, the synapse becomes more resistant to synaptic depression (Taschenberger and von Gersdorff 2000), which is the result of several mechanisms. Faster Ca^{2+} clearance will decrease the impact of Ca-dependent inactivation of Ca channels on synaptic depression (Taschenberger et al. 2002; Nakamura et al. 2008). Similar mechanisms are probably responsible for the reduction of PTP, which depends on a long-lasting increase in residual calcium at more mature synapses (Korogod et al. 2005). The lower release probability, the faster recovery from desensitization of the AMPA receptors, and the faster clearance of glutamate from the synaptic cleft together strongly reduce the contribution of glutamate receptor desensitization to short-term depression after hearing onset (Joshi and Wang 2002; Taschenberger et al. 2002, 2005; Koike-Tani et al. 2008). These same factors reduce the contribution of activation of the presynaptic metabotropic glutamate receptor to short-term depression (Iwasaki and Takahashi 2001).

These effects all contribute to the transformation of the calyx synapse into a relay synapse that is more resistant to fatigue. In fact, little evidence for synaptic depression was observed in vivo in the adult calyx synapse (Lorteije et al. 2009). These changes are accompanied by a down-regulation of the effects of group II/III metabotropic glutamate receptors (Elezgarai et al. 1999; Renden et al. 2005), adenosine (Kimura et al. 2003), serotonin (Mizutani et al. 2006), and noradrenalin (Leão and von Gersdorff 2002). The developmental decline in the stimulatory effects of cAMP on transmitter release is consistent with the reduction in the effects of these modulators (Kaneko and Takahashi 2004).

3.3.3 The Role of Activity in the Maturation of the Calyx Synapse

The role of activity in the maturation of the calyx synapse has been studied using mouse mutants that are deaf because of a cochlear mutation. The effects of these changes have been surprisingly mild. For example, in a Ca^{2+} channel mutant where release from hair cells was completely absent, transmission at P14–17, an age where many of the changes that are thought to be the result of hearing onset have already taken place, was in many respects indistinguishable from mice with normal hearing (Erazo-Fischer et al. 2007). These mice had somewhat larger release probability, which may have been due to broader presynaptic APs, leading to larger Ca^{2+} influx and stronger synaptic depression. In addition, the NMDA-EPSCs were clearly larger. Studies on a spontaneously deaf mouse mutant showed no difference in calyx morphology (Youssoufian et al. 2008), little change in synaptic transmission shortly after hearing onset (Oleskevich et al. 2004; Youssoufian et al. 2005), an increase in the hyperpolarization-activated cation currents (Leão et al. 2005b) but reduced low-threshold K^+ channels, a condition leading to increased postsynaptic excitability (Leão et al. 2004). In these mice, the normal tonotopic gradients of K^+ currents and of the hyperpolarization-activated cation currents are no longer present (Leão et al. 2006). From these studies in deaf mice it has become clear that development of the calyx synapse can proceed to a large extent normally in the absence of auditory nerve activity, but a possible instructive role of spontaneous activity in the AVCN still awaits further study (Youssoufian et al. 2008).

4 Outlook

The unique accessibility of the calyx to direct electrophysiological recordings has made it possible to study the electrical and biochemical steps that control transmitter release at a mammalian synapse. In the past 15 years it has functioned as a model system to study the biophysical properties of transmitter release and has provided a wealth of information about the electrical and biochemical processes that control that transmitter release and allow the calyx synapse to act as an auditory relay. Much still remains to be investigated. In the next section, some outstanding questions are summarized, which may be answered in the next 15 years.

4.1 Technical Hurdles

Technical advances have driven research on the calyx, but these studies have mostly been dominated by cell physiological questions, whereas its advantages for studying molecular mechanisms have been less clear. Although it is possible to use the calyx as an expression system by using viral vectors (Wimmer et al. 2004), this is not an easy procedure. Some *cre* lines have been reported that allow a relatively specific expression at the calyx synapse (Voiculescu et al. 2000; Saul et al. 2008), but an extension in the repertoire of genetic tools is clearly desirable. Another hurdle is the lack of a culture system in which the properties of the calyx synapse can be studied under defined conditions. Although it has been possible to culture the principal neurons, it has not yet been possible to preserve the calyx synapse or to induce its formation in vitro (Lohmann et al. 1998; Lou et al. 2008; Tong et al. 2010).

4.2 Development

A conspicuous feature of the calyx synapse is the fact that each principal cell is contacted by only a single giant axosomatic terminal. The developmental mechanisms that ensure this one-to-one relation are almost fully unexplored. Expression studies may result in the identification of pre- and postsynaptic proteins that pave the first steps of the road to a relay synapse.

4.3 Anatomical Specializations and Heterogeneity

Despite progress, there is still much to be discovered about the way vesicles recycle and which processes control replenishment, release, and endocytosis. A combination of cell physiology, electron microscopy, optical imaging, and molecular biological techniques will hopefully allow identification of the physical vesicle pools at the calyx and the molecular mechanisms that eventually control transmitter release.

Most previous studies have been aimed at the identification of general mechanisms for neurotransmitter release at mammalian synapses. However, although the calyx synapse has functioned as a model synapse to study synaptic transmission, it does contain a number of anatomical specializations, the function of which still remains to be explored. Another interesting question is the degree of specialization within the MNTB and what the role of afferent activity is in setting up functional differences between calyces. Even though one of its main attractions has been that the variability between different principal neurons is much smaller than in, for example, the cerebral cortex, there are differences between calyces as well. Apart from the well-known medial-to-lateral tonotopic gradient within the MNTB, which is correlated with difference in the expression of K^+ channels, among others (e.g., Gazula et al. 2010), one may ask whether there are functional gradients in rostro-caudal direction as well, because the inputs from the PVCN terminate more caudally in the MNTB (Harrison and Irving 1964; Vater and Feng 1990). How are the projection patterns of principal neurons related to their functional properties? What is the significance of structural differences between calyces (Spirou et al. 2008)? More generally, to what extent do functional properties of the calyx synapse depend on activity and what are the mechanisms that allow long-term changes in synaptic transmission at this synapse (Walmsley et al. 2006)?

4.4   *Physiology*

Because of its unique properties, most of the experimental efforts have focused on the calyx synapse, whereas the functional role of the dendritic excitatory inputs to the principal cells is unknown. The same holds true for the large, somatic inhibitory inputs on principal neurons. There is consensus that the calyx synapse acts as a fast relay synapse, enabling the principal neurons to provide inhibition to many ipsilateral auditory brainstem nuclei. Its monaural function in providing tonic inhibition to the SPON and its binaural function in the high-frequency ILD pathway, involving the projection to the LSO, are well established. The function of the projection to the MSO is less well established. How important is precise timing in this projection? These questions illustrate some of the possible directions for future research. In combination with its unique technical advantages, the many unanswered questions about synaptic transmission at the calyx of Held will ensure that it remains popular among neuroscientists in the future.

Acknowledgments   We thank Dr. Larry Trussell for helpful comments on an earlier version of this chapter. The research was supported by ALW-NWO (Moving vesicles, 814.02.004).

References

Altman, J., & Bayer, S. A. (1980). Development of the brain stem in the rat. III. Thymidine-radiographic study of the time of origin of neurons of the vestibular and auditory nuclei of the upper medulla. *Journal of Comparative Neurology, 194*(4), 877–904.

Awatramani, G. B., Turecek, R., & Trussell, L. O. (2004). Inhibitory control at a synaptic relay. *Journal of Neuroscience, 24*(11), 2643–2647.

Awatramani, G. B., Price, G. D., & Trussell, L. O. (2005). Modulation of transmitter release by presynaptic resting potential and background calcium levels. *Neuron, 48*(1), 109–121.

Banks, M. I., & Smith, P. H. (1992). Intracellular recordings from neurobiotin-labeled cells in brain slices of the rat medial nucleus of the trapezoid body. *Journal of Neuroscience, 12*(7), 2819–2837.

Barnes-Davies, M., & Forsythe, I. D. (1995). Pre- and postsynaptic glutamate receptors at a giant excitatory synapse in rat auditory brainstem slices. *Journal of Physiology, 488*(2), 387–406.

Bergsman, J. B., De Camilli, P., & McCormick, D. A. (2004). Multiple large inputs to principal cells in the mouse medial nucleus of the trapezoid body. *Journal of Neurophysiology, 92*(1), 545–552.

Billups, B., & Forsythe, I. D. (2002). Presynaptic mitochondrial calcium sequestration influences transmission at mammalian central synapses. *Journal of Neuroscience, 22*(14), 5840–5847.

Billups, B., Graham, B. P., Wong, A. Y. C., & Forsythe, I. D. (2005). Unmasking group III metabotropic glutamate autoreceptor function at excitatory synapses in the rat CNS. *Journal of Physiology, 565*(Pt. 3), 885–896.

Bollmann, J. H., & Sakmann, B. (2005). Control of synaptic strength and timing by the release-site Ca^{2+} signal. *Nature Neuroscience, 8*(4), 426–434.

Bollmann, J. H., Sakmann, B., & Borst, J. G. G. (2000). Calcium sensitivity of glutamate release in a calyx-type terminal. *Science, 289*(5481), 953–957.

Borst, J. G. G., & Sakmann, B. (1996). Calcium influx and transmitter release in a fast CNS synapse. *Nature, 383*(6599), 431–434.

Borst, J. G. G., & Sakmann, B. (1998a). Calcium current during a single action potential in a large presynaptic terminal of the rat brainstem. *Journal of Physiology, 506*, 143–157.

Borst, J. G. G., & Sakmann, B. (1998b). Facilitation of presynaptic calcium currents in the rat brainstem. *Journal of Physiology, 513*, 149–155.

Borst, J. G. G., & Sakmann, B. (1999a). Depletion of calcium in the synaptic cleft of a calyx-type synapse in the rat brainstem. *Journal of Physiology, 521*, 123–133.

Borst, J. G. G., & Sakmann, B. (1999b). Effect of changes in action potential shape on calcium currents and transmitter release in a calyx-type synapse of the rat auditory brainstem. *Philosophical Transactions of the Royal Society of London. Series B, Biological Sciences, 354*(1381), 347–355.

Borst, J. G. G., Helmchen, F., & Sakmann, B. (1995). Pre- and postsynaptic whole-cell recordings in the medial nucleus of the trapezoid body of the rat. *Journal of Physiology, 489*(Pt. 3), 825–840.

Boudreau, J. C., & Tsuchitani, C. (1968). Binaural interaction in the cat superior olive S segment. *Journal of Neurophysiology, 31*(3), 442–454.

Brew, H. M., & Forsythe, I. D. (1995). Two voltage-dependent K^+ conductances with complementary functions in postsynaptic integration at a central auditory synapse. *Journal of Neuroscience, 15*(12), 8011–8022.

Brew, H. M., & Forsythe, I. D. (2005). Systematic variation of potassium current amplitudes across the tonotopic axis of the rat medial nucleus of the trapezoid body. *Hearing Research, 206*(1–2), 116–132.

Brown, M. R., Kronengold, J., Gazula, V.-R., Chen, Y., Strumbos, J. G., Sigworth, F. J., Navaratnam, D., & Kaczmarek, L. K. (2010). Fragile X mental retardation protein controls gating of the sodium-activated potassium channel Slack. *Nature Neuroscience, 13*(7), 819–821.

Caicedo, A., & Eybalin, M. (1999). Glutamate receptor phenotypes in the auditory brainstem and mid-brain of the developing rat. *European Journal of Neuroscience, 11*(1), 51–74.

Casey, M. A., & Feldman, M. L. (1982). Aging in the rat medial nucleus of the trapezoid body. I. Light microscopy. *Neurobiology of Aging, 3*(3), 187–195.

Casey, M. A., & Feldman, M. L. (1985). Aging in the rat medial nucleus of the trapezoid body. II. Electron microscopy. *Journal of Comparative Neurology, 232*(3), 401–413.

Casey, M. A., & Feldman, M. L. (1988). Age-related loss of synaptic terminals in the rat medial nucleus of the trapezoid body. *Neuroscience, 24*(1), 189–194.

Chuhma, N., & Ohmori, H. (1998). Postnatal development of phase-locked high-fidelity synaptic transmission in the medial nucleus of the trapezoid body of the rat. *Journal of Neuroscience, 18*(1), 512–520.

Chuhma, N., & Ohmori, H. (2001). Differential development of Ca^{2+} dynamics in presynaptic terminal and postsynaptic neuron of the rat auditory synapse. *Brain Research, 904*(2), 341–344.

Cuttle, M. F., Tsujimoto, T., Forsythe, I. D., & Takahashi, T. (1998). Facilitation of the presynaptic calcium current at an auditory synapse in rat brainstem. *Journal of Physiology, 512*(3), 723–729.

Cuttle, M. F., Rusznák, Z., Wong, A. Y., Owens, S., & Forsythe, I. D. (2001). Modulation of a presynaptic hyperpolarization-activated cationic current (I_h) at an excitatory synaptic terminal in the rat auditory brainstem. *Journal of Physiology, 534*(Pt. 3), 733–744.

de Lange, R. P. J., de Roos, A. D. G., & Borst, J. G. G. (2003). Two modes of vesicle recycling in the rat calyx of Held. *Journal of Neuroscience, 23*(31), 10164–10173.

Dittman, J., & Ryan, T. A. (2009). Molecular circuitry of endocytosis at nerve terminals. *Annual Review of Cell and Developmental Biology, 25*, 133–160.

Dodson, P. D., Barker, M. C., & Forsythe, I. D. (2002). Two heteromeric Kv1 potassium channels differentially regulate action potential firing. *Journal of Neuroscience, 22*(16), 6953–6961.

Dodson, P. D., Billups, B., Rusznák, Z., Szũcs, G., Barker, M. C., & Forsythe, I. D. (2003). Presynaptic rat Kv1.2 channels suppress synaptic terminal hyperexcitability following action potential invasion. *Journal of Physiology, 550*(Pt. 1), 27–33.

Dondzillo, A., Sätzler, K., Horstmann, H., Altrock, W. D., Gundelfinger, E. D., & Kuner, T. (2010). Targeted three-dimensional immunohistochemistry reveals localization of presynaptic proteins Bassoon and Piccolo in the rat calyx of Held before and after the onset of hearing. *Journal of Comparative Neurology, 518*(7), 1008–1029.

Elezgarai, I., Benítez, R., Mateos, J. M., Lázaro, E., Osorio, A., Azkue, J. J., Bilbao, A., Lingenhoehl, K., Van Der Putten, H., Hampson, D. R., Kuhn, R., Knöpfel, T., & Grandes, P. (1999). Developmental expression of the group III metabotropic glutamate receptor mGluR4a in the medial nucleus of the trapezoid body of the rat. *Journal of Comparative Neurology, 411*(3), 431–440.

Elezgarai, I., Bilbao, A., Mateos, J. M., Azkue, J. J., Benítez, R., Osorio, A., Diez, J., Puente, N., Doñate-Oliver, F., & Grandes, P. (2001). Group II metabotropic glutamate receptors are differentially expressed in the medial nucleus of the trapezoid body in the developing and adult rat. *Neuroscience, 104*(2), 487–498.

Elezgarai, I., Díez, J., Puente, N., Azkue, J. J., Benítez, R., Bilbao, A., Knöpfel, T., Doñate-Oliver, F., & Grandes, P. (2003). Subcellular localization of the voltage-dependent potassium channel Kv3.1b in postnatal and adult rat medial nucleus of the trapezoid body. *Neuroscience, 118*(4), 889–898.

Englitz, B., Tolnai, S., Typlt, M., Jost, J., & Rübsamen, R. (2009). Reliability of synaptic transmission at the synapses of Held in vivo under acoustic stimulation. *PLoS One, 4*(10), e7014.

Erazo-Fischer, E., Striessnig, J., & Taschenberger, H. (2007). The role of physiological afferent nerve activity during *in vivo* maturation of the calyx of Held synapse. *Journal of Neuroscience, 27*(7), 1725–1737.

Fedchyshyn, M. J., & Wang, L.-Y. (2005). Developmental transformation of the release modality at the calyx of Held synapse. *Journal of Neuroscience, 25*(16), 4131–4140.

Fedchyshyn, M. J., & Wang, L.-Y. (2007). Activity-dependent changes in temporal components of neurotransmission at the juvenile mouse calyx of Held synapse. *Journal of Physiology, 581*(Pt. 2), 581–602.

Felmy, F., & Schneggenburger, R. (2004). Developmental expression of the Ca^{2+}-binding proteins calretinin and parvalbumin at the calyx of Held of rats and mice. *European Journal of Neuroscience, 20*(6), 1473–1482.

Felmy, F., Neher, E., & Schneggenburger, R. (2003). Probing the intracellular calcium sensitivity of transmitter release during synaptic facilitation. *Neuron, 37*(5), 801–811.

Ford, M. C., Grothe, B., & Klug, A. (2009). Fenestration of the calyx of Held occurs sequentially along the tonotopic axis, is influenced by afferent activity, and facilitates glutamate clearance. *Journal of Comparative Neurology, 514*(1), 92–106.

Forsythe, I. D. (1994). Direct patch recording from identified presynaptic terminals mediating glutamatergic EPSCs in the rat CNS, *in vitro. Journal of Physiology, 479*(3), 381–387.

Forsythe, I. D., & Barnes-Davies, M. (1993). The binaural auditory pathway: Excitatory amino acid receptors mediate dual timecourse excitatory postsynaptic currents in the rat medial nucleus of the trapezoid body. *Proceedings of the Royal Society B, 251*(1331), 151–157.

Forsythe, I. D., Tsujimoto, T., Barnes-Davies, M., Cuttle, M. F., & Takahashi, T. (1998). Inactivation of presynaptic calcium current contributes to synaptic depression at a fast central synapse. *Neuron, 20*(4), 797–807.

Futai, K., Okada, M., Matsuyama, K., & Takahashi, T. (2001). High-fidelity transmission acquired via a developmental decrease in NMDA receptor expression at an auditory synapse. *Journal of Neuroscience, 21*(10), 3342–3349.

Galambos, R., Schwartzkopff, J., & Rupert, A. (1959). Microelectrode study of superior olivary nuclei. *American Journal of Physiology, 197*, 527–536.

Gazula, V.-R., Strumbos, J. G., Mei, X., Chen, H., Rahner, C., & Kaczmarek, L. K. (2010). Localization of Kv1.3 channels in presynaptic terminals of brainstem auditory neurons. *Journal of Comparative Neurology, 518*(16), 3205–3220.

Geal-Dor, M., Freeman, S., Li, G., & Sohmer, H. (1993). Development of hearing in neonatal rats: Air and bone conducted ABR thresholds. *Hearing Research, 69*(1–2), 236–242.

Geiger, J. R. P., Melcher, T., Koh, D.-S., Sakmann, B., Seeburg, P. H., Jonas, P., & Monyer, H. (1995). Relative abundance of subunit mRNAs determines gating and Ca^{2+} permeability of AMPA receptors in principal neurons and interneurons in rat CNS. *Neuron, 15*(1), 193–204.

Glendenning, K. K., Brunso-Bechtold, J. K., Thompson, G. C., & Masterton, R. B. (1981). Ascending auditory afferents to the nuclei of the lateral lemniscus. *Journal of Comparative Neurology, 197*(4), 673–703.

Goldberg, J. M., & Brown, P. B. (1968). Functional organization of the dog superior olivary complex: An anatomical and electrophysiological study. *Journal of Neurophysiology, 31*(4), 639–656.

Green, J. S., & Sanes, D. H. (2005). Early appearance of inhibitory input to the MNTB supports binaural processing during development. *Journal of Neurophysiology, 94*(6), 3826–3835.

Groffen, A. J., Martens, S., Arazola, R. D., Cornelisse, L. N., Lozovaya, N., de Jong, A. P. H., Goriounova, N. A., Habets, R. L. P., Takai, Y., Borst, J. G. G., Brose, N., McMahon, H. T., & Verhage, M. (2010). Doc2b is a high-affinity Ca^{2+} sensor for spontaneous neurotransmitter release. *Science, 327*(5973), 1614–1618.

Grothe, B., Pecka, M., & McAlpine, D. (2010). Mechanisms of sound localization in mammals. *Physiological Reviews, 90*(3), 983–1012.

Guinan, J. J. Jr., & Li, R. Y.-S. (1990). Signal processing in brainstem auditory neurons which receive giant endings (calyces of Held) in the medial nucleus of the trapezoid body of the cat. *Hearing Research, 49*(1–3), 321–334.

Guinan, J. J. Jr., Guinan, S. S., & Norris, B. E. (1972a). Single auditory units in the superior olive complex I: Responses to sounds and classifications based on physiological properties. *International Journal of Neuroscience, 4*, 101–120.

Guinan, J. J. Jr., Norris, B. E., & Guinan, S. S. (1972b). Single auditory units in the superior olive complex II: Tonotopic organization and locations of unit categories. *International Journal of Neuroscience, 4*, 147–166.

Habets, R. L. P., & Borst, J. G. G. (2005). Post-tetanic potentiation in the rat calyx of Held synapse. *Journal of Physiology, 564*(Pt. 1), 173–187.

Habets, R. L. P., & Borst, J. G. G. (2006). An increase in calcium influx contributes to post-tetanic potentiation at the rat calyx of Held synapse. *Journal of Neurophysiology, 96*(6), 2868–2876.

Habets, R. L. P., & Borst, J. G. G. (2007). Dynamics of the readily releasable pool during post-tetanic potentiation in the rat calyx of Held synapse. *Journal of Physiology, 581*(Pt. 2), 467–478.

Hamann, M., Billups, B., & Forsythe, I. D. (2003). Non-calyceal excitatory inputs mediate low fidelity synaptic transmission in rat auditory brainstem slices. *European Journal of Neuroscience, 18*(10), 2899–2902.

Hardman, R. M., & Forsythe, I. D. (2009). *Ether-à-go-go*–related gene K^+ channels contribute to threshold excitability of mouse auditory brainstem neurons. *Journal of Physiology, 587*(Pt. 11), 2487–2497.

Harrison, J. M., & Irving, R. (1964). Nucleus of the trapezoid body: Dual afferent innervation. *Science, 143*, 473–474.

Harrison, J. M., & Warr, W. B. (1962). A study of the cochlear nuclei and ascending auditory pathways of the medulla. *Journal of Comparative Neurology, 119*, 341–378.

Härtig, W., Singer, A., Grosche, J., Brauer, K., Ottersen, O. P., & Brückner, G. (2001). Perineuronal nets in the rat medial nucleus of the trapezoid body surround neurons immunoreactive for various amino acids, calcium-binding proteins and the potassium channel subunit Kv3.1b. *Brain Research, 899*(1–2), 123–133.

He, L., & Wu, L.-G. (2007). The debate on the kiss-and-run fusion at synapses. *Trends in Neurosciences, 30*(9), 447–455.

He, L., Xue, L., Xu, J., McNeil, B. D., Bai, L., Melicoff, E., Adachi, R., & Wu, L.-G. (2009). Compound vesicle fusion increases quantal size and potentiates synaptic transmission. *Nature, 459*(7243), 93–97.

Held, H. (1893). Die centrale Gehorleitung. *Archiv für Anatomie und Physiologie, Anatomie Abtheil, 201*–248.

Helmchen, F., Borst, J. G. G., & Sakmann, B. (1997). Calcium dynamics associated with a single action potential in a CNS presynaptic terminal. *Biophysical Journal, 72*(3), 1458–1471.

Hermida, D., Elezgarai, I., Puente, N., Alonso, V., Anabitarte, N., Bilbao, A., Doñate-Oliver, F., & Grandes, P. (2006). Developmental increase in postsynaptic alpha-amino-3-hydroxy-5-methyl-4 isoxazolepropionic acid receptor compartmentalization at the calyx of Held synapse. *Journal of Comparative Neurology, 495*(5), 624–634.

Hermida, D., Mateos, J. M., Elezgarai, I., Puente, N., Bilbao, A., Bueno-López, J. L., Streit, P., & Grandes, P. (2010). Spatial compartmentalization of AMPA glutamate receptor subunits at the calyx of Held synapse. *Journal of Comparative Neurology, 518*(2), 163–174.

Hilbig, H., Beil, B., Hilbig, H., Call, J., & Bidmon, H.-J. (2009). Superior olivary complex organization and cytoarchitecture may be correlated with function and catarrhine primate phylogeny. *Brain Structure and Function, 213*(4–5), 489–497.

Hoffpauir, B. K., Grimes, J. L., Mathers, P. H., & Spirou, G. A. (2006). Synaptogenesis of the calyx of Held: Rapid onset of function and one-to-one morphological innervation. *Journal of Neuroscience, 26*(20), 5511–5523.

Hori, T., Takai, Y., & Takahashi, T. (1999). Presynaptic mechanism for phorbol ester–induced synaptic potentiation. *Journal of Neuroscience, 19*(17), 7262–7267.

Hosoi, N., Sakaba, T., & Neher, E. (2007). Quantitative analysis of calcium-dependent vesicle recruitment and its functional role at the calyx of Held synapse. *Journal of Neuroscience, 27*(52), 14286–14298.

Hosoi, N., Holt, M., & Sakaba, T. (2009). Calcium dependence of exo- and endocytotic coupling at a glutamatergic synapse. *Neuron, 63*(2), 216–229.

Hsieh, C. Y., Nakamura, P. A., Luk, S. O., Miko, I. J., Henkemeyer, M., & Cramer, K. S. (2010). Ephrin-B reverse signaling is required for formation of strictly contralateral auditory brainstem pathways. *Journal of Neuroscience, 30*(29), 9840–9849.

Huang, H., & Trussell, L. O. (2008). Control of presynaptic function by a persistent Na$^+$ current. *Neuron, 60*(6), 975–979.

Inchauspe, C. G., Martini, F. J., Forsythe, I. D., & Uchitel, O. D. (2004). Functional compensation of P/Q by N-type channels blocks short-term plasticity at the calyx of Held presynaptic terminal. *Journal of Neuroscience, 24*(46), 10379–10383.

Inchauspe, C. G., Forsythe, I. D., & Uchitel, O. D. (2007). Changes in synaptic transmission properties due to the expression of N-type calcium channels at the calyx of Held synapse of mice lacking P/Q-type calcium channels. *Journal of Physiology, 584*(Pt. 3), 835–851.

Irving, R., & Harrison, J. M. (1967). The superior olivary complex and audition: A comparative study. *Journal of Comparative Neurology, 130*(1), 77–86.

Isaacson, J. S. (1998). GABA$_B$ receptor-mediated modulation of presynaptic currents and excitatory transmission at a fast central synapse. *Journal of Neurophysiology, 80*(3), 1571–1576.

Ishikawa, T., Sahara, Y., & Takahashi, T. (2002). A single packet of transmitter does not saturate postsynaptic glutamate receptors. *Neuron, 34*(4), 613–621.

Ishikawa, T., Nakamura, Y., Saitoh, N., Li, W.-B., Iwasaki, S., & Takahashi, T. (2003). Distinct roles of Kv1 and Kv3 potassium channels at the calyx of Held presynaptic terminal. *Journal of Neuroscience, 23*(32), 10445–10453.

Ishikawa, T., Kaneko, M., Shin, H. S., & Takahashi, T. (2005). Presynaptic N-type and P/Q-type Ca^{2+} channels mediating synaptic transmission at the calyx of Held of mice. *Journal of Physiology, 568*(Pt. 1), 199–209.

Iwasaki, S., & Takahashi, T. (2001). Developmental regulation of transmitter release at the calyx of Held in rat auditory brainstem. *Journal of Physiology, 534*(Pt. 3), 861–871.

Iwasaki, S., Momiyama, A., Uchitel, O. D., & Takahashi, T. (2000). Developmental changes in calcium channel types mediating central synaptic transmission. *Journal of Neuroscience, 20*(1), 59–65.

Jeffress, L. A. (1948). A place theory of sound localization. *Journal of Comparative and Physiological Psychology, 41*(1), 35–39.

Johnston, J., Griffin, S. J., Baker, C., Skrzypiec, A., Chernova, T., & Forsythe, I. D. (2008). Initial segment Kv2.2 channels mediate a slow delayed rectifier and maintain high frequency action potential firing in medial nucleus of the trapezoid body neurons. *Journal of Physiology, 586*(14), 3493–3509.

Joris, P. X., & Yin, T. C. T. (1995). Envelope coding in the lateral superior olive. I. Sensitivity to interaural time differences. *Journal of Neurophysiology, 73*(3), 1043–1062.

Joris, P., & Yin, T. C. T. (2007). A matter of time: Internal delays in binaural processing. *Trends in Neurosciences, 30*(2), 70–78.

Joshi, I., & Wang, L.-Y. (2002). Developmental profiles of glutamate receptors and synaptic transmission at a single synapse in the mouse auditory brainstem. *Journal of Physiology, 540*(Pt. 3), 861–873.

Joshi, I., Shokralla, S., Titis, P., & Wang, L.-Y. (2004). The role of AMPA receptor gating in the development of high-fidelity neurotransmission at the calyx of Held synapse. *Journal of Neuroscience, 24*(1), 183–196.

Kadner, A., & Berrebi, A. S. (2008). Encoding of temporal features of auditory stimuli in the medial nucleus of the trapezoid body and superior paraolivary nucleus of the rat. *Neuroscience, 151*(3), 868–887.

Kadner, A., Kulesza, R. J. Jr., & Berrebi, A. S. (2006). Neurons in the medial nucleus of the trapezoid body and superior paraolivary nucleus of the rat may play a role in sound duration coding. *Journal of Neurophysiology, 95*(3), 1499–1508.

Kajikawa, Y., Saitoh, N., & Takahashi, T. (2001). GTP-binding protein βγ subunits mediate presynaptic calcium current inhibition by GABA$_B$ receptor. *Proceedings of the National Academy of Sciences of the United States of America, 98*(14), 8054–8058.

Kandler, K., & Friauf, E. (1993). Pre- and postnatal development of efferent connections of the cochlear nucleus in the rat. *Journal of Comparative Neurology, 328*(2), 161–184.

Kaneko, M., & Takahashi, T. (2004). Presynaptic mechanism underlying cAMP-dependent synaptic potentiation. *Journal of Neuroscience, 24*(22), 5202–5208.

Katz, B., & Miledi, R. (1968). The role of calcium in neuromuscular facilitation. *Journal of Physiology, 195*(2), 481–492.

Kay, A. R., Alfonso, A., Alford, S., Cline, H. T., Holgado, A. M., Sakmann, B., Snitsarev, V. A., Stricker, T. P., Takahashi, M., & Wu, L.-G. (1999). Imaging synaptic activity in intact brain and slices with FM1-43 in C. elegans, lamprey, and rat. *Neuron, 24*(4), 809–817.

Kil, J., Kageyama, G. H., Semple, M. N., & Kitzes, L. M. (1995). Development of ventral cochlear nucleus projections to the superior olivary complex in gerbil. *Journal of Comparative Neurology, 353*(3), 317–340.

Kim, M.-H., Korogod, N., Schneggenburger, R., Ho, W.-K., & Lee, S.-H. (2005). Interplay between Na$^+$/Ca^{2+} exchangers and mitochondria in Ca^{2+} clearance at the calyx of Held. *Journal of Neuroscience, 25*(26), 6057–6065.

Kim, J. H., Sizov, I., Dobretsov, M., & von Gersdorff, H. (2007). Presynaptic Ca^{2+} buffers control the strength of a fast post-tetanic hyperpolarization mediated by the α3 Na$^+$/K$^+$-ATPase. *Nature Neuroscience, 10*(2), 196–205.

Kimura, M., Saitoh, N., & Takahashi, T. (2003). Adenosine A₁ receptor-mediated presynaptic inhibition at the calyx of Held of immature rats. *Journal of Physiology, 553*(Pt. 2), 415–426.

Klug, A., & Trussell, L. O. (2006). Activation and deactivation of voltage-dependent K⁺ channels during synaptically driven action potentials in the MNTB. *Journal of Neurophysiology, 96*(3), 1547–1555.

Kochubey, O., Han, Y., & Schneggenburger, R. (2009). Developmental regulation of the intracellular Ca^{2+} sensitivity of vesicle fusion and Ca^{2+}-secretion coupling at the rat calyx of Held. *Journal of Physiology, 587*(Pt. 12), 3009–3023.

Koike-Tani, M., Saitoh, N., & Takahashi, T. (2005). Mechanisms underlying developmental speeding in AMPA-EPSC decay time at the calyx of Held. *Journal of Neuroscience, 25*(1), 199–207.

Koike-Tani, M., Kanda, T., Saitoh, N., Yamashita, T., & Takahashi, T. (2008). Involvement of AMPA receptor desensitization in short-term synaptic depression at the calyx of Held in developing rats. *Journal of Physiology, 586*(9), 2263–2275.

Kopp-Scheinpflug, C., Fuchs, K., Lippe, W. R., Tempel, B. L., & Rübsamen, R. (2003a). Decreased temporal precision of auditory signaling in *Kcna1*-null mice: An electrophysiological study *in vivo. Journal of Neuroscience, 23*(27), 9199–9207.

Kopp-Scheinpflug, C., Lippe, W. R., Dorrscheidt, G. J., & Rübsamen, R. (2003b). The medial nucleus of the trapezoid body in the gerbil is more than a relay: Comparison of pre- and postsynaptic activity. *Journal of the Association for Research in Otolaryngology, 4*(1), 1–23.

Kopp-Scheinpflug, C., Tolnai, S., Malmierca, M. S., & Rübsamen, R. (2008). The medial nucleus of the trapezoid body: Comparative physiology. *Neuroscience, 154*(1), 160–170.

Korogod, N., Lou, X., & Schneggenburger, R. (2005). Presynaptic Ca^{2+} requirements and developmental regulation of posttetanic potentiation at the calyx of Held. *Journal of Neuroscience, 25*(21), 5127–5137.

Korogod, N., Lou, X., & Schneggenburger, R. (2007). Posttetanic potentiation critically depends on an enhanced Ca^{2+} sensitivity of vesicle fusion mediated by presynaptic PKC. *Proceedings of the National Academy of Sciences of the United States of America, 104*(40), 15923–15928.

Kulesza, R. J. Jr. (2008). Cytoarchitecture of the human superior olivary complex: Nuclei of the trapezoid body and posterior tier. *Hearing Research, 241*(1–2), 52–63.

Kulesza, R. J. Jr., Viñuela, A., Saldaña, E., & Berrebi, A. S. (2002). Unbiased stereological estimates of neuron number in subcortical auditory nuclei of the rat. *Hearing Research, 168*(1–2), 12–24.

Kulesza, R. J. Jr., Spirou, G. A., & Berrebi, A. S. (2003). Physiological response properties of neurons in the superior paraolivary nucleus of the rat. *Journal of Neurophysiology, 89*(4), 2299–2312.

Kulesza, R. J. Jr., Kadner, A., & Berrebi, A. S. (2007). Distinct roles for glycine and GABA in shaping the response properties of neurons in the superior paraolivary nucleus of the rat. *Journal of Neurophysiology, 97*(2), 1610–1620.

Kuwabara, N., & Zook, J. M. (1991). Classification of the principal cells of the medial nucleus of the trapezoid body. *Journal of Comparative Neurology, 314*(4), 707–720.

Kuwabara, N., DiCaprio, R. A., & Zook, J. M. (1991). Afferents to the medial nucleus of the trapezoid body and their collateral projections. *Journal of Comparative Neurology, 314*(4), 684–706.

Leão, R. M., & von Gersdorff, H. (2002). Noradrenaline increases high-frequency firing at the calyx of Held synapse during development by inhibiting glutamate release. *Journal of Neurophysiology, 87*(5), 2297–2306.

Leão, R. N., Berntson, A., Forsythe, I. D., & Walmsley, B. (2004). Reduced low-voltage activated K⁺ conductances and enhanced central excitability in a congenitally deaf (*dn/dn*) mouse. *Journal of Physiology, 559*(Pt. 1), 25–33.

Leão, R. M., Kushmerick, C., Pinaud, R., Renden, R., Li, G.-L., Taschenberger, H., Spirou, G., Levinson, S. R., & von Gersdorff, H. (2005a). Presynaptic Na⁺ channels: Locus, development, and recovery from inactivation at a high-fidelity synapse. *Journal of Neuroscience, 25*(14), 3724–3738.

Leão, R. N., Svahn, K., Berntson, A., & Walmsley, B. (2005b). Hyperpolarization-activated (I_h) currents in auditory brainstem neurons of normal and congenitally deaf mice. *European Journal of Neuroscience, 22*(1), 147–157.

Leão, R. N., Sun, H., Svahn, K., Berntson, A., Youssoufian, M., Paolini, A. G., Fyffe, R. E. W., & Walmsley, B. (2006). Topographic organization in the auditory brainstem of juvenile mice is disrupted in congenital deafness. *Journal of Physiology, 571*(Pt. 3), 563–578.

Leão, R. N., Leão, R. M., da Costa, L. F., Rock Levinson, S., & Walmsley, B. (2008). A novel role for MNTB neuron dendrites in regulating action potential amplitude and cell excitability during repetitive firing. *European Journal of Neuroscience, 27*(12), 3095–3108.

Lee, J. S., Kim, M.-H., Ho, W.-K., & Lee, S.-H. (2008). Presynaptic release probability and readily releasable pool size are regulated by two independent mechanisms during posttetanic potentiation at the calyx of Held synapse. *Journal of Neuroscience, 28*(32), 7945–7953.

Lee, J. S., Ho, W.-K., & Lee, S.-H. (2010). Post-tetanic increase in the fast-releasing synaptic vesicle pool at the expense of the slowly releasing pool. *Journal of General Physiology, 136*(3), 259–272.

Lenn, N. J., & Reese, T. S. (1966). The fine structure of nerve endings in the nucleus of the trapezoid body and the ventral cochlear nucleus. *American Journal of Anatomy, 118*(2), 375–390.

Li, W., Kaczmarek, L. K., & Perney, T. M. (2001). Localization of two high-threshold potassium channel subunits in the rat central auditory system. *Journal of Comparative Neurology, 437*(2), 196–218.

Lohmann, C., & Friauf, E. (1996). Distribution of the calcium-binding proteins parvalbumin and calretinin in the auditory brainstem of adult and developing rats. *Journal of Comparative Neurology, 367*(1), 90–109.

Lohmann, C., Ilic, V., & Friauf, E. (1998). Development of a topographically organized auditory network in slice culture is calcium dependent. *Journal of Neurobiology, 34*(2), 97–112.

Lorteije, J. A. M., Rusu, S. I., Kushmerick, C., & Borst, J. G. G. (2009). Reliability and precision of the mouse calyx of Held synapse. *Journal of Neuroscience, 29*(44), 13770–13784.

Lou, X., Scheuss, V., & Schneggenburger, R. (2005). Allosteric modulation of the presynaptic Ca^{2+} sensor for vesicle fusion. *Nature, 435*(7041), 497–501.

Lou, X., Korogod, N., Brose, N., & Schneggenburger, R. (2008a). Phorbol esters modulate spontaneous and Ca^{2+}-evoked transmitter release via acting on both Munc13 and protein kinase C. *Journal of Neuroscience, 28*(33), 8257–8267.

Lou, X., Paradise, S., Ferguson, S. M., & De Camilli, P. (2008b). Selective saturation of slow endocytosis at a giant glutamatergic central synapse lacking dynamin 1. *Proceedings of the National Academy of Sciences of the United States of America, 105*(45), 17555–17560.

Matveev, V., Zucker, R. S., & Sherman, A. (2004). Facilitation through buffer saturation: Constraints on endogenous buffering properties. *Biophysical Journal, 86*(5), 2691–2709.

McLaughlin, M., van der Heijden, M., & Joris, P. X. (2008). How secure is *in vivo* synaptic transmission at the calyx of Held? *Journal of Neuroscience, 28*(41), 10206–10219.

Meinrenken, C. J., Borst, J. G. G., & Sakmann, B. (2002). Calcium secretion coupling at calyx of Held governed by nonuniform channel-vesicle topography. *Journal of Neuroscience, 22*(5), 1648–1667.

Mizutani, H., Hori, T., & Takahashi, T. (2006). 5-HT$_{1B}$ receptor-mediated presynaptic inhibition at the calyx of Held of immature rats. *European Journal of Neuroscience, 24*(7), 1946–1954.

Moore, J. K. (1987). The human auditory brain stem: A comparative view. *Hearing Research, 29*(1), 1–32.

Moore, M. J., & Caspary, D. M. (1983). Strychnine blocks binaural inhibition in lateral superior olivary neurons. *Journal of Neuroscience, 3*(1), 237–242.

Moore, J. K., & Moore, R. Y. (1987). Glutamic acid decarboxylase-like immunoreactivity in brainstem auditory nuclei of the rat. *Journal of Comparative Neurology, 260*(2), 157–174.

Morest, D. K. (1968a). The collateral system of the medial nucleus of the trapezoid body of the cat, its neuronal architecture and relation to the olivo-cochlear bundle. *Brain Research, 9*(2), 288–311.

Morest, D. K. (1968b). The growth of synaptic endings in the mammalian brain: A study of the calyces of the trapezoid body. *Zeitschrift fur Anatomie und Entwicklungsgeschichte, 127*(3), 201–220.

Müller, M., Felmy, F., Schwaller, B., & Schneggenburger, R. (2007). Parvalbumin is a mobile presynaptic Ca^{2+} buffer in the calyx of Held that accelerates the decay of Ca^{2+} and short-term facilitation. *Journal of Neuroscience, 27*(9), 2261–2271.

Müller, M., Felmy, F., & Schneggenburger, R. (2008). A limited contribution of Ca^{2+} current facilitation to paired-pulse facilitation of transmitter release at the rat calyx of Held. *Journal of Physiology, 586*(Pt. 22), 5503–5520.

Müller, J., Reyes-Haro, D., Pivneva, T., Nolte, C., Schaette, R., Lübke, J., & Kettenmann, H. (2009). The principal neurons of the medial nucleus of the trapezoid body and $NG2^+$ glial cells receive coordinated excitatory synaptic input. *Journal of General Physiology, 134*(2), 115–127.

Müller, M., Goutman, J. D., Kochubey, O., & Schneggenburger, R. (2010). Interaction between facilitation and depression at a large CNS synapse reveals mechanisms of short-term plasticity. *Journal of Neuroscience, 30*(6), 2007–2016.

Nakajima, Y. (1971). Fine structure of the medial nucleus of the trapezoid body of the bat with special reference to two types of synaptic endings. *Journal of Cell Biology, 50*(1), 121–134.

Nakamura, Y., & Takahashi, T. (2007). Developmental changes in potassium currents at the rat calyx of Held presynaptic terminal. *Journal of Physiology, 581*(Pt. 3), 1101–1112.

Nakamura, T., Yamashita, T., Saitoh, N., & Takahashi, T. (2008). Developmental changes in calcium/calmodulin-dependent inactivation of calcium currents at the rat calyx of Held. *Journal of Physiology, 586*(9), 2253–2261.

Neher, E., & Sakaba, T. (2001). Combining deconvolution and noise analysis for the estimation of transmitter release rates at the calyx of Held. *Journal of Neuroscience, 21*(2), 444–461.

Neher, E., & Sakaba, T. (2008). Multiple roles of calcium ions in the regulation of neurotransmitter release. *Neuron, 59*(6), 861–872.

Oleskevich, S., Youssoufian, M., & Walmsley, B. (2004). Presynaptic plasticity at two giant auditory synapses in normal and deaf mice. *Journal of Physiology, 560*(Pt. 3), 709–719.

Palmer, M. J., Taschenberger, H., Hull, C., Tremere, L., & von Gersdorff, H. (2003). Synaptic activation of presynaptic glutamate transporter currents in nerve terminals. *Journal of Neuroscience, 23*(12), 4831–4841.

Pan, B., & Zucker, R. S. (2009). A general model of synaptic transmission and short-term plasticity. *Neuron, 62*(4), 539–554.

Paolini, A. G., FitzGerald, J. V., Burkitt, A. N., & Clark, G. M. (2001). Temporal processing from the auditory nerve to the medial nucleus of the trapezoid body in the rat. *Hearing Research, 159*(1–2), 101–116.

Perkins, G. A., Tjong, J., Brown, J. M., Poquiz, P. H., Scott, R. T., Kolson, D. R., Ellisman, M. H., & Spirou, G. A. (2010). The micro-architecture of mitochondria at active zones: Electron tomography reveals novel anchoring scaffolds and cristae structured for high-rate metabolism. *Journal of Neuroscience, 30*(3), 1015–1026.

Price, G. D., & Trussell, L. O. (2006). Estimate of the chloride concentration in a central glutamatergic terminal: A gramicidin perforated-patch study on the calyx of Held. *Journal of Neuroscience, 26*(44), 11432–11436.

Ramón y Cajal, S. (1899, 1904). *Textura del sistema nervioso del hombre y los vertebrados.* Madrid: Imprenta y Liberia de Nicolas Moya.

Renden, R., Taschenberger, H., Puente, N., Rusakov, D. A., Duvoisin, R., Wang, L.-Y., Lehre, K. P., & von Gersdorff, H. (2005). Glutamate transporter studies reveal the pruning of metabotropic glutamate receptors and absence of AMPA receptor desensitization at mature calyx of Held synapses. *Journal of Neuroscience, 25*(37), 8482–8497.

Reyes-Haro, D., Müller, J., Boresch, M., Pivneva, T., Benedetti, B., Scheller, A., Nolte, C., & Kettenmann, H. (2010). Neuron-astrocyte interactions in the medial nucleus of the trapezoid body. *Journal of General Physiology.*

Richter, E. A., Norris, B. E., Fullerton, B. C., Levine, R. A., & Kiang, N. Y. S. (1983). Is there a medial nucleus of the trapezoid body in humans? *American Journal of Anatomy, 168*(2), 157–166.

Riemann, R., & Reuss, S. (1998). Projection neurons in the superior olivary complex of the rat auditory brainstem: A double retrograde tracing study. *ORL: Journal of Oto-Rhino-Laryngology and Its Related Specialties, 60*(5), 278–282.

Robertson, D. (1985). Brainstem location of efferent neurones projecting to the guinea pig cochlea. *Hearing Research, 20*(1), 79–84.

Rodriguez-Contreras, A., de Lange, R. P. J., Lucassen, P. J., & Borst, J. G. G. (2006). Branching of calyceal afferents during postnatal development in the rat auditory brainstem. *Journal of Comparative Neurology, 496*(2), 214–228.

Rodriguez-Contreras, A., van Hoeve, J. S., Habets, R. L. P., Locher, H., & Borst, J. G. G. (2008). Dynamic development of the calyx of Held synapse. *Proceedings of the National Academy of Sciences of the United States of America, 105*(14), 5603–5608.

Rowland, K. C., Irby, N. K., & Spirou, G. A. (2000). Specialized synapse-associated structures within the calyx of Held. *Journal of Neuroscience, 20*(24), 9135–9144.

Sakaba, T. (2006). Roles of the fast-releasing and the slowly releasing vesicles in synaptic transmission at the calyx of Held. *Journal of Neuroscience, 26*(22), 5863–5871.

Sakaba, T., & Neher, E. (2001a). Calmodulin mediates rapid recruitment of fast-releasing synaptic vesicles at a calyx-type synapse. *Neuron, 32*, 1–13.

Sakaba, T., & Neher, E. (2001b). Preferential potentiation of fast-releasing synaptic vesicles by cAMP at the calyx of Held. *Proceedings of the National Academy of Sciences of the United States of America, 98*(1), 331–336.

Sakaba, T., & Neher, E. (2001c). Quantitative relationship between transmitter release and calcium current at the calyx of Held synapse. *Journal of Neuroscience, 21*(2), 462–476.

Sakaba, T., & Neher, E. (2003a). Direct modulation of synaptic vesicle priming by $GABA_B$ receptor activation at a glutamatergic synapse. *Nature, 424*(6950), 775–778.

Sakaba, T., & Neher, E. (2003b). Involvement of actin polymerization in vesicle recruitment at the calyx of Held synapse. *Journal of Neuroscience, 23*(3), 837–846.

Sakaba, T., Stein, A., Jahn, R., & Neher, E. (2005). Distinct kinetic changes in neurotransmitter release after SNARE protein cleavage. *Science, 309*(5733), 491–494.

Saldaña, E., Aparicio, M.-A., Fuentes-Santamaria, V., & Berrebi, A. S. (2009). Connections of the superior paraolivary nucleus of the rat: Projections to the inferior colliculus. *Neuroscience, 163*(1), 372–387.

Sanes, J. R., & Yamagata, M. (2009). Many paths to synaptic specificity. *Annual Review of Cell and Developmental Biology, 25*, 161–195.

Sätzler, K., Söhl, L. F., Bollmann, J. H., Borst, J. G. G., Frotscher, M., Sakmann, B., & Lübke, J. H. R. (2002). Three-dimensional reconstruction of a calyx of Held and its postsynaptic principal neuron in the medial nucleus of the trapezoid body. *Journal of Neuroscience, 22*(24), 10567–10579.

Saul, S. M., Brzezinski, J. A., Altschuler, R. A., Shore, S. E., Rudolph, D. D., Kabara, L. L., Halsey, K. E., Hufnagel, R. B., Zhou, J., Dolan, D. F., & Glaser, T. (2008). *Math5* expression and function in the central auditory system. *Molecular and Cellular Neurosciences, 37*(1), 153–169.

Schlüter, O. M., Basu, J., Südhof, T. C., & Rosenmund, C. (2006). Rab3 superprimes synaptic vesicles for release: implications for short-term synaptic plasticity. *Journal of Neuroscience, 26*(4), 1239–1246.

Schneggenburger, R., & Neher, E. (2000). Intracellular calcium dependence of transmitter release rates at a fast central synapse. *Nature, 406*(6798), 889–893.

Schneggenburger, R., Meyer, A. C., & Neher, E. (1999). Released fraction and total size of a pool of immediately available transmitter quanta at a calyx synapse. *Neuron, 23*(2), 399–409.

Smith, P. H., Joris, P. X., & Yin, T. C. T. (1998). Anatomy and physiology of principal cells of the medial nucleus of the trapezoid body (MNTB) of the cat. *Journal of Neurophysiology, 79*(6), 3127–3142.

Sommer, I., Lingenhöhl, K., & Friauf, E. (1993). Principal cells of the rat medial nucleus of the trapezoid body: An intracellular in vivo study of their physiology and morphology. *Experimental Brain Research, 95*(2), 223–239.

Song, P., Yang, Y., Barnes-Davies, M., Bhattacharjee, A., Hamann, M., Forsythe, I. D., Oliver, D. L., & Kaczmarek, L. K. (2005). Acoustic environment determines phosphorylation state of the Kv3.1 potassium channel in auditory neurons. *Nature Neuroscience, 8*(10), 1335–1342.

Sonntag, M., Englitz, B., Kopp-Scheinpflug, C., & Rübsamen, R. (2009). Early postnatal development of spontaneous and acoustically evoked discharge activity of principal cells of the medial nucleus of the trapezoid body: An *in vivo* study in mice. *Journal of Neuroscience, 29*(30), 9510–9520.

Spangler, K. M., Warr, W. B., & Henkel, C. K. (1985). The projections of principal cells of the medial nucleus of the trapezoid body in the cat. *Journal of Comparative Neurology, 238*(3), 249–262.

Spirou, G. A., Brownell, W. E., & Zidanic, M. (1990). Recordings from cat trapezoid body and HRP labeling of globular bushy cell axons. *Journal of Neurophysiology, 63*(5), 1169–1190.

Spirou, G. A., Chirila, F. V., von Gersdorff, H., & Manis, P. B. (2008). Heterogeneous Ca^{2+} influx along the adult calyx of Held: A structural and computational study. *Neuroscience, 154*(1), 171–185.

Srinivasan, G., Kim, J. H., & von Gersdorff, H. (2008). The pool of fast releasing vesicles is augmented by myosin light chain kinase inhibition at the calyx of Held synapse. *Journal of Neurophysiology, 99*(4), 1810–1824.

Steinert, J. R., Kopp-Scheinpflug, C., Baker, C., Challiss, R. A. J., Mistry, R., Haustein, M. D., Griffin, S. J., Tong, H., Graham, B. P., & Forsythe, I. D. (2008). Nitric oxide is a volume transmitter regulating postsynaptic excitability at a glutamatergic synapse. *Neuron, 60*(4), 642–656.

Steinert, J. R., Postlethwaite, M., Jordan, M. D., Chernova, T., Robinson, S. W., & Forsythe, I. D. (2010). NMDAR-mediated EPSCs are maintained and accelerate in time course during maturation of mouse and rat auditory brainstem *in vitro. Journal of Physiology, 588*(Pt. 3), 447–463.

Strumbos, J. G., Brown, M. R., Kronengold, J., Polley, D. B., & Kaczmarek, L. K. (2010a). Fragile X mental retardation protein is required for rapid experience-dependent regulation of the potassium channel Kv3.1b. *Journal of Neuroscience, 30*(31), 10263–10271.

Strumbos, J. G., Polley, D. B., & Kaczmarek, L. K. (2010b). Specific and rapid effects of acoustic stimulation on the tonotopic distribution of Kv3.1b potassium channels in the adult rat. *Neuroscience, 167*(3), 567–572.

Südhof, T. C., & Rothman, J. E. (2009). Membrane fusion: Grappling with SNARE and SM proteins. *Science, 323*(5913), 474–477.

Sun, J.-Y., & Wu, L.-G. (2001). Fast kinetics of exocytosis revealed by simultaneous measurements of presynaptic capacitance and postsynaptic currents at a central synapse. *Neuron, 30*(1), 171–182.

Sun, J.-Y., Wu, X. S., Wu, W., Jin, S. X., Dondzillo, A., & Wu, L.-G. (2004). Capacitance measurements at the calyx of Held in the medial nucleus of the trapezoid body. *Journal of Neuroscience Methods, 134*(2), 121–131.

Sun, J., Pang, Z. P., Qin, D., Fahim, A. T., Adachi, R., & Südhof, T. C. (2007). A dual-Ca^{2+}-sensor model for neurotransmitter release in a central synapse. *Nature, 450*(7170), 676–682.

Takahashi, T., Forsythe, I. D., Tsujimoto, T., Barnes-Davies, M., & Onodera, K. (1996). Presynaptic calcium current modulation by a metabotropic glutamate receptor. *Science, 274*(5287), 594–597.

Takahashi, T., Kajikawa, Y., & Tsujimoto, T. (1998). G-protein-coupled modulation of presynaptic calcium currents and transmitter release by a $GABA_B$ receptor. *Journal of Neuroscience, 18*(9), 3138–3146.

Taschenberger, H., & von Gersdorff, H. (2000). Fine-tuning an auditory synapse for speed and fidelity: Developmental changes in presynaptic waveform, EPSC kinetics, and synaptic plasticity. *Journal of Neuroscience, 20*(24), 9162–F9173.

Taschenberger, H., Leão, R. M., Rowland, K. C., Spirou, G. A., & von Gersdorff, H. (2002). Optimizing synaptic architecture and efficiency for high-frequency transmission. *Neuron, 36*(6), 1127–1143.

Taschenberger, H., Scheuss, V., & Neher, E. (2005). Release kinetics, quantal parameters and their modulation during short-term depression at a developing synapse in the rat CNS. *Journal of Physiology, 568*(Pt. 2), 513–537.

Tatsuoka, H., & Reese, T. S. (1989). New structural features of synapses in the anteroventral cochlear nucleus prepared by direct freezing and freeze-substitution. *Journal of Comparative Neurology, 290*(3), 343–357.

Thompson, A. M., & Schofield, B. R. (2000). Afferent projections of the superior olivary complex. *Microscopy Research and Technique, 51*(4), 330–354.

Tollin, D. J., & Yin, T. C. T. (2005). Interaural phase and level difference sensitivity in low-frequency neurons in the lateral superior olive. *Journal of Neuroscience, 25*(46), 10648–10657.

Tolnai, S., Englitz, B., Scholbach, J., Jost, J., & Rübsamen, R. (2009). Spike transmission delay at the calyx of Held in vivo: Rate dependence, phenomenological modeling, and relevance for sound localization. *Journal of Neurophysiology, 102*(2), 1206–1217.

Tong, H., Steinert, J. R., Robinson, S. W., Chernova, T., Read, D. J., Oliver, D. L., & Forsythe, I. D. (2010). Regulation of Kv channel expression and neuronal excitability in rat medial nucleus of the trapezoid body maintained in organotypic culture. *Journal of Physiology, 588*(Pt. 9), 1451–1468.

Tritsch, N. X., Rodríguez-Contreras, A., Crins, T. T. H., Wang, H. C., Borst, J. G. G., & Bergles, D. E. (2010). Calcium action potentials in hair cells pattern auditory neuron activity before hearing onset. *Nature Neuroscience, 13*(9), 1050–1052.

Tsuchitani, C., & Boudreau, J. C. (1966). Single unit analysis of cat superior olive S segment with tonal stimuli. *Journal of Neurophysiology, 29*(4), 684–697.

Tsujimoto, T., Jeromin, A., Saitoh, N., Roder, J. C., & Takahashi, T. (2002). Neuronal calcium sensor 1 and activity-dependent facilitation of P/Q-type calcium currents at presynaptic nerve terminals. *Science, 295*(5563), 2276–2279.

Turecek, R., & Trussell, L. O. (2001). Presynaptic glycine receptors enhance transmitter release at a mammalian central synapse. *Nature, 411*(6837), 587–590.

Turecek, R., & Trussell, L. O. (2002). Reciprocal developmental regulation of presynaptic ionotropic receptors. *Proceedings of the National Academy of Sciences of the United States of America, 99*(21), 13884–13889.

Vater, M., & Feng, A. S. (1990). Functional organization of ascending and descending connections of the cochlear nucleus of horseshoe bats. *Journal of Comparative Neurology, 292*(3), 373–395.

Voiculescu, O., Charnay, P., & Schneider-Maunoury, S. (2000). Expression pattern of a *Krox-20/Cre* knock-in allele in the developing hindbrain, bones, and peripheral nervous system. *Genesis, 26*(2), 123–126.

von Gersdorff, H., Schneggenburger, R., Weis, S., & Neher, E. (1997). Presynaptic depression at a calyx synapse: The small contribution of metabotropic glutamate receptors. *Journal of Neuroscience, 17*(21), 8137–8146.

von Hehn, C. A., Bhattacharjee, A., & Kaczmarek, L. K. (2004). Loss of Kv3.1 tonotopicity and alterations in cAMP response element-binding protein signaling in central auditory neurons of hearing impaired mice. *Journal of Neuroscience, 24*(8), 1936–1940.

Wadel, K., Neher, E., & Sakaba, T. (2007). The coupling between synaptic vesicles and Ca^{2+} channels determines fast neurotransmitter release. *Neuron, 53*(4), 563–575.

Walmsley, B., Berntson, A., Leão, R. N., & Fyffe, R. E. W. (2006). Activity-dependent regulation of synaptic strength and neuronal excitability in central auditory pathways. *Journal of Physiology, 572*(Pt. 2), 313–321.

Wang, L.-Y., & Kaczmarek, L. K. (1998). High-frequency firing helps replenish the readily releasable pool of synaptic vesicles. *Nature, 394*(6691), 384–388.

Wang, L.-Y., Gan, L., Forsythe, I. D., & Kaczmarek, L. K. (1998). Contribution of the Kv3.1 potassium channel to high-frequency firing in mouse auditory neurones. *Journal of Physiology, 509*(Pt. 1), 183–194.

Wang, L.-Y., Neher, E., & Taschenberger, H. (2008). Synaptic vesicles in mature calyx of Held synapses sense higher nanodomain calcium concentrations during action potential-evoked glutamate release. *Journal of Neuroscience, 28*(53), 14450–14458.

Wenthold, R. J., Huie, D., Altschuler, R. A., & Reeks, K. A. (1987). Glycine immunoreactivity localized in the cochlear nucleus and superior olivary complex. *Neuroscience, 22*(3), 897–912.

Wimmer, V. C., Nevian, T., & Kuner, T. (2004). Targeted in vivo expression of proteins in the calyx of Held. *Pflugers Archiv: European Journal of Physiology, 449*(3), 319–333.

Wimmer, V. C., Horstmann, H., Groh, A., & Kuner, T. (2006). Donut-like topology of synaptic vesicles with a central cluster of mitochondria wrapped into membrane protrusions: A novel structure-function module of the adult calyx of Held. *Journal of Neuroscience, 26*(1), 109–116.

Wölfel, M., Lou, X., & Schneggenburger, R. (2007). A mechanism intrinsic to the vesicle fusion machinery determines fast and slow transmitter release at a large CNS synapse. *Journal of Neuroscience, 27*(12), 3198–3210.

Wong, A. Y., Graham, B. P., Billups, B., & Forsythe, I. D. (2003). Distinguishing between presynaptic and postsynaptic mechanisms of short-term depression during action potential trains. *Journal of Neuroscience, 23*(12), 4868–4877.

Wong, A. Y. C., Billups, B., Johnston, J., Evans, R. J., & Forsythe, I. D. (2006). Endogenous activation of adenosine A1 receptors, but not P2X receptors, during high-frequency synaptic transmission at the calyx of Held. *Journal of Neurophysiology, 95*(6), 3336–3342.

Wu, L. G., & Borst, J. G. G. (1999). The reduced release probability of releasable vesicles during recovery from short-term synaptic depression. *Neuron, 23*(4), 821–832.

Wu, S. H., & Kelly, J. B. (1991). Physiological properties of neurons in the mouse superior olive: Membrane characteristics and postsynaptic responses studied in vitro. *Journal of Neurophysiology, 65*(2), 230–246.

Wu, S. H., & Kelly, J. B. (1995). Inhibition in the superior olivary complex: Pharmacological evidence from mouse brain slice. *Journal of Neurophysiology, 73*(1), 256–269.

Wu, W., & Wu, L.-G. (2007). Rapid bulk endocytosis and its kinetics of fission pore closure at a central synapse. *Proceedings of the National Academy of Sciences of the United States of America, 104*(24), 10234–10239.

Wu, X.-S., & Wu, L.-G. (2009). Rapid endocytosis does not recycle vesicles within the readily releasable pool. *Journal of Neuroscience, 29*(35), 11038–11042.

Wu, L.-G., Westenbroek, R. E., Borst, J. G. G., Catterall, W. E., & Sakmann, B. (1999). Calcium channel types with distinct presynaptic localization couple differentially to transmitter release in single calyx-type synapses. *Journal of Neuroscience, 19*, 726–736.

Wu, L.-G., Ryan, T. A., & Lagnado, L. (2007). Modes of vesicle retrieval at ribbon synapses, calyx-type synapses, and small central synapses. *Journal of Neuroscience, 27*(44), 11793–11802.

Wu, X. S., McNeil, B. D., Xu, J., Fan, J., Xue, L., Melicoff, E., Adachi, R., Bai, L., & Wu, L.-G. (2009). Ca^{2+} and calmodulin initiate all forms of endocytosis during depolarization at a nerve terminal. *Nature Neuroscience, 12*(8), 1003–1010.

Xiao, L., Han, Y., Runne, H., Murray, H., Kochubey, O., Luthi-Carter, R., & Schneggenburger, R. (2010). Developmental expression of Synaptotagmin isoforms in single calyx of Held–generating neurons. *Molecular and Cellular Neurosciences.*

Xu, J., & Wu, L.-G. (2005). The decrease in the presynaptic calcium current is a major cause of short-term depression at a calyx-type synapse. *Neuron, 46*(4), 633–645.

Xu, J., Mashimo, T., & Südhof, T. C. (2007). Synaptotagmin-1, -2, and −9: Ca^{2+} sensors for fast release that specify distinct presynaptic properties in subsets of neurons. *Neuron, 54*(4), 567–581.

Xu, J., McNeil, B., Wu, W., Nees, D., Bai, L., & Wu, L.-G. (2008). GTP-independent rapid and slow endocytosis at a central synapse. *Nature Neuroscience, 11*(1), 45–53.

Yamashita, T., Hige, T., & Takahashi, T. (2005). Vesicle endocytosis requires dynamin-dependent GTP hydrolysis at a fast CNS synapse. *Science, 307*(5706), 124–127.

Yamashita, T., Eguchi, K., Saitoh, N., von Gersdorff, H., & Takahashi, T. (2010). Developmental shift to a mechanism of synaptic vesicle endocytosis requiring nanodomain Ca²⁺. *Nature Neuroscience, 13*(7), 838–844.

Yang, Y.-M., & Wang, L.-Y. (2006). Amplitude and kinetics of action potential-evoked Ca²⁺ current and its efficacy in triggering transmitter release at the developing calyx of Held synapse. *Journal of Neuroscience, 26*(21), 5698–5708.

Yang, B., Desai, R., & Kaczmarek, L. K. (2007). Slack and slick K_{Na} channels regulate the accuracy of timing of auditory neurons. *Journal of Neuroscience, 27*(10), 2617–2627.

Yang, Y.-M., Fedchyshyn, M. J., Grande, G., Aitoubah, J., Tsang, C. W., Xie, H., Ackerley, C. A., Trimble, W. S., & Wang, L.-Y. (2010). Septins regulate developmental switching from microdomain to nanodomain coupling of Ca²⁺ influx to neurotransmitter release at a central synapse. *Neuron, 67*(1), 100–115.

Young, S. M. Jr., & Neher, E. (2009). Synaptotagmin has an essential function in synaptic vesicle positioning for synchronous release in addition to its role as a calcium sensor. *Neuron, 63*(4), 482–496.

Youssoufian, M., Oleskevich, S., & Walmsley, B. (2005). Development of a robust central auditory synapse in congenital deafness. *Journal of Neurophysiology, 94*(5), 3168–3180.

Youssoufian, M., Couchman, K., Shivdasani, M. N., Paolini, A. G., & Walmsley, B. (2008). Maturation of auditory brainstem projections and calyces in the congenitally deaf (*dn/dn*) mouse. *Journal of Comparative Neurology, 506*(3), 442–451.

Zhou, Y., Carney, L. H., & Colburn, H. S. (2005). A model for interaural time difference sensitivity in the medial superior olive: Interaction of excitatory and inhibitory synaptic inputs, channel dynamics, and cellular morphology. *Journal of Neuroscience, 25*(12), 3046–3058.

Chapter 6
Synaptic Mechanisms of Coincidence Detection

Katrina M. MacLeod and Catherine E. Carr

1 Introduction

Localization of sounds in space is a capability crucial to an animal's survival in a world full of predators, scarce of prey, and with heavy selection pressures for mates. For neuroscientists, sound localization offers an opportunity to ask precise questions relating sensory stimuli to their neural representation and the computation of sensory percepts. Both monaural and binaural cues are used to generate a sense of auditory spatial location, but the most thorough analysis of localization has focused on the use of binaural temporal cues. The study of the binaural cues allows the investigation of the neural mechanisms of sensory integration as the brain combines information from the left and right ears. The field has been enriched by studying animals that have highly developed capabilities to localize sound, such as the barn owl (*Tyto alba*), which hunts its prey in complete darkness (Payne 1971; Konishi 1973a, b). When combined with cellular and anatomical studies in the chicken and several other avian species, a remarkable confluence of evidence has emerged to reveal the synaptic and biophysical mechanisms that combine to create specialized brainstem neural circuits that perform coincidence detection on the temporal information and encode sound location, as described in this chapter.

K.M. MacLeod (✉)
Department of Biology, University of Maryland, College Park, MD 20742–4415, USA
e-mail: macleod@umd.edu

L.O. Trussell et al. (eds.), *Synaptic Mechanisms in the Auditory System*,
Springer Handbook of Auditory Research 41, DOI 10.1007/978-1-4419-9517-9_6,
© Springer Science+Business Media, LLC 2012

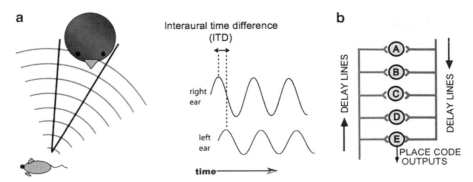

Fig. 6.1 Animals use differences in the time of arrival in the two ears to localize sound in space. (a) When sound arises on one side of an animal, the arrival time at the far ear is delayed relative to the near ear due to the difference in the path length between the source and the ears. The delay is termed the interaural time difference (ITD). For a sinusoidal pure tone, this delay introduces an ongoing phase difference. (b) The Jeffress model for explaining the computation of ITDs in the brainstem. Monaural channels act as delay lines and project to an array of coincidence detectors that each tap the signal at a different ITD. The cells for which the internal (axonal) delay is equal but opposite to the acoustic ITD are maximally active. The delay lines create a map of ITD whereby the temporal code is transformed into a place code

1.1 Using the Timing Cues in Acoustic Stimuli for Sound Localization

Studies in barn owls and other species have revealed two principal binaural cues for localization: Timing cues and intensity, or sound level, cues. These cues arise from the different paths the sound waves take to reach the ears when the sound source originates to one side in the horizontal (azimuthal) plane (Fig. 6.1). Because the path length to the more distant ear is longer, the arrival time is delayed relative to the near ear, resulting in an interaural time difference (ITD). The head also shadows sound to some degree, resulting in an interaural level difference (ILD), which will not be discussed further here (but see Konishi 1993; Klump 2000; Kubke and Carr 2005). The range of time differences available to the barn owl and other avian species depends on the head size and frequency but is generally limited to about ±200 μs (Woodworth 1954; Klump 2000). Behavioral experiments have shown that barn owls use interaural time differences to localize sounds with an angular accuracy of up to ~2° (as measured as the minimal resolvable angle in free-field conditions using broad-band noise stimuli; Knudsen et al. 1979; Bala and Takahashi 2000). Such accuracy using time difference would require the detection of ITDs of tens of microseconds – much smaller than a typical action potential. This chapter reviews the synaptic and cellular specializations needed to accomplish this feat.

1.2 Tuning for ITD Using Neural Coincidence Detection

1.2.1 The Jeffress Model for Sound Localization

In 1948 Jeffress put forth a model describing how a neural system might be set up to accomplish sound localization using the timing information in the stimulus waveform. The model circuit consists of two elements, delay lines and coincidence detectors (Fig. 6.1; Jeffress 1948). The delay lines are created from variations in axonal path lengths, and the coincidence detectors are units that respond most vigorously when they receive inputs simultaneously from the opposing afferents, which can only occur when the external time difference is exactly compensated by the delay introduced by the axonal travel time. The Jeffress model elegantly explains both the measurement and the encoding of ITDs. The position of coincidence detectors in the array determines the precise combination of delayed inputs, such that each neuron responds only to sound coming from a particular direction. Thus the location of the sound is encoded in the anatomical *place* of the neuron. A new variable, time difference, is calculated and transforms the time code into a place code.

1.2.2 Evidence for Jeffress Model Coincidence Detectors in the Avian Brainstem

The sound localization circuits in avian brainstem circuits conform to the Jeffress model, and the major components of the model are outlined in this chapter. Mammalian circuits share many similar features, although the sources of delay may be more variable and the coincidence detectors may not be organized to form a map of ITD (see McAlpine and Grothe 2003). The rest of the chapter focuses on features of the synaptic inputs and integration that contribute to making this circuit work.

For coincidence detection and delay lines to be useful in detecting very small external differences in timing, sound must first be encoded very precisely in the periphery. Phase-locked spikes encode the timing of the sinusoidal waveform of a sound stimulus within the coincidence detector's frequency sensitivity. The phase-locked activity of the auditory nerve fibers are relayed by the cochlear nucleus concerned with timing. In birds, this is the cochlear *nucleus magnocellularis* (NM). Time differences from the two ears converge on the neurons of the *nucleus laminaris* (NL) (Fig. 6.2). Much has been written about the specializations required to create precision phase-locking in NM as well as in their mammalian counterpart, the spherical bushy neurons in the anteroventral cochlear nucleus (Oertel 1999), and there are many similarities with the intrinsic properties of NM and NL neurons that are reviewed elsewhere (Trussell 1999; Carr and Code 2000; Carr et al. 2001).

In the Jeffress model, the targets of the delay lines act as coincidence detectors and are sensitive to binaural disparities or ITDs. In mammals, Goldberg and Brown (1969) showed that dog MSO neurons fit the predictions of the Jeffress model,

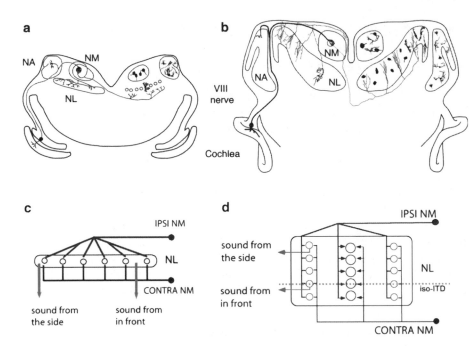

Fig. 6.2 Jeffress-like circuits in the avian auditory brainstem. Diagrams of transverse sections from the brainstems of (**a**) chicken and (**b**) barn owl. The eighth nerve projects from the cochlea to the cochlear nuclei, nucleus angularis (NA), and nucleus magnocellularis (NM). NM afferents project bilaterally to the third-order nucleus laminaris (NL). (**c, d**) The Jeffress model as applied to avian auditory brainstem. In chicken, delays are formed as the contralateral axons run along the mediolateral length of the NL. Ipsilateral axons do not form delay lines. In barn owl, the NL is expanded in the dorsoventral axis. Delay lines are formed in both the ipsilateral and contralateral afferent as the axons traverse the dorsoventral axis. The iso-ITD line delineates the axis of equal ITD tuning. (Modified with permission from Kubke and Carr 2000)

as did cat MSO neurons (Yin and Chan 1990) and barn owl NL neurons (Sullivan and Konishi 1986). NL's projections to the inferior colliculus contribute to the maps of spatial location (Knudsen and Konishi 1978a, 1978b; Moiseff and Konishi 1983). Both NL and MSO neurons respond maximally at a favorable ITD and minimally at an unfavorable ITD. When pure tones are used, the tuning curves have multiple peaks 2π apart, revealing their dependence on interaural phase differences (Fig. 6.3). NL and MSO neurons also phase-lock to both binaural and monaural stimuli. If the precise phase at which the firing occurs to the left stimulus versus the right stimulus is different, the most favorable ITD is the one that compensates for this difference. Thus, observing the phase-locking of the NL or MSO neuron to the stimulus, one can see that maximal firing occurs when phase-locked spikes from each side arrive simultaneously, that is, when the difference in the internal conduction delay is nullified by the ITD (Goldberg and Brown 1969; Carr and Konishi 1990; Yin and Chan 1990). They respond minimally when the phase-locked spikes arrive 180° out of phase.

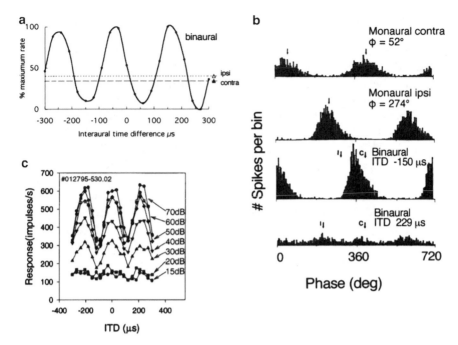

Fig. 6.3 ITD tuning curves in the barn owl show characteristics of coincidence detection. (**a**) In vivo recordings from NL neurons in barn owl show modulation of firing rate to binaural stimulation with changes in ITD (*thick line*). Responses to monaural stimulation of either the ipsilateral ear or the contralateral ear (*markers*) had firing rates that were less than half the peak binaural response. Note the troughs of the tuning curve are well below the monaural activity levels. (**b**) Period histograms show NL responses phase-lock to the sound stimulus. Responses to monaural stimulation of the contralateral ear or ipsilateral ear (top two plots) and binaural stimulation at favorable and unfavorable ITDs (bottom two plots). The binaural response is robust when the ITD is opposite and equal to the difference in best phase between the two monaural phase preferences. Fourth PSTH shows the weak response when the ITD corresponds to an unfavorable phase difference. (Panels **a**, **b** modified with permission from Carr and Konishi 1990). (**c**) Overlaid ITD curves for different average binaural intensities, indicated on the right in dB SPL, shows that ITD tuning persists with increasing sound levels. Note how the peaks and troughs of the tuning curves line up, indicating that the best ITD is generally insensitive to level. (Panel **c** modified with permission from Pena et al. 1996)

Are these coincidence detectors simply adding their inputs? Analysis of the NL responses shows that simple linear summation cannot account for the data. There is a nonlinear transformation: The peak firing rates of coincidence detectors when their inputs are in phase are sometimes greater than the sum of the responses to the stimulus presented monaurally (Moiseff and Konishi 1983; Carr and Konishi 1990). Most significantly, the minimal firing rate (in the trough of the tuning curve) during out-of-phase binaural stimulation is *less* than the monaural firing rate. These data suggest that the NL neurons behave more like true coincidence detectors, like biophysical AND-gates, and not like simple summators. Numerous mechanisms contribute to coincidence detection, including those that suppress monaural coincidences (Colburn et al. 1990), such as the segregation of the ipsilateral and contralateral inputs onto different dendrites, coupled with shunting of the postsynaptic current (Sect. 2.2; Agmon-Snir et al. 1998; Grau-Serrat et al. 2003; Zhou et al. 2005). Another recently discovered mechanism is inhibition at "bad" or out-of-phase ITDs, counteracting the dispersive effects of monaural coincidences (Nishino et al. 2008).

Several other features of ITD tuning suggest a refinement beyond simple summation. ITD tuning must be robust to variations in intensity, in terms of both the average binaural intensity and interaural differences. Increasing sound level increases the firing rates of the afferents. This means that on each cycle, a larger number of afferents from the same side are firing in synchrony. If the firing of the NL neurons depends on the coincidence of inputs, the larger number of same-side firings should increase the likelihood of the monaural coincidences, thus increasing the firing rate during unfavorable ITDs, leading to a flattening of the tuning curve. However, NL neurons continue to show sensitivity to ITD with no change in best ITD well above threshold levels (Fig. 6.3). The synaptic and biophysical mechanisms that contribute to maintaining ITD sensitivity in the face of changing sound pressure levels are the major focus of Sect. 2.2.

1.2.3 Evidence for Jeffress Model Delay Lines

The Jeffress model circuit depends not only on coincidence detection, but also on delay lines. These have an anatomical correlate in the axonal projections from NM to NL. Reconstructions of the NM axons that provide input to NL show orderly patterns of axonal projections, where the length of the axon can generate conduction delays that compensate for external ITDS. In the barn owl, the bilateral projections from NM to NL innervate dorsoventral arrays of neurons in NL in a sequential fashion (Carr and Konishi 1990; Pena et al. 2001). Measurements of time delays from these fibers show an orderly progression of phase relationships along the length of these delay lines (Carr and Konishi 1988, 1990).

In the chicken, the delay line circuit is similar to the barn owl circuit, except that NL is a monolayer of bipolar cells oriented in the mediolateral dimension of the brain stem (Fig. 6.2; Ramon y Cajal 1908; Parks and Rubel 1975; Jhaveri and Morest 1982). Chicken NL cells receive input from the ipsilateral NM onto their dorsal dendrites and input from the contralateral NM onto their ventral dendrites. The ipsilateral axons may not act as delay lines, but the contralateral axons do; estimates of conduction velocity suggest that the contralateral delay line could encode 200 µs or more of delay (Overholt et al. 1992; Köppl and Carr 2008). Patterns of axonal projections in the cat MSO are very similar to those of the chicken. Axons from the contralateral cochlear nucleus form a ladder-like projection across the rostro-caudal axis of MSO, while the ipsilateral axons form a less organized projection (Smith et al. 1993; Joris et al. 1998; Beckius et al. 1999). Whether or not these axons act as delay lines, in the sense that they form an ordered representation of ITD, remains controversial (see Sect. 3.2).

In summary, the avian system follows Jeffress's original conception to a remarkable degree. Synaptic and cellular mechanisms coordinate to create the coincidence detectors crucial to the model's function. Section 2 focuses on how synaptic properties contribute to coincidence detection. NL neurons fire most when inputs coincide, but many are tuned to nonzero ITDs. They do this precisely by following the Jeffress model of introducing internal delays, which is described in more detail in Sect. 3. Interestingly, the mammalian system provides an alternative model, with important differences evident in the synaptic inputs and anatomical mapping.

2 Building a Coincidence Detector Neuron

An "ITD discriminator" neuron should fire when inputs from two independent (binaural) neural sources coincide (or almost coincide), but not when two inputs from the same neural source (almost) coincide. A neuron that sums its inputs linearly would not be able to distinguish between these two scenarios, and in this section the biophysical and morphological adaptations that allow for ITD discrimination are reviewed.

Neurons in NL and MSO achieve coincidence detection with phenomenally fine temporal resolution. Both MSO and NL neurons exhibit similar anatomical and biophysical specializations, such as stereotyped bipolar dendrites, with inputs from each ear segregated onto each set of dendrites, allowing nonlinear integration between the inputs from left and right ears (Agmon-Snir et al. 1998; Scott et al. 2005). Coincidence detectors appear to work by "cross-correlating" sinusoidal synaptic conductances, which mirror the stimulus waveform, as conveyed through their phase-locked inputs. These neurons have a high density of low voltage-activated potassium channels, which speed up their synaptic dynamics, yielding excitatory postsynaptic potentials (EPSPs) that are typically around 400 μs wide at half amplitude (Reyes et al. 1996).

2.1 Excitatory Synaptic Time Course

2.1.1 Fast Glutamatergic Synaptic Conductances Contribute to Temporal Integration

Synaptic transmission in the auditory brainstem is characterized by AMPA receptor (AMPA-R) mediated glutamatergic conductances with some of the fastest kinetics known (Levin et al. 1997; Parks 2000; Sugden et al. 2002). Coincidence detectors that must deal with very small time differences in their inputs are able to do this in part because the AMPA-R EPSPs they receive are very brief, with rapid rises and decays (Fig. 6.4; Reyes et al. 1996; Smith et al. 2000; Kuba et al. 2003). The brevity of the EPSPs is due to fast synaptic conductances and intrinsic properties that minimize the low-pass filtering of the synaptic signal (Reyes et al. 1994, 1996). The AMPA receptors in the cochlear nuclei and nucleus laminaris have conductances

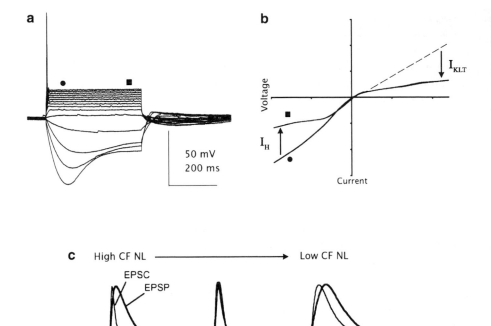

Fig. 6.4 Synaptic properties contribute to coincidence detection. (**a**) Voltage responses of an NL neuron to current steps. NL neurons fire only a single spike at the onset. Hyperpolarizing current steps evoke a "sag" current due to activation of I_h. (**b**) Current-voltage plots show double rectification. Outward rectification is due to I_h; inward rectification is due to I_{KLT}. Markers indicate the time at which IV is taken in A. Dashed line: linear current of membrane lacking I_{KLT} (**c**) Excitatory postsynaptic currents (EPSCs) and potentials (EPSPs) have rapid rises and decays. While the EPSPs are typically longer than the very rapid EPSCs, in the middle CF range the EPSPs are nearly as rapid due to a preponderance of low-threshold potassium channels (see Sect. 3.3) (Panel **c** modified with permission from Kuba et al. 2005)

characterized by very fast rise times and decay times that are partially due to rapid desensitization (how quickly the receptors enter a desensitization state during continued presence of the neurotransmitter), in addition to rapid deactivation (how quickly the receptors close after withdrawal of the neurotransmitter) (Raman et al. 1994).

Regional variations in the subunits that comprise the functional AMPA receptors contribute to the synaptic diversity in the nervous system. In the auditory system, where timing issues are crucial, the time course of these AMPA-R excitatory postsynaptic currents (EPSCs) is achieved through the presence of glutamate receptor subunits GluR3 and GluR4 and through the absence of subunit GluR2 (Parks 2000; Sugden et al. 2002). These combinations of subunits confer faster desensitization kinetics, as well as permeability to calcium ions (Mosbacher et al. 1994; Geiger et al. 1995; Dingledine et al. 1999). These subunits can further exist in different variants due to posttranslational modification; auditory brainstem neurons are distinguished by expressing high levels of the "flop" splice variants, which have faster kinetics than the "flip" variant. Direct measurements of the EPSCs in NL neurons in chick brainstem slice preparation under voltage clamp show very narrow half-times, as low as 400 µs (Kuba et al. 2005; Reyes et al. 1996). This estimate of the EPSC time course is likely to be an overestimate for the conductances close to the synaptic site in the dendrite, given the limitations of voltage clamping the dendritic tree and that the subunit profiles are nearly identical in NL and NM (Parks 2000), the latter of which can achieve near-ideal voltage clamp and have EPSCs as rapid as 200 µs (Zhang and Trussell 1994; Brenowitz and Trussell 2001).

2.1.2 Interactions Between Synaptic and Intrinsic Properties Lead to Fast EPSPs

Although fast EPSCs underlie the rapid synaptic potential changes in coincidence detector neurons, the intrinsic electrical properties also shape the synaptic response. Voltage-sensitive potassium channels play an important role throughout the timing pathways of the auditory circuits. Two types of potassium channels are critical for precise phase-locking in the neurons involved in the timing pathways: a low-threshold conductance (G_{KLT}) and a high-threshold conductance (G_{KHT}) (Manis and Marx 1991; Brew and Forsythe 1995; Trussell 1999). Similar conductances are found in NL neurons and serve to enhance coincidence detection (Parameshwaran et al. 2001; Kuba et al. 2002a; Parameshwaran-Iyer et al. 2003). These potassium conductances distinguish the NL, NM, and mammalian bushy neurons from the multipolar cells of the avian cochlear nucleus angularis (Soares et al. 2002; Fukui and Ohmori 2003) and the mammalian ventral cochlear nucleus (Wu and Oertel 1984; Manis and Marx 1991; Oertel 1991). The G_{KHT} is associated with the Kv3 family of potassium channels and helps maintain phase-locked output firing by keeping the action potentials narrow (Wang et al. 1998). G_{KHT} appears to have little impact on EPSC time course, however, and therefore we focus on the G_{KLT} conductance.

G_{KLT} has a large impact on both the firing responses and the time course of synaptic inputs. The G_{KLT} is so-called because its activation threshold is near the resting potential of the neuron and is largely responsible for the outward rectification and nonlinear current voltage relationship near rest (Rathouz and Trussell 1998; see also Fig. 6.4). It is dendrotoxin-sensitive and associated with the Kv1 family. G_{KLT} activation is crucial because even very fast currents, once injected into the neuron, will result in voltage changes governed by the intrinsic properties of the neuronal membrane, which acts like a low-pass filter. The time constant of the membrane (τ_m) is proportional to the membrane capacitance (C_m) and membrane resistance (R_m), so reducing R_m (by increasing the membrane conductance) will reduce τ_m, leading to a faster voltage response. Because of the G_{KLT}, NL neurons typically have an R_m of 10–24 MΩ and a τ_m of 0.5–1.5 ms (Reyes et al. 1996; Cook et al. 2003; Kuba et al. 2005). The large conductance and steep activation curve between rest and the spike threshold suppress firing in response to current injections, so that a flat step current injection will elicit only a single spike at the onset. Blocking the G_{KLT} with dendrotoxin leads to the generation of multiple spikes in response to a flat current step (Grigg et al. 2000; Kuba et al. 2002a).

The short membrane time constant results in EPSPs that are very brief, and under certain conditions even as brief as the underlying EPSC (Fig. 6.4c; Kuba et al. 2005). Depolarizing the membrane potential further activates the G_{KLT}, while repolarizing the membrane generates a post-EPSP hyperpolarization, like the effect of a delayed rectifier after an action potential. Blocking G_{KLT} with dendrotoxin prolongs the EPSPs. The narrow EPSPs, combined with the intrinsic properties of the NL neurons, lead to effective coincidence detection. Bilateral activation of synaptic currents further shows that the temporal window for synaptic integration is very narrow, with half-widths of less than 400 μs (Funabiki et al. 1998; Kuba et al. 2003). The integration window further narrows with maturity and at physiological temperatures (Kuba et al. 2002a). The EPSP time course has a strong positive correlation with the sharpness of coincidence detection. The limiting value of the time window (0.16 ms), calculated from the estimated EPSP time course, is narrow enough to explain the acuity of ITD detection in NL in vivo (Kuba et al. 2003). Using conductance clamp and recording from NL neurons in vitro, Reyes and coworkers systematically altered the size of the simulated EPSPs and the underlying phase relationship between the right and left inputs (Reyes et al. 1996). When left and right inputs were delivered in phase, EPSPs summed to generate large, rapid, and well-separated fluctuations in membrane potential that could effectively trigger a spike. When left and right inputs were delivered out of phase, the EPSPs summed to generate a flatter, low-variability plateau potential, which failed to generate spikes even though it reached or exceeded the nominal spiking threshold. Furthermore, out-of-phase inputs generated fewer spikes than when the inputs from only a single side were applied; that is, the out-of-phase "binaural" condition suppressed spiking relative to the "monaural" condition. Thus voltage-gated potassium channels in the coincidence detection neurons assist in limiting the temporal integration and allow spike generation only in response to the steeply fluctuating inputs that occur when phase-locked inputs from both sides coincide. Using simple injected current stimuli with noisy waveforms shows that NL neurons closely approximate ideal "differentiators" (Higgs et al. 2006).

2.2   Dendritic Computation in Nucleus Laminaris

2.2.1   Bitufted Morphology and the Dendritic Length Gradient

The receiving elements proposed by Jeffress were hypothetical coincidence detectors, not based on morphological studies of the MSO (he incorrectly assumed that the MSO had been ruled out by the presence of phase-locked responses in the lemniscal nuclei). Nevertheless, although Jeffress designed his famous figure of bipolar neurons innervated by delay lines "merely to illustrate a principle," he was remarkably prescient. Anatomical studies of both NL and MSO describe the strikingly

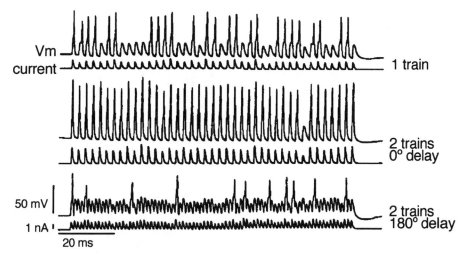

Fig. 6.5 Dynamic clamp experiments with NL neurons in brainstem slices used computer-controlled intracellular current injection to investigate the effects of simulated postsynaptic current trains. Simulated inputs were generated to mimic the firing of NM neurons during acoustic stimulation. Voltage traces (top of each pair) due to current waveforms (bottom of each pair) in response to input of a single train (as if inputs were from only one NM, top), inputs of two trains with no delay between them (as if from binaural stimulation, middle), and inputs with a 180° delay (bottom) (Modified with permission from Reyes et al. 1996)

bipolar, or bitufted, nature of the dendritic arborization of the real coincidence detectors (Figs. 6.5 and 6.6; Parks and Rubel 1975; Jhaveri and Morest 1982; Smith 1995). Typically NL is composed of a compact horizontal monolayer of neurons, ventral and rostral to nucleus magnocellularis, that extend their polarized dendritic fields into the dorsal and ventral neuropil (see Fig. 6.2). Synaptic inputs carrying information from either ear (via the cochlear nucleus) are segregated onto opposite dendritic arbors (Rubel and Parks 1975; Smith and Rubel 1979). Originally demonstrated in the chicken *Gallus gallus*, the polarization of the dendritic arbors is a common feature of the NL/MSO across many vertebrate species, including other birds such as emu (*Dromaius novaehollandiae*), barn owl (though in the latter, bipolar dendrites are restricted to the lateral low-characteristic frequency (CF) NL, as discussed later) (Carr et al. 1997; Kubke et al. 1999; MacLeod et al. 2006), and alligators (Carr and Soares 2002), as well as in mammals like guinea pig (Smith 1995) and cat (Stotler 1953). In the barn owl, the medial NL consists of a thickened neuropil, expanded in the dorsoventral dimension with neurons distributed throughout (Schwartzkopff and Winter 1960; Takahashi et al. 1987; Carr and Boudreau 1993b). These high-CF NL neurons have short unpolarized dendrites; this feature, along with the neuropil hypertrophy, is thought to be a derived feature (Carr et al. 2001; Kubke et al. 2002) that contributes to the barn owl's accurate sound localization (Schwartzkopff and Winter 1960; Carr and Konishi 1988, 1990).

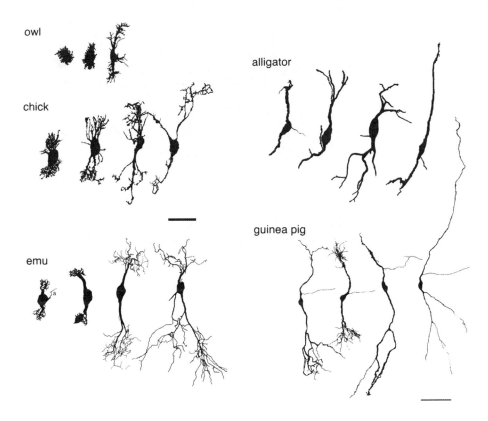

Fig. 6.6 Variations in bipolar dendritic morphology in NL and MSO. Both avian and mammalian coincidence detectors share bitufted morphology. Barn owl NLs were labeled with the Golgi technique (Carr et al. 1997). Chicken NL neurons were labeled with the Golgi technique (Jhaveri and Morest 1982). Emu NL neurons were labeled with biocytin (MacLeod et al. 2006). Alligator NL neurons were labeled with the Golgi technique (Soares and Carr, unpublished). Guinea pig medial superior olive (MSO) neurons were labeled with neurobiotin (Smith 1995). Dendritic length increases from left to right along the presumed high to low best-frequency axis, except in the principal cells of the MSO from the guinea pig, where a frequency gradient is not apparent. The bipolar architecture and the segregation of the inputs arriving from both ears are common to both mammalian and avian coincidence detectors. In the barn owl, coincidence detectors have lost this bipolar organization (except in low best-frequency regions), and their short dendrites radiate around the cell body. The bar on the left applies to bird and alligator reconstructions, 40 μm; the right bottom bar applies to MSO, 100 μm (Modified with permission from MacLeod et al. 2006; Carr et al. 1997; Smith 1995; and Jhaveri and Morest 1982)

A second striking feature that was not predicted by Jeffress is the correlation of the length of the dendrites with the tonotopic mapping of sound frequency in NL. First shown in the chicken (*Gallus gallus*) NL neurons responsive to low frequencies are located caudolaterally, with progressively higher-frequency tuned neurons encountered as one moves rostromedially (Rubel and Parks 1975; Lippe and Rubel 1983).

The caudolateral, low-CF NL neurons have few primary dendrites and long dendritic trees. Dendritic length becomes progressively shorter as one moves toward the rostromedial, high-CF neurons; in chickens these neurons also have a larger number of these short dendrites (Smith and Rubel 1979; Smith 1981; Jhaveri and Morest 1982). A gradient in dendritic length appears in other archosaurs such as the emu (MacLeod et al. 2006) and the alligator *Alligator mississippiensis* (Carr and Soares 2002). The preponderance of short versus long dendrites may also depend on the relative importance of different frequencies to the animal. In the barn owl, very short, unpolarized dendrites in the expanded high-CF region predominate, while bipolar dendrites are confined to the low-CF region (Carr and Boudreau 1993b). In the emu, which has an overlapping hearing range with the chicken but a greater representation of low CFs, the entire field of NL neurons contains neurons that have two primary dendrites that end in dendritic tufts. *A. mississippiensis* is limited to low-CF hearing, and have longer relatively unelaborated dendrites. How dendritic variations emerge in each of these species is an interesting question, as well as why such a gradient is not evident in MSO. In the following section, we offer hypotheses about why such dendritic gradients exist and about their computational value for coincidence detection.

2.2.2 Dendritic Computation in the Nucleus Laminaris Improves ITD Discrimination

The beautiful organization of NL leads one to the hypothesis that there is a compelling relationship between dendritic structure and function. The copious data on the function of coincidence detector neurons provide a unique opportunity for investigating fundamental questions about the computational capabilities of dendrites. Modeling studies not only show how dendrites enhance ITD detection, but also offer a rationale for the existence of a dendritic gradient.

The presence of dendrites results in two biophysical nonlinearities that aid coincidence detection (Fig. 6.7; Agmon-Snir et al. 1998; Simon et al. 1999; Grau-Serrat et al. 2003). These nonlinearities have been studied using compartmental cable theory models that expand on point neuron models by adding dendritic compartments. First, the segregation of the synaptic inputs onto the dendritic compartments improves ITD tuning due to a saturation nonlinearity on the summation of monaural input. Synaptic inputs arriving at the same dendritic compartment add nonlinearly because the driving force decreases with depolarization. The net synaptic current from several inputs arriving simultaneously near one another on the same dendrite is therefore smaller than the net current generated if these inputs are distributed on different dendrites. If the "conductance threshold" is defined as the minimum synaptic conductance needed to trigger a somatic action potential, one can show that this threshold is higher when the synaptic events are on the same dendrite, compared to when they are split between the bipolar dendrites, that is, there is a bias toward binaural coincidences over monaural coincidences. Second, each dendrite acts as a current sink for inputs on the other dendrite, consequently increasing the voltage change needed to trigger a spike at the soma when inputs arrive on only one side. This effect is boosted by the presence of G_{KLT}, such that out-of-phase inputs are subtractively inhibited (Grau-Serrat et al. 2003). With only monaural input, the G_{KLT} in the opposite dendrite is somewhat activated, producing a mild current sink. When there are recent EPSPs in the opposite dendrite due to out-of-phase inputs, however, the G_{KLT} is strongly activated and acts as a large current sink suppressing spike initiation. Thus, the model predicts the experimental finding that the monaural firing rate, although lower than the binaural in-phase rate, is higher than the binaural out-of-phase rate (Goldberg and Brown 1969; Carr and Konishi 1990; Yin and Chan 1990).

Because longer dendrites increase the nonlinearities due to greater segregation from the opposite dendrite, it would seem the longer the better. So why don't all NL neurons have long dendrites? Modeling results show that the improvement of ITD detection with longer dendrites is most pronounced at lower sound frequencies. At a given stimulus frequency, there is a dendritic length beyond which performance no longer increases (Fig. 6.7; Grau-Serrat et al. 2003). A dendritic length gradient is predicted under the assumption that dendritic length is minimized while optimizing for ITD discrimination. This ideal dendritic length decreases with increasing frequency, in agreement with the observation of dendritic length gradients described earlier. In addition, at higher sound frequencies, the dendritic saturation nonlinearity may actually begin to impose a cost (Agmon-Snir et al. 1998). At higher sound frequencies, phase locking is poorer and the afferent spike train contains more jitter. This results in greater overlap in time between inputs from both ears, even when there was a 180° phase shift between them. The effect of this overlap on the conductance threshold is comparably large when the inputs were on the dendrites; even a small "erroneous" input from one ear significantly lowers the conductance threshold for the inputs from the other ear.

In summary, when typical chick-like parameters are used, sublinear summation in the dendrites only improves coincidence detection below 2 kHz, after which discrimination between in-phase and out-of-phase inputs is poor. This is consistent with observations from rabbit MSO neurons, where ITD sensitivity has only been observed for sounds at or below 2 kHz (Batra et al. 1997). The second dendritic nonlinearity, subtractive inhibition of out-of-phase inputs, improves coincidence detection at all frequencies (Grau-Serrat et al. 2003) and might therefore be most significant in avian coincidence detectors between 2 and 8 kHz. It is also clear that the quality of phase-locked inputs has some bearing on coincidence detection: typical chick-like parameters, but with barn owl–like phase locking, allow ITD discrimination up to 4–6 kHz (Fig. 6.7). The benefits conveyed by the neuronal structure of the coincidence detectors allow one to argue for the convergent or parallel evolution of coincidence detectors in the bird NL and mammalian MSO.

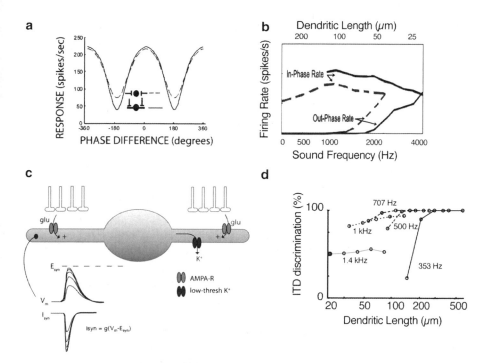

Fig. 6.7 Dendritic computation enhances coincidence detection in NL. (**a**) Comparison of ITD coding between a point neuron and a bipolar-neuron compartmental model, which is composed of a soma and two cylindrical dendrites that receive inputs. Model ITD curves for a 500-Hz stimulus showed an improvement in ITD coding with dendrites: the contrast between the maximum spike rate (0° delay) and the minimum spike rate (180° delay) was larger when the inputs were segregated on the dendrites (*solid line*) compared with the point-neuron model (*dashed line*) (Modified with permission from Agmon-Snir et al. 1998) (**b**) Schematic of nonlinear dendritic effects. Each dendrite receives multiple afferent inputs via glutamatergic synaptic inputs. With each additional input, the voltage in the dendrites (V_m) gets closer to the reversal potential for the AMPA receptor (E_{syn}), reducing the synaptic drive and effectively limiting the efficacy of multiple inputs into the dendritic compartment. A low-threshold K^+ current in the opposite dendrite acts like a current sink. (**c**) ITD discrimination index, as a function of best frequency location (along the tonotopic axis). Stimuli presented at CF with the corresponding NM vector strength input. The upper plots are for chick, as is the lower plot, but with input vector strength that of barn owl. (**d**) ITD discrimination index as a function of dendritic length for five stimulus frequencies. There is a dendritic length beyond which any increase in length no longer aids ITD discrimination. That length grows with decreasing frequency, predicting a dendritic length gradient (**b** and **d** are modified with permission from Grau-Serrat et al. 2003)

2.3   Contributions of Synaptic Inhibition to Coincidence Detection

Synaptic inhibition plays several key roles in coincidence detection. Known functions of inhibitory inputs in avian NL include suppression of responses at unfavorable ITDs and maintenance of sensitivity to ITDs when sounds are loud, while in mammals, glycinergic inputs provide precisely timed synaptic inputs to shift ITD sensitivity. The roles of inhibition in ITD processing have recently been the subject of a cogent review by Burger and Rubel (2008), which is recommended for those interested in learning more about this topic.

In birds, the coincidence detectors in NL receive GABAergic inputs (Carr et al. 1989), mostly from the superior olive (Lachica et al. 1994; Yang et al. 1999; Burger et al. 2005). The superior olive receives excitatory inputs from NL and NA, and the projection back to NL forms an inhibitory feedback loop, perhaps to offset the effects of bilateral differences in sound intensity (Lachica et al. 1994; Burger et al. 2005) and to decrease or eliminate the effects of intensity on ITD computations (Grun et al. 1992; Pena et al. 1996; Viete et al. 1997). GABAergic synapses generally exert a depolarizing effect on NL neurons. Unusually, intracellular chloride concentrations in NL remain high even after the onset of hearing (Hyson et al. 1995; Monsivais et al. 2000; Lu and Trussell 2001). GABAergic IPSPs are therefore depolarizing, activating the G_{KLT} conductance, decreasing the input resistance, and improving coincidence detection (Funabiki et al. 1998). Lesions of the superior olive also eliminate the suppression of firing activity at the worst ITD in the low best-frequency neurons (Nishino et al. 2008). An additional potentially inhibitory component has recently been discovered; there is a descending projection from the ventral nucleus of the lateral lemniscus to NA, NM, and NL (Wild et al. 2009) that might be the source of the fast glycinergic IPSCs observed in NA and NL (Kuo et al. 2009).

In the mammalian MSO, glycinergic inputs from the medial nucleus of the trapezoid body provide a precisely timed synaptic input that shifts the ITD curve so that the slope of the ITD function falls within the physiological range of ITDs available to the gerbil (Brand et al. 2002; McAlpine and Grothe 2003; Pecka et al. 2008). Modeling studies suggest these adjustments of best ITD may be achieved by interplay of somatic sodium currents and synaptic inhibitory currents (Zhou et al. 2005). The differences in inhibitory synaptic transmission in birds (slow and GABAergic) and mammals (temporally precise and glycinergic) underlie a major difference in ITD coding in birds and small mammals where, at least for low best frequencies, the change in firing rate (the slope) of the coincidence detectors in the MSO is used to determine sound location (McAlpine et al. 2001; Hancock and Delgutte 2004; Joris and Yin 2007).

2.4   Short-Term Depression Improves Coincidence Detection

Short-term synaptic plasticity is defined as a rapid activity-dependent alteration in synaptic efficacy that occurs on the time scale of tens of milliseconds to seconds. This form of synaptic plasticity has been proposed to contribute to neural coding by influencing what type of information is passed across the synapse, acting like a type of filter (for reviews, see Zucker and Regehr 2002, and Abbott and Regehr 2004). Synapses in the timing circuits can express the form of synaptic plasticity termed short-term depression, in which the responses are reduced with increasing presynaptic firing rates (Trussell 1999; Kuba et al. 2002b; Schneggenburger and Forsythe 2006). In contrast, more complex, less-depressing forms of short-term plasticity characterize synapses onto cochlear nucleus neurons involved in sound intensity or other non-ITD processing (MacLeod and Carr 2007; MacLeod et al. 2007; Cao et al. 2008). The short-term depression observed at synapses onto NL neurons has a similar dependence on the presynaptic firing rate as those in NM (Zhang and Trussell 1994; Kuba et al. 2002b; Cook et al. 2003). This is a presynaptic effect, perhaps due to the depletion of synaptic vesicles in the presynaptic terminal, as there appears to be little contribution by postsynaptic mechanisms such as AMPA-receptor desensitization (Kuba et al. 2002a), despite expressing receptor subtypes with rapid desensitization properties (Parks 2000; Raman et al. 1994). Increased presynaptic firing rates lead to greater depression of the synaptic response, scaling the synaptic response inversely with the firing rate. This effectively maintains the synaptic amplitudes within a range suitable for coincidence detection (Fig. 6.8; Cook et al. 2003; Kuba et al. 2002b). Because firing rates of the auditory nerve afferents and NM increase with sound intensity, a scaling mechanism could counteract the problems that arise with high-intensity inputs. Alternatively, scaling could equalize the inputs across the two ears when the sound is louder in one than the other. When inputs from one side are driven at a higher rate than inputs from the other side, due to interaural intensity differences, short-term depression results in a greater reduction in the amplitude of the high-rate inputs (from the loud side) than the relatively lower-rate inputs (from the softer side). Because short-term plasticity is synapse-specific in its effects, it can differentially affect inputs arising from different afferents to NL, a property that distinguishes such plasticity from a more generic inhibitory feedback mechanism. ITD tuning in NL is in fact relatively invariant to both overall sound intensity and interaural intensity differences (Pena et al. 1996; Viete et al. 1997). Simply put, short-term depression can improve coincidence detection by removing rate (intensity) information. Thus, synaptic depression acts like an adaptive mechanism to remain within the range for effective coincidence detection.

2.5   Are Coincidence Detectors Performing Cross-Correlation?

Studies of cross-correlation are key to understanding the synaptic mechanisms of coincidence detection, because incoming brief phase-locked EPSCs arrive at the coincidence detectors, where they represent the stimulus in a cycle-by-cycle fashion (Reyes et al. 1996; Marsalek et al. 1997; Ashida et al. 2007). Coincidence detectors should fire when the inputs from two ears coincide (or almost coincide), but not when two inputs from the same ear coincide (also see Sect. 2.2). In other words, they should distinguish between ITD-dependent binaural inputs and ITD-independent monaural inputs, which is what a cross-correlator does. Neurons that sum their inputs linearly cannot do this.

For cross-correlation, a representation of the effective monaural input must emerge in the membrane potential of the coincidence detector. The large number of

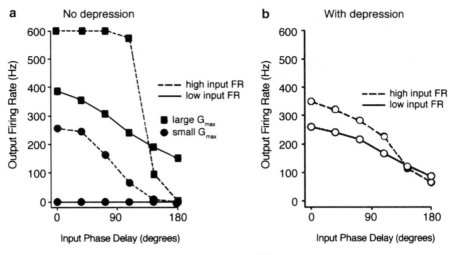

Fig. 6.8 Model of the effects of short-term synaptic depression on NL ITD tuning. Short-term depression buffers coincidence detection against changing sound intensity. (**a**) Without synaptic depression, using a low value for the synaptic conductance (low G_{max}) leads to NL output firing that is modulated at high sound intensity (high-input FR; circles with dashed line) but shows poor modulation with phase delay at low sound intensity (low-input FR; *circles with solid line*). Increasing the synaptic conductance (high G_{max}) improves tuning at low sound intensity (squares with solid line), but firing rates saturate at high sound intensity for phase delays less than 90° and decline sharply to near zero for phase delays near 180° (*squares with dashed line*). A single conductance value fails to code ITD over the range of sound intensities likely to be encountered. (**b**) With synaptic depression, coincidence detection curves encode ITD well at low (*solid*) and high (*dashed*) sound intensities (Modified with permission from Cook et al. 2003)

NM axons connecting to each NL cell (Carr and Boudreau 1993b), coupled with the high spontaneous rate of NM neurons (Köppl 1997), is consistent with sufficient convergence to re-create an unrectified representation of the stimulus in NL neurons. This key assumption is not yet confirmed by in vivo intracellular recordings from coincidence detectors but has been predicted by models of the owl's NL as a sound analog synaptic input formed by phase-locked spikes of NM axons (Ashida et al. 2007). This assumption has been tested in vitro, to some extent. Reyes et al. (1996) used computer-controlled intracellular current injection to mimic postsynaptic currents generated in NL neurons by the firing of NM neurons during acoustic stimulation. They found that when they summed two stimulus trains at different delays, the NL firing rate dropped below the monaural rate with a delay of half the stimulus period, showing that inputs delivered at twice the in-phase frequency (2 F) did not effectively stimulate the NL neuron.

How close is coincidence detection to cross-correlation? It has been proposed that cross-correlation provides a more general description of coincidence detection (Colburn et al. 1990; Yin and Chan 1990; Han and Colburn 1993). Careful testing of this hypothesis in the MSO has shown, however, that these coincidence detectors are not ideal cross correlators (Batra and Yin 2004). They respond to monaural inputs, and their tuning is broader than expected by cross-correlation. Batra and Yin (2004) describe MSO output as a degraded version of cross-correlation, which raises the question of whether coincidence detectors are adapted to approximate cross-correlation. Are MSO neurons as good as they can get, given biophysical constraints on phase locking, or are there other significant constraints? Interestingly, barn owl NL neurons conform more closely to cross-correlation-based models of binaural interaction (Fischer et al. 2008). Are they better adapted to act as cross-correlators, or do MSO neurons do more than one thing?

3 Building an Array of Coincidence Detectors

The magnocellular-laminaris circuit conforms to the predictions of the Jeffress model, which posits that systematically varying axonal conduction delays along the length of the nucleus serve to offset ITDs, so that each neuron is "tuned" to a best ITD value. This scheme turns systematic variations in ITD into a topographic map of sound source location, or place map, in which the location of maximal activity corresponds to a particular azimuthal location. In owls, the place map in the nucleus laminaris is preserved at higher stations of the auditory system and is finally aligned with the visual space map in the auditory midbrain and the optic tectum (for review, see Knudsen 2002).

3.1 Jeffress Model: Expansion to Multiple Maps Across Different Frequencies

Within birds, the details of delay line circuit organization vary, as discussed earlier (see Sect. 1.2). In the chicken, NL is composed of a monolayer of bipolar neurons that receive input from ipsi- and contralateral cochlear nuclei onto their dorsal and ventral dendrites, respectively (Parks and Rubel 1975). Evidence from brain slices suggests that only the projection from the contralateral cochlear nucleus acts as a delay line, while inputs from the ipsilateral cochlear nucleus arrive simultaneously at all neurons (Overholt et al. 1992). This pattern of inputs creates a single map of ITD in any tonotopic band in the mediolateral dimension of NL (Fig. 6.8; Overholt et al. 1992; Köppl and Carr 2008).

The barn owl NL is both much larger and reorganized when compared to the plesiomorphic condition exemplified by the chicken and alligator (Carr et al. 2009; Kubke et al. 2002, 2004). Magnocellular axons from both cochlear nuclei act as delay lines (Carr and Konishi 1988). They convey the phase of the auditory stimulus to NL such that axons from the ipsilateral NM enter the NL from the dorsal side, while axons from the contralateral NM enter from the ventral side. Recordings from these interdigitating ipsilateral and contralateral axons show regular changes in delay with depth in NL (Carr and Konishi 1990). Thus, these afferents interdigitate to innervate dorso-ventral arrays of neurons in NL in a sequential fashion, producing multiple representations of ITD within NL. Despite the differences in organization of NL in owls and chickens, ITDs are detected by neurons that act as coincidence detectors in both species (Sullivan and Konishi 1984; Joseph and Hyson 1993; Pena et al. 1996).

3.2 Place Codes and Current Theories of Local Versus Population Codes in NL and MSO

Jeffress model circuits form both a place map, otherwise known as a "labeled line" or local population code, where position in an array encodes location, and a rate code, where change in firing rate with location provides capacity for fine temporal discrimination (Takahashi et al. 2003). The existence of a place map has been demonstrated in the avian NL (Köppl and Carr 2008), but in the MSO, direct confirmation of a such a map is lacking. Precisely timed inputs do arrive at the MSO, and their organization is not inconsistent with that posited for delay lines. Reconstruction of labeled contralateral anterior ventral cochlear nucleus (AVCN) axons in cat revealed that the projections to the contralateral, but not the ipsilateral MSO, show a rostral-to-caudal delay line configuration (Smith et al. 1993). This was consistent with coarse but apparently topographic maps of ITDs reported by Yin and Chan (1990). Labeled contralateral AVCN axons form ladder-like collaterals, with shorter collaterals innervating more rostral parts of MSO and longer collaterals innervating more caudal parts of MSO. Other axons had restricted terminal fields comparable to the size of a single dendritic tree in the MSO. In the ipsilateral MSO, some axons had a reverse, but less steep, gradient in axonal length with greater axonal length associated with more rostral locations; others had restricted terminal fields (Beckius et al. 1999; Smith et al. 1993).

Given these data, until recently it was assumed that the AVCN inputs to the MSO formed a map, as proposed by Jeffress and as observed in birds. Then, work in guinea pigs and gerbils revealed a tendency for the steepest region of the slope of the ITD tuning curve to fall close to midline, irrespective of best frequency (Brand et al. 2002; McAlpine and Grothe 2003). McAlpine and his colleagues proposed that sound source location could be encoded by activity in two broad hemispheric spatial channels, especially in animals with small heads and/or low best-frequency hearing (Fig. 6.9; McAlpine et al. 2001). The two-channel scheme would thus encode ITD using a rate code alone (Pecka et al. 2008).

The emergence of different ITD coding strategies in birds and mammals is interesting. Birds and mammals have similar constraints, yet birds and their close relatives, the crocodilians, have maps of ITD, while the mammals so far examined do not appear to (Köppl 2009). Chickens and gerbils have similar head sizes and ability to encode temporal information, but chickens have a place code of ITD in NL (Köppl and Carr 2008). Data from chickens and barn owls suggest instead that evolutionary history

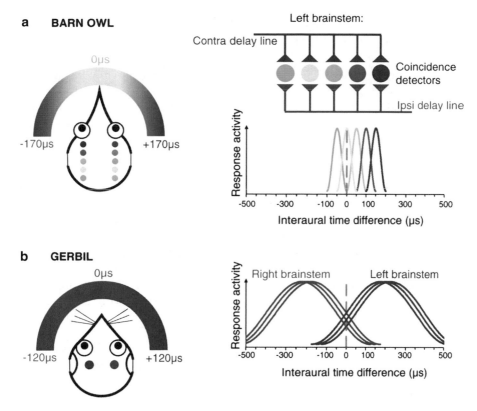

Fig. 6.9 Comparison of avian and mammalian ITD representations. Two coding schemes have been proposed for the encoding of ITDs. (**a**) In the barn owl, each half of the brainstem contains a map of interaural time differences corresponding to sound sources mostly in the contralateral acoustic hemifield. This map is created by the basic circuit illustrated for the left brainstem. Auditory nerve input is relayed from both ears to cells (*colored circles*), which act as coincidence detectors, as in the Jeffress model. Typical responses of barn owl coincidence detector neurons are shown below the circuit diagram as a function of ITD. Together they cover the interaural time difference range experienced by the owl and form a place map. (**b**) In the gerbil, the responses of neurons in half of the brainstem are all similar and show only broad selectivity compared to the animal's ITD range. It is thought that higher-order brain areas compare the relative activity levels of both brainstem sides to derive a correlate of sound source location (Modified with permission from Köppl 2009)

may constrain neural circuits. This is borne out by recordings from the alligator NL. Crocodilians are birds' closest living relatives, and their NL displays the major features of a place code of ITD (Carr et al. 2009). The differences between birds and mammals may indicate either that the MSO circuit is simply another good enough solution to the same problem or that MSO organization reflects different constraints. Movable external ears and sensitivity to higher frequencies than most frogs and reptiles set mammalian hearing apart, and both features could impact sound localization strategies.

3.3 Tonotopy and How Synaptic Mechanisms Change in a Gradient Across NL

Gradients in expression level of gene products are not uncommon in the auditory system. Frequency selectivity in nonmammalian hair cells is enhanced by the systematic variation in hair cell membrane properties along the tonotopic axis, for example (for review, see Fettiplace and Fuchs 1999). Coincidence detection circuits are notable for gradients in morphology, synaptic properties. and membrane properties along the tonotopic axis. Dendritic length and number change with best frequency, as does the expression of potassium channels. Gradients in the low threshold potassium current (I_{KLT}) expression sculpt the EPSP time course, while gradients in the high threshold potassium current (I_{KHT}) channel expression sculpt presynaptic action potentials – a high-frequency specialization. Generally, these gradients appear focused on tuning the synaptic mechanisms of coincidence detection.

Elegant work using the chick brainstem slice preparation has quantified systematic tonotopic changes in the electrophysiological properties of NL neurons. These properties include changes in input capacitance with change in the length of dendrites, changes in the conductance of hyperpolarization-activated conductance (I_H), and changes in the EPSC time course (Reyes et al. 1996; Kuba et al. 2005; Kuba 2007). Gradients in both I_{KLT} and I_{KHT} underlie systematic tonotopic changes in electrophysiological properties (Manis and Marx 1991; Brew and Forsythe 1995; Rathouz and Trussell 1998). These cellular specializations are hypothesized to be frequency-dependent adaptations for ITD coding.

Not all changes are continuous. In both owls and chickens, there is a binary change in axonal morphology, in which neurons with low best frequencies and long dendrites had normal initial segments, while neurons in the rostral higher best-frequency portion of NL had myelinated axon initial segments (Carr and Boudreau 1993a). Thus for high best-frequency neurons, the site of spike initiation is in the axon, distant from the soma (Kuba et al. 2006). Sodium channels are not found in the soma of high best-frequency (2.5–3.3 kHz) and middle best-frequency (1.0–2.5 kHz) neurons, while in low best-frequency (0.4–1.0 kHz) neurons they are clustered in a normal initial segment closer to the soma (Kuba et al. 2006). Thus, neurons initiate spikes at a more remote site as the best frequency of neurons increases. Consequently, the somatic amplitudes of action potentials are smaller in high-CF and middle-CF neurons and are large in low-CF neurons. Computer simulations suggest that the geometry of the initiation site is optimized to reduce the threshold of spike generation and thus increase the ITD sensitivity. Especially in high best-frequency neurons, a distant localization of the spike initiation site improves the ITD sensitivity because of electrical isolation of the initiation site from the soma and dendrites, and because of reduction of Na-channel inactivation by attenuating the temporal summation of synaptic potentials through the low-pass filtering along the axon. The importance of initial segment geometry is borne out by auditory deprivation studies that showed increases in axon initial segment length (Kuba et al. 2010).

It is clear that the higher the frequency, the more difficult ITD detection becomes. The phase-locking of the inputs falls off with frequency, while the intrinsic properties of the neuronal membrane act like a low-pass filter (see Sect. 2.1). Changes in I_{KLT} conductance and expression of Kv1.2 channel protein reflect the increasing biophysical demands on coincidence detection (Kuba et al. 2005). Kv1.2 expression is maximal in the central middle best-frequency region of NL. Thus, the middle-frequency neurons have the smallest input resistance and membrane time constant and consequently the fastest EPSPs.

4 Summary and Comparison with Other Coincidence-Detection Systems

In 1998, Joris et al. (1998) wrote a minireview on coincidence detection entitled "50 Years after Jeffress" in which they discussed how well Jeffress's simple model held up. Jeffress's three predictions were that first, delay lines converge on a binaural nucleus and contain temporal information in their discharge pattern about the waveform of the acoustic stimulus. Second, cells in the binaural nucleus only discharge when receiving coincident spikes. Third, the delay lines form a place map of the ITD. It is clear that this simple model remains a useful focus of both experimental and modeling studies. Section 1 focused on the synaptic mechanisms of coincidence detection in the auditory brainstem and provided brief discussions of phase-locking, delay lines, and place maps. Section 2 focused on the key feature of coincidence detection, which is that these neurons should fire when inputs from two independent sources coincide (or almost coincide), but not when two monaural inputs coincide. This discrimination is a remarkable example of the computational role of dendrites and the tonotopic variation displayed in morphology, biophysics, and synaptic properties, especially in birds, and allows one to argue that selective pressures have driven the convergent evolution of coincidence detectors in bird NL and mammalian MSO.

Precise coincidence detection characterizes other systems as well, including weakly electric fish that can distinguish phase differences as small as 0.5 μs between different parts of their body surface (Carr et al. 1986b). Temporal hyperacuity has evolved independently in both of the electrosensory fish groups, Mormyriformes and Gymnotiformes (Kawasaki 1993). Remarkably, these unrelated fish have evolved the same computational algorithms, and both use coincidence detectors to detect microsecond time differences. In the African Gymnarchus niloticus, large hindbrain neurons called ovoidal cells receive two types of phase-locking inputs, segregated onto the soma and dendrite, respectively (Matsushita and Kawasaki 2004). In the South American Gymnotiformes, midbrain coincidence detectors also receive inputs segregated onto cell body and dendrites (Carr et al. 1986a). The parallel emergence of segregated inputs permitting detection of inputs from two independent sources points to the importance of dendritic compartments.

Escape circuits achieve similar precision, although their output is naturally more binary. In the crayfish giant fiber system, precise coincidence detection emerges from the normal function of rectifying electrical synapses (Edwards et al. 1998). Such synapses allow bidirectional current flow when presynaptic cells depolarize relative to their postsynaptic targets and remain open until well after completion of presynaptic spikes. When multiple input neurons fire simultaneously, the synaptic currents sum effectively and produce a large EPSP. When some inputs are delayed relative to the rest, however, their effect is reduced because the early EPSP retards the opening of additional voltage-sensitive synapses, and the late synaptic currents are shunted by already opened junctions. These mechanisms account for the ability of the lateral giant neurons of crayfish to sum synchronous inputs but not inputs separated by only 100 µs. This is similar to the shunting effect of the opposite dendrite in nucleus laminaris, as proposed by Grau-Serrat et al. (2003).

Inhibition can also play a pivotal role. In another well-understood escape system, the Mauthner neuron, the inhibitory electric field around the axon cap can coincide with the electrotonic excitatory drive to the Mauthner cell to regulate the threshold of the acoustic startle with high temporal precision (Weiss et al. 2008). Similarly, Mittmann et al. (2005) found that feed-forward inhibition shapes the spike output of cerebellar Purkinje cells. This temporally precise inhibition can be achieved by the interplay of somatic currents and synaptic inhibitory currents, reminiscent of the role of precisely timed glycinergic inhibition in the MSO (Brand et al. 2002; Zhou et al. 2005).

Acknowledgments The authors received support for their work from the National Institutes of Health grants R01DC000436 (C.E.C.) and R03DC007972 (K.M.M.) and a grant from the National Organization for Hearing Research (K.M.M.) The authors thank C. Köppl and H. Kuba for helpful comments on the manuscript.

References

Abbott, L. F., & Regehr, W. G. (2004). Synaptic computation. *Nature, 431*(7010), 796–803.

Agmon-Snir, H., Carr, C. E., & Rinzel, J. (1998). The role of dendrites in auditory coincidence detection. *Nature, 393*(6682), 268–272.

Ashida, G., Abe, K., Funabiki, K., & Konishi, M. (2007). Passive soma facilitates submillisecond coincidence detection in the owl's auditory system. *Journal of Neurophysiology, 97*(3), 2267–2282.

Bala, A. D., & Takahashi, T. T. (2000). Pupillary dilation response as an indicator of auditory discrimination in the barn owl. *Journal of Comparative Physiology [A], 186*(5), 425–434.

Batra, R., & Yin, T. C. (2004). Cross correlation by neurons of the medial superior olive: A reexamination. *JARO: Journal of the Association for Research Otolaryngology, 5*(3), 238–252.

Batra, R., Kuwada, S., & Fitzpatrick, D. C. (1997). Sensitivity to interaural temporal disparities of low- and high-frequency neurons in the superior olivary complex. I. Heterogeneity of responses. *Journal of Neurophysiology, 78*(3), 1222–1236.

Beckius, G. E., Batra, R., & Oliver, D. L. (1999). Axons from anteroventral cochlear nucleus that terminate in medial superior olive of cat: Observations related to delay lines. *Journal of Neuroscience, 19*(8), 3146–3161.

Brand, A., Behrend, O., Marquardt, T., McAlpine, D., & Grothe, B. (2002). Precise inhibition is essential for microsecond interaural time difference coding. *Nature, 417*(6888), 543–547.

Brenowitz, S., & Trussell, L. O. (2001). Maturation of synaptic transmission at end-bulb synapses of the cochlear nucleus. *Journal of Neuroscience, 21*(23), 9487–9498.

Brew, H. M., & Forsythe, I. D. (1995). Two voltage-dependent K + conductances with complementary functions in postsynaptic integration at a central auditory synapse. *Journal of Neuroscience, 15*(12), 8011–8022.

Burger, R. M., & Rubel, E. W. (2008). Encoding of interaural timing for binaural hearing. In P. Dallos & D. Oertel (Eds.), *The Senses: A Comprehensive Reference* (pp 613–630). San Diego: Academic Press.

Burger, R. M., Cramer, K. S., Pfeiffer, J. D., & Rubel, E. W. (2005). Avian superior olivary nucleus provides divergent inhibitory input to parallel auditory pathways. *Journal of Comparative Neurology, 481*(1), 6–18.

Cao, X. J., McGinley, M. J., & Oertel, D. (2008). Connections and synaptic function in the posteroventral cochlear nucleus of deaf jerker mice. *Journal of Comparative Neurology, 510*(3), 297–308.

Carr, C. E., & Boudreau, R. E. (1993a). An axon with a myelinated initial segment in the bird auditory system. *Brain Research, 628*, 330–334.

Carr, C. E., & Boudreau, R. E. (1993b). Organization of the nucleus magnocellularis and the nucleus laminaris in the barn owl: Encoding and measuring interaural time differences. *Journal of Comparative Neurology, 334*(3), 337–355.

Carr, C. E., & Code, R. A. (2000). The central auditory system of reptiles and birds. In R. J. Dooling, R. R. Fay, & A. N. Popper (Eds.), *Comparative Hearing: Birds and Reptiles* (pp 197–248). New York: Springer.

Carr, C. E., & Konishi, M. (1988). Axonal delay lines for time measurement in the owl's brainstem. *Proceedings of the National Academy of Sciences of the United States of America, 85*(21), 8311–8315.

Carr, C. E., & Konishi, M. (1990). A circuit for detection of interaural time differences in the brainstem of the barn owl. *Journal of Neuroscience, 10*, 3227–3246.

Carr, C. E., & Soares, D. (2002). Evolutionary convergence and shared computational principles in the auditory system. *Brain, Behavior and Evolution, 59*(5–6), 294–311.

Carr, C., Heiligenberg, W., & Rose, G. (1986a). A time-comparison circuit in the electric fish midbrain. I. Behavior and physiology. *Journal of Neuroscience, 6*, 107–119.

Carr, C. E., Maler, L., & Taylor, B. (1986b). A time comparison circuit in the electric fish midbrain. II. Functional morphology. *Journal of Neuroscience, 6*, 1372–1383.

Carr, C. E., Fujita, I., & Konishi, M. (1989). Distribution of GABAergic neurons and terminals in the auditory system of the barn owl. *Journal of Comparative Neurology, 286*(2), 190–207.

Carr, C. E., Kubke, M. F., Massoglia, D. P., Cheng, S. M., Rigby, L. L., & Moiseff, A. (1997). Development of temporal coding circuits in the barn owl. In A. R. Palmer, A. Rees, Q. Summerfield, & R. Meddis (Eds.), *Psychophysical and Physiological Advances in Hearing* (pp 344–351). London: Whurr.

Carr, C. E., Soares, D., Parameshwaran, S., & Perney, T. (2001). Evolution and development of time coding systems. *Current Opinion in Neurobiology, 11*(6), 727–733.

Carr, C., Soares, D., Simon, J., & Smolders, J. (2009). Detection of interaural time differences in the alligator. *Journal of Neuroscience, 29*(25), 7978–7990.

Colburn, H. S., Han, Y. A., & Culotta, C. P. (1990). Coincidence model of MSO responses. *Hearing Research, 49*(1–3), 335–346.

Cook, D. L., Schwindt, P. C., Grande, L. A., & Spain, W. J. (2003). Synaptic depression in the localization of sound. *Nature, 421*(6918), 66–70.

Dingledine, R., Borges, K., Bowie, D., & Traynelis, S. F. (1999). The glutamate receptor ion channels. *Pharmacology Reviews and Communications, 51*(1), 7–61.

Edwards, D. H., Yeh, S. R., & Krasne, F. B. (1998). Neuronal coincidence detection by voltage-sensitive electrical synapses. *Proceedings of the National Academy of Sciences of the United States of America, 95*(12), 7145–7150.

Fettiplace, R., & Fuchs, P. A. (1999). Mechanisms of hair cell tuning. *Annual Review of Physiology, 61*, 809–834.

Fischer, B. J., Christianson, G. B., & Pena, J. L. (2008). Cross-correlation in the auditory coincidence detectors of owls. *Journal of Neuroscience, 28*(32), 8107–8115.

Fukui, I., & Ohmori, H. (2003). Developmental changes in membrane excitability and morphology of neurons in the nucleus angularis of the chicken. *Journal of Physiology (London), 548*(Pt. 1), 219–232.

Funabiki, K., Koyano, K., & Ohmori, H. (1998). The role of GABAergic inputs for coincidence detection in the neurones of nucleus laminaris of the chick. *Journal of Physiology (London)*, *508*(3), 851–869.

Geiger, J. R., Melcher, T., Koh, D. S., Sakmann, B., Seeburg, P. H., Jonas, P., & Monyer, H. (1995). Relative abundance of subunit mRNAs determines gating and Ca2+ permeability of AMPA receptors in principal neurons and interneurons in rat CNS. *Neuron*, *15*(1), 193–204.

Goldberg, J. M., & Brown, P. B. (1969). Response of binaural neurons of dog superior olivary complex to dichotic tonal stimuli: Some physiological mechanisms of sound localization. *Journal of Neurophysiology*, *32*, 613–636.

Grau-Serrat, V., Carr, C. E., & Simon, J. Z. (2003). Modeling coincidence detection in nucleus laminaris. *Biological Cybernetics*, *89*(5), 388–396.

Grigg, J. J., Brew, H. M., & Tempel, B. L. (2000). Differential expression of voltage-gated potassium channel genes in auditory nuclei of the mouse brainstem. *Hearing Research*, *140*(1–2), 77–90.

Grun, S., Aertsen, A., Wagner, H., & Carr, C. (1992). Binaural interaction in the nucleus laminaris of the barn owl: A quantitative model. *BrainWorks* v1991-01, http://www.brainworks.uni-freiburg.de, Albert-Ludwigs-University, Freiburg.

Han, Y., & Colburn, H. S. (1993). Point-neuron model for binaural interaction in MSO. *Hearing Research*, *68*(1), 115–130.

Hancock, K. E., & Delgutte, B. (2004). A physiologically based model of interaural time difference discrimination. *Journal of Neuroscience*, *24*(32), 7110–7117.

Higgs, M. H., Slee, S. J., & Spain, W. J. (2006). Diversity of gain modulation by noise in neocortical neurons: Regulation by the slow afterhyperpolarization conductance. *Journal of Neuroscience*, *26*(34), 8787–8799.

Hyson, R. L., Reyes, A. D., & Rubel, E. W. (1995). A depolarizing inhibitory response to GABA in brainstem auditory neurons of the chick. *Brain Research*, *677*(1), 117–126.

Jeffress, L. (1948). A place theory of sound localization. *Journal of Comparative Physiology and Psychology*, *41*, 35–39.

Jhaveri, S., & Morest, D. K. (1982). Sequential alterations of neuronal architecture in nucleus magnocellularis of the developing chicken: A Golgi study. *Neuroscience*, *7*(4), 837–853.

Joris, P., & Yin, T. C. (2007). A matter of time: Internal delays in binaural processing. *Trends in Neurosciences*, *30*(2), 70–78.

Joris, P. X., Smith, P. H., & Yin, T. C. (1998). Coincidence detection in the auditory system: 50 years after Jeffress. *Neuron*, *21*(6), 1235–1238.

Joseph, A. W., & Hyson, R. L. (1993). Coincidence detection by binaural neurons in the chick brain stem. *Journal of Neurophysiology*, *69*(4), 1197–1211.

Kawasaki, M. (1993). Independently evolved jamming avoidance responses employ identical computational algorithms: A behavioral study of the African electric fish, Gymnarchus niloticus. *Journal of Comparative Physiology [A]*, *173*(1), 9–22.

Klump, G. M. (2000). Sound localization in birds. In R. J. Dooling, R. R. Fay, & A. N. Popper (Eds.), *Comparative Hearing: Birds and Reptiles* (pp. 249–307). New York: Springer.

Knudsen, E. I. (2002). Instructed learning in the auditory localization pathway of the barn owl. *Nature*, *417*(6886), 322–328.

Knudsen, E. I., & Konishi, M. (1978a). A neural map of auditory space in the owl. *Science*, *200*, 795–797.

Knudsen, E. I., & Konishi, M. (1978b). Space and frequency are represented separately in the auditory midbrain of the owl. *Journal of Neurophysiology*, *41*, 870–884.

Knudsen, E. I., Blasdel, G. G., & Konishi, M. (1979). Sound localization by the barn owl (Tyto alba) measured with the search coil technique. *Journal of Comparative Physiology*, *133*, 1–11.

Konishi, M. (1973a). How the owl tracks its prey. *American Scientist*, *61*, 414–424.

Konishi, M. (1973b). Locatable and nonlocatable acoustic signals for barn owls. *American Naturalist*, *107*, 775–785.

Konishi, M. (1993). Listening with two ears. *Scientific American*, *268*(4), 66–73.

Köppl, C. (1997). Phase locking to high frequencies in the auditory nerve and cochlear nucleus magnocellularis of the barn owl, Tyto alba. *Journal of Neuroscience, 17*(9), 3312–3321.

Köppl, C. (2009). Evolution of sound localisation in land vertebrates. *Current Biology, 19*(15), R635–639.

Köppl, C., & Carr, C. E. (2008). Maps of interaural time difference in the chicken's brainstem nucleus laminaris. *Biological Cybernetics, 98*(6), 541–559.

Kuba, H. (2007). Cellular and molecular mechanisms of avian auditory coincidence detection. *Neuroscience Research, 59*(4), 370–376.

Kuba, H., Koyano, K., & Ohmori, H. (2002a). Development of membrane conductance improves coincidence detection in the nucleus laminaris of the chicken. *Journal of Physiology (London), 540*(Pt. 2), 529–542.

Kuba, H., Koyano, K., & Ohmori, H. (2002b). Synaptic depression improves coincidence detection in the nucleus laminaris in brainstem slices of the chick embryo. *European Journal of Neuroscience, 15*(6), 984–990.

Kuba, H., Yamada, R., & Ohmori, H. (2003). Evaluation of the limiting acuity of coincidence detection in nucleus laminaris of the chicken. *Journal of Physiology (London), 552*(Pt. 2), 611–620.

Kuba, H., Yamada, R., Fukui, I., & Ohmori, H. (2005). Tonotopic specialization of auditory coincidence detection in nucleus laminaris of the chick. *Journal of Neuroscience, 25*(8), 1924–1934.

Kuba, H., Ishii, T., & Ohmori, H. (2006). Axonal site of spike initiation enhances auditory coincidence detection. *Nature, 444*, 1069–1072.

Kuba, H., Oichi, Y., & Ohmori, H. (2010). Presynaptic activity regulates Na(+) channel distribution at the axon initial segment. *Nature, 465*(7301), 1075–1078.

Kubke, M. F., & Carr, C. E. (2000). Development of the auditory brainstem of birds: Comparison between barn owls and chickens. *Hearing Research 147*(1–2), 1–20.

Kubke, M. F., & Carr, C. E. (2005). Development of sound localization. In A. N. Popper & R. Fay (Eds.), *Sound Source Localization,* 179–237 New York: Springer.

Kubke, M. F., Gauger, B., Basu, L., Wagner, H., & Carr, C. E. (1999). Development of calretinin immunoreactivity in the brainstem auditory nuclei of the barn owl (Tyto alba). *Journal of Comparative Neurology, 415*(2), 189–203.

Kubke, M. F., Massoglia, D. P., & Carr, C. E. (2002). Developmental changes underlying the formation of the specialized time coding circuits in barn owls (Tyto alba). *Journal of Neuroscience, 22*(17), 7671–7679.

Kubke, M. F., Massoglia, D. P., & Carr, C. E. (2004). Bigger brains or bigger nuclei? Regulating the size of auditory structures in birds. *Brain, Behavior and Evolution, 63*(3), 169–180.

Kuo, S. P., Bradley, L. A., & Trussell, L. O. (2009). Heterogeneous kinetics and pharmacology of synaptic inhibition in the chick auditory brainstem. *Journal of Neuroscience, 29*(30), 9625–9634.

Lachica, E. A., Rubsamen, R., & Rubel, E. W. (1994). GABAergic terminals in nucleus magnocellularis and laminaris originate from the superior olivary nucleus. *Journal of Comparative Neurology, 348*(3), 403–418.

Levin, M. D., Kubke, M. F., Schneider, M., Wenthold, R., & Carr, C. E. (1997). Localization of AMPA-selective glutamate receptors in the auditory brainstem of the barn owl. *Journal of Comparative Neurology, 378*(2), 239–253.

Lippe, W., & Rubel, E. W. (1983). Development of the place principle: Tonotopic organization. *Science, 219*(4584), 514–516.

Lu, T., & Trussell, L. O. (2001). Mixed excitatory and inhibitory GABA-mediated transmission in chick cochlear nucleus. *Journal of Physiology (London), 535*(Pt. 1), 125–131.

MacLeod, K. M., & Carr, C. E. (2007). Beyond timing in the auditory brainstem: Intensity coding in the avian cochlear nucleus angularis. *Progress in Brain Research, 165*, 123–133.

MacLeod, K. M., Soares, D., & Carr, C. E. (2006). Interaural timing difference circuits in the auditory brainstem of the emu (Dromaius novaehollandiae). *Journal of Comparative Neurology, 495*(2), 185–201.

MacLeod, K. M., Horiuchi, T. K., & Carr, C. E. (2007). A role for short-term synaptic facilitation and depression in the processing of intensity information in the auditory brain stem. *Journal of Neurophysiology, 97*(4), 2863–2874.

Manis, P. B., & Marx, S. O. (1991). Outward currents in isolated ventral cochlear nucleus neurons. *Journal of Neuroscience, 11*(9), 2865–2880.

Marsalek, P., Koch, C., & Maunsell, J. (1997). On the relationship between synaptic input and spike output jitter in individual neurons. *Proceedings of the National Academy of Sciences of the United States of America, 94*(2), 735–740.

Matsushita, A., & Kawasaki, M. (2004). Unitary giant synapses embracing a single neuron at the convergent site of time-coding pathways of an electric fish, Gymnarchus niloticus. *Journal of Comparative Neurology, 472*, 140–155.

McAlpine, D., & Grothe, B. (2003). Sound localization and delay lines – do mammals fit the model? *Trends in Neurosciences, 26*(7), 347–350.

McAlpine, D., Jiang, D., & Palmer, A. R. (2001). A neural code for low-frequency sound localization in mammals. *Nature Neuroscience, 4*(4), 396–401.

Mittmann, W., Koch, U., & Hausser, M. (2005). Feed-forward inhibition shapes the spike output of cerebellar Purkinje cells. *Journal of Physiology (London), 563*(Pt. 2), 369–378.

Moiseff, A., & Konishi, M. (1983). Binaural characteristics of units in the owl's brainstem auditory pathway: Precursors of restricted spatial receptive fields *Journal of Neuroscience, 3*, 2553–2562.

Monsivais, P., Yang, L., & Rubel, E. W. (2000). GABAergic inhibition in nucleus magnocellularis: Implications for phase locking in the avian auditory brainstem. *Journal of Neuroscience, 20*(8), 2954–2963.

Mosbacher, J., Schoepfer, R., Monyer, H., Burnashev, N., Seeburg, P. H., & Ruppersberg, J. P. (1994). A molecular determinant for submillisecond desensitization in glutamate receptors. *Science, 266*(5187), 1059–1062.

Nishino, E., Yamada, R., Kuba, H., Hioki, H., Furuta, T., Kaneko, T., & Ohmori, H. (2008). Sound-intensity-dependent compensation for the small interaural time difference cue for sound source localization. *Journal of Neuroscience, 28*(28), 7153–7164.

Oertel, D. (1991). The role of intrinsic neuronal properties in the encoding of auditory information in the cochlear nuclei. *Current Opinion in Neurobiology, 1*(2), 221–228.

Oertel, D. (1999). The role of timing in the brain stem auditory nuclei of vertebrates. *Annual Review of Physiology, 61*, 497–519.

Overholt, E. M., Rubel, E. W., & Hyson, R. L. (1992). A circuit for coding interaural time differences in the chick brainstem. *Journal of Neuroscience, 12*(5), 1698–1708.

Parameshwaran, S., Carr, C. E., & Perney, T. M. (2001). Expression of the Kv3.1 potassium channel in the avian auditory brainstem. *Journal of Neuroscience, 21*(2), 485–494.

Parameshwaran-Iyer, S., Carr, C. E., & Perney, T. M. (2003). Localization of KCNC1 (Kv3.1) potassium channel subunits in the avian auditory nucleus magnocellularis and nucleus laminaris during development. *Journal of Neurobiology, 55*(2), 165–178.

Parks, T. N. (2000). The AMPA receptors of auditory neurons. *Hearing Research, 147*(1–2), 77–91.

Parks, T. N., & Rubel, E. W. (1975). Organization and development of brain stem auditory nucleus of the chicken: Organization of projections from N. magnocellularis to N. laminaris. *Journal of Comparative Neurology, 164*, 435–448.

Payne, R. (1971). Acoustic localization of prey by barn owls (Tyto alba). *Journal of Experimental Biology, 54*, 535–573.

Pecka, M., Brand, A., Behrend, O., & Grothe, B. (2008). Interaural time difference processing in the mammalian medial superior olive: The role of glycinergic inhibition. *Journal of Neuroscience, 28*(27), 6914–6925.

Pena, J. L., Viete, S., Albeck, Y., & Konishi, M. (1996). Tolerance to sound intensity of binaural coincidence detection in the nucleus laminaris of the owl. *Journal of Neuroscience, 16*(21), 7046–7054.

Pena, J. L., Viete, S., Funabiki, K., Saberi, K., & Konishi, M. (2001). Cochlear and neural delays for coincidence detection in owls. *Journal of Neuroscience, 21*(23), 9455–9459.

Raman, I. M., Zhang, S., & Trussell, L. O. (1994). Pathway-specific variants of AMPA receptors and their contribution to neuronal signaling. *Journal of Neuroscience, 14*(8), 4998–5010.

Ramon y Cajal, S. (1908). Les ganlions terminaux du nerf acoustique des oiseaux.*Trabajos del Instituto Cajal de investigaciones biológicas, 6,* 195–225.

Rathouz, M., & Trussell, L. (1998). Characterization of outward currents in neurons of the avian nucleus magnocellularis. *Journal of Neurophysiology, 80*(6), 2824–2835.

Reyes, A. D., Rubel, E. W., & Spain, W. J. (1994). Membrane properties underlying the firing of neurons in the avian cochlear nucleus. *Journal of Neuroscience, 14*(9), 5352–5364.

Reyes, A. D., Rubel, E. W., & Spain, W. J. (1996). In vitro analysis of optimal stimuli for phase-locking and time-delayed modulation of firing in avian nucleus laminaris neurons. *Journal of Neuroscience, 16*(3), 993–1007.

Rubel, E. W., & Parks, T. N. (1975). Organization and development of brain stem auditory nuclei of the chicken: Tonotopic organization of n. magnocellularis and n. laminaris. *Journal of Comparative Neurology, 164*(4), 411–434.

Schneggenburger, R., & Forsythe, I. D. (2006). The calyx of Held. *Cell and Tissue Research, 326*(2), 311–337.

Schwartzkopff, J., & Winter, P. (1960). Zur Anatomie der Vogel-Cochlea unter naturlichen Bedingungen. *Biologisches Zentralblatt, 79,* 607–625.

Scott, L. L., Mathews, P. J., & Golding, N. L. (2005). Posthearing developmental refinement of temporal processing in principal neurons of the medial superior olive. *Journal of Neuroscience, 25*(35), 7887–7895.

Simon, J. Z., Carr, C. E., & Shamma, S. A. (1999). A dendritic model of coincidence detection in the avian brainstem. *Neurocomputing, 26–27,* 263–269.

Smith, Z. D. (1981). Organization and development of brain stem auditory nuclei of the chicken: Dendritic development in N. laminaris. *Journal of Comparative Neurology, 203*(3), 309–333.

Smith, P. H. (1995). Structural and functional differences distinguish principal from nonprincipal cells in the guinea pig MSO slice. *Journal of Neurophysiology, 73*(4), 1653–1667.

Smith, D. J., & Rubel, E. W. (1979). Organization and development of brain stem auditory nuclei of the chicken: Dendritic gradients in nucleus laminaris. *Journal of Comparative Neurology, 186*(2), 213–239.

Smith, P. H., Joris, P. X., & Yin, T. C. (1993). Projections of physiologically characterized spherical bushy cell axons from the cochlear nucleus of the cat: evidence for delay lines to the medial superior olive. *Journal of Comparative Neurology, 331*(2), 245–260.

Smith, A. J., Owens, S., & Forsythe, I. D. (2000). Characterisation of inhibitory and excitatory postsynaptic currents of the rat medial superior olive. *Journal of Physiology, 529*(Pt. 3), 681–698.

Soares, D., Chitwood, R. A., Hyson, R. L., & Carr, C. E. (2002). Intrinsic neuronal properties of the chick nucleus angularis. *Journal of Neurophysiology, 88*(1), 152–162.

Stotler, W. A. (1953). An experimental study of the cells and connections of the superior olivary complex of the cat. *Journal of Comparative Neurology, 98,* 401–432.

Sugden, S. G., Zirpel, L., Dietrich, C. J., & Parks, T. N. (2002). Development of the specialized AMPA receptors of auditory neurons. *Journal of Neurobiology, 52*(3), 189–202.

Sullivan, W. E., & Konishi, M. (1984). Segregation of stimulus phase and intensity coding in the cochlear nucleus of the barn owl. *Journal of Neuroscience, 4*(7), 1787–1799.

Sullivan, W. E., & Konishi, M. (1986). Neural map of interaural phase difference in the owl's brainstem. *Proceedings of the National Academy of Science of the United States of America, 83,* 8400–8404.

Takahashi, T. T., Carr, C. E., Brecha, N., & Konishi, M. (1987). Calcium binding protein-like immunoreactivity labels the terminal field of nucleus laminaris of the barn owl. *Journal of Neuroscience, 7*(6), 1843–1856.

Takahashi, T. T., Bala, A. D., Spitzer, M. W., Euston, D. R., Spezio, M. L., & Keller, C. H. (2003). The synthesis and use of the owl's auditory space map. *Biological Cybernetics, 89*(5), 378–387.

Trussell, L. O. (1999). Synaptic mechanisms for coding timing in auditory neurons. *Annual Review of Physiology, 61*, 477–496.

Viete, S., Pena, J. L., & Konishi, M. (1997). Effects of interaural intensity difference on the processing of interaural time difference in the owl's nucleus laminaris. *Journal of Neuroscience, 17*(5), 1815–1824.

Wang, L. Y., Gan, L., Forsythe, I. D., & Kaczmarek, L. K. (1998). Contribution of the Kv3.1 potassium channel to high-frequency firing in mouse auditory neurones. *Journal of Physiology (London), 509*(Pt. 1), 183–194.

Weiss, S. A., Preuss, T., & Faber, D. S. (2008). A role of electrical inhibition in sensorimotor integration. *Proceedings of the National Academy of Sciences of the United States of America, 105*(46), 18047–18052.

Wild, J. M., Krutzfeldt, N. O., & Kubke, M. F. (2009). Afferents to the cochlear nuclei and nucleus laminaris from the ventral nucleus of the lateral lemniscus in the zebra finch (Taeniopygia guttata). *Hearing Research, 257*(1–2), 1–7.

Woodworth, R. S. (1954). *Experimental Psychology.* New York: Holt, Rinehart and Winston.

Wu, S. H., & Oertel, D. (1984). Intracellular injection with horseradish peroxidase of physiologically characterized stellate and bushy cells in slices of mouse anteroventral cochlear nucleus. *Journal of Neuroscience, 4*(6), 1577–1588.

Yang, L., Monsivais, P., & Rubel, E. W. (1999). The superior olivary nucleus and its influence on nucleus laminaris: A source of inhibitory feedback for coincidence detection in the avian auditory brainstem. *Journal of Neuroscience, 19*(6), 2313–2325.

Yin, T. C., & Chan, J. C. (1990). Interaural time sensitivity in medial superior olive of cat. *Journal of Neurophysiology, 64*(2), 465–488.

Zhang, S., & Trussell, L. O. (1994). Voltage clamp analysis of excitatory synaptic transmission in the avian nucleus magnocellularis. *Journal of Physiology (London), 480*(1), 123–136.

Zhou, Y., Carney, L. H., & Colburn, H. S. (2005). A model for interaural time difference sensitivity in the medial superior olive: Interaction of excitatory and inhibitory synaptic inputs, channel dynamics, and cellular morphology. *Journal of Neuroscience, 25*(12), 3046–3058.

Zucker, R. S., & Regehr, W. G. (2002). Short-term synaptic plasticity. *Annual Review of Physiology, 64*, 355–405.

Chapter 7
Inhibitory Neurons in the Auditory Brainstem

Laurence O. Trussell

1 Introduction

Most chapters in this volume address the function of the major excitatory synapses in the lower auditory pathways, with respect to coincidence detection, the ion channels that determine neuronal firing, and the control of excitation through modulation and plasticity. However, none of these processes can be understood at a functional level without considering synaptic inhibition. Indeed, inhibition through GABAergic and glycinergic interneurons is likely to play an essential role in controlling the excitation at every level of central auditory processing. The present chapter examines inhibitory interneurons in several contexts in order to illustrate the diversity of their cellular mechanisms and circuit-level function, with a focus on the auditory brainstem. The term "interneuron" is used loosely; in fact, inhibitory cells are so fundamental to auditory processing that individual neurons may act as both proper interneurons (intrinsic neurons, i.e., those inhibiting within a local circuit) and inhibitory projection neurons (inhibiting across brainstem nuclei or regions). After introducing the study of interneurons and their general function, the chapter examines two prominent examples from the cochlear nucleus and superior olivary complex, rather than provide an exhaustive summary of all known auditory interneurons. Then four aspects of interneuron physiology are explored: the control of the reversal potential for Cl^-, the gating properties of the receptor-channel complex, the role of corelease of the transmitters GABA and glycine from interneuronal synapses; and, lastly, mechanisms for prolonging the action of the transmitter.

L.O. Trussell (✉)
Vollum Institute, Oregon Hearing Research Center,
Oregon Health and Science University, 3181 Southwest
Sam Jackson Park Road, L335A, Portland, OR 97239, USA
e-mail: trussell@ohsu.edu

L.O. Trussell et al. (eds.), *Synaptic Mechanisms in the Auditory System*,
Springer Handbook of Auditory Research 41, DOI 10.1007/978-1-4419-9517-9_7,
© Springer Science+Business Media, LLC 2012

1.1   The Study of Inhibitory Neurons in the Auditory System

Anatomical, physiological, and combined approaches have been essential to defining many of the classes of inhibitory interneurons in the auditory brainstem. Nissl and Golgi studies identified major cell types in diverse species, grouping cells on the basis of their somatic size and location and the layout of their dendritic arbors (Osen 1969; Brawer et al. 1974; Lorente de No 1981). Intracellular recordings followed by anatomical reconstruction defined more completely the axonal projections of neurons and were able to correlate neuronal morphology with physiological firing patterns and synaptic inputs (Rhode et al. 1983a, b; Zhang and Oertel 1994).

Immunohistochemical labeling for GABA and glycine was an essential first step in confirming the inhibitory function of cell types in the auditory system (Mugnaini 1985; Wenthold et al. 1987; Kolston et al. 1992; Moore et al. 1996). Finally, in vivo studies, in which blockers of GABA and glycine receptors were micro-iontophoresed near sites of electrophysiological recording, were able to determine the role of particular transmitter systems in shaping the response profile to acoustic stimuli (Palombi and Caspary 1992; Backoff et al. 1997, 1999).

As vital as these approaches have been, newer approaches are required in order to clarify the subtypes and roles of inhibitory neurons in auditory pathways. For example, some cell types, such as the superficial stellate cells of the outer layer of the dorsal cochlear nucleus (DCN), are too small for routine extracellular or sharp intracellular recording and are relatively difficult to identify in brain slice studies. As a result, their function remains largely unknown (Wouterlood et al. 1984; Zhang and Oertel 1993). Moreover, it is likely that the diversity of interneuron subtypes is much greater than has been revealed through the approaches described earlier. Immunohistochemical colabeling for multiple antigens has the power to resolve molecular differences among cells of similar morphology. This approach is especially useful when paired with electrophysiology and/or the use of mice with targeted expression of green fluorescent protein (GFP) in subtypes of cells. When these approaches were applied to the cerebellar cortex, it became apparent that Golgi cells, a common interneuron that modified incoming mossy fiber input, may actually be comprised of five different cell types (Simat et al. 2007). Moreover, the use of diverse electrophysiological and molecular tools have pointed to a staggering diversity of GABAergic neurons in the cerebral cortex and hippocampus (McBain and Fisahn 2001; Butt et al. 2005). It is expected that a similar interneuron complexity will become recognized within the auditory system.

Despite these rather intimidating possibilities, there are some general principles regarding how interneurons modify the electrical activity of their neighbors, and these should be appreciated before investigating the body of this review. Readers more familiar with cellular neurophysiology may skip the next section.

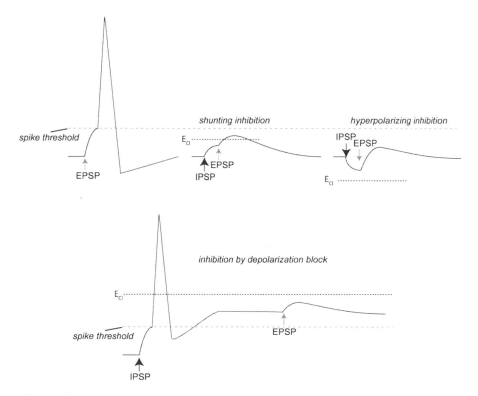

Fig. 7.1 Three cellular mechanisms of inhibition. In shunting inhibition, the excitatory postsynaptic potential (EPSP) is prevented from triggering a spike because the inhibitory Cl⁻ current activated by the transmitter has a reversal potential negative to spike threshold. The IPSP remains inhibitory even though it may be depolarizing because the Cl⁻ conductance shunts the excitatory current. In hyperpolarizing inhibition the IPSP simply brings the resting membrane potential away from spike threshold. Depolarization block may occur when a long-lasting IPSP has a reversal potential above spike threshold, leading to Na⁺ channel inactivation

1.2   Synaptic Inhibition

Synaptic inhibition is generally mediated by one or both of two neurotransmitters, GABA and glycine. While GABAergic transmission is the predominant inhibitory transmitter of the midbrain and forebrain, glycine shares a major role with GABA in inhibition in the spinal cord and brainstem. In the auditory system, glycine is probably the major inhibitory transmitter, and glycinergic neurons are found in the cochlear nucleus, superior olivary complex, and the nuclei of the lateral lemniscus (Wenthold et al. 1987). Glycine acts on an ionotropic glycine receptor while GABA acts on $GABA_A$ and $GABA_B$ receptors, which are ionotropic and metabotropic receptors, respectively. Both ionotropic receptors are Cl⁻ channels and, as such, will depolarize or hyperpolarize postsynaptic neurons according to the transmembrane Cl⁻ gradient. Figure 7.1 illustrates the relation between the Nernst potential for Cl⁻

(defined by the transmembrane Cl⁻ gradient) and the type of inhibition mediated by GABA$_A$ or glycine receptors. Except as in some cases discussed later, the Nernst potential for Cl⁻ is more negative than spike threshold, indicating that the transmitters oppose excitation. However, there are a number of important subtleties in how this inhibition is mediated. Strongly hyperpolarizing responses to GABA or glycine simply bring the membrane potential far from spike threshold; stronger depolarizing excitation would then be required to trigger action potentials. However, following a hyperpolarizing inhibitory postsynaptic potential (IPSP) the cell is potentially in a state of enhanced excitability, in part because Na⁺ channels are partially recovered from inactivation. This effect may account for enhancement of spike timing by IPSPs in principal cells of the dorsal cochlear nucleus (Street and Manis 2007).

Another common form of inhibition in the auditory system occurs where the Nernst potential for Cl⁻ is between threshold and the resting potential (Golding and Oertel 1996; Kim and Trussell 2009). In this case, the conductance increase produced by the activated receptors electrically shunts the excitatory stimuli (Fatt and Katz 1953). If the Nernst potential is somewhat depolarized relative to rest, then a remarkable effect is possible in which a short period of enhanced excitability exists after the GABA/glycine channels have closed but the weak depolarization from the inhibitory transmitter is still ongoing (Gulledge and Stuart 2003). This occurs because the cells are transiently depolarized but there is no longer an inhibitory shunt. Lastly, auditory neurons in the avian brainstem utilize a strongly depolarizing form of inhibition, discussed in Sect. 7.3.1.

Several general circuit-level aspects of inhibition are relevant to an understanding of the computational role of interneurons. Generally, three types of circuit are considered; these are illustrated in Fig. 7.2. Feed-forward inhibition describes the case of afferent fibers terminating both on a principal cell and on the neuron that inhibits the principal cell. Two examples of feed-forward inhibition are described in the next section. In feedback inhibition, the inhibitory neuron is excited by the principal cell itself. An example of this is found in Chap. 6 by MacLeod and Carr, in which GABAergic neurons of the avian superior olive feed back onto neurons of the cochlear nucleus and nucleus laminaris (Burger et al. 2005). Finally, in lateral inhibition, common in auditory cortex (de la Rocha et al. 2008), afferent fibers may excite both a principal cell and interneurons that terminate on neighboring principal cells. Thus, inhibition reduces activity in principal cells surrounding the excited principal cell. In some cases, as in the parallel fiber system of the cochlear nucleus, the nature of inhibition, whether feed-forward or lateral, is ambiguous because the population of afferent fibers carries diverse sensory signals, and each fiber makes only sparse synaptic contact onto nearby interneurons and principal cells (Roberts and Trussell 2010).

Inhibition may be phasic or tonic, that is, it may consist of discrete, brief inhibitory events or a summated, ongoing increase in inhibitory conductance. The latter may arise from either temporal summation of small IPSPs or the action of elevated ambient transmitter levels in brain tissue (Farrant and Nusser 2005). Moreover, forms of inhibition can be distinguished on the basis of their effects on input-output relations of neurons, as shown in Fig. 7.3. Increase in the frequency or strength of

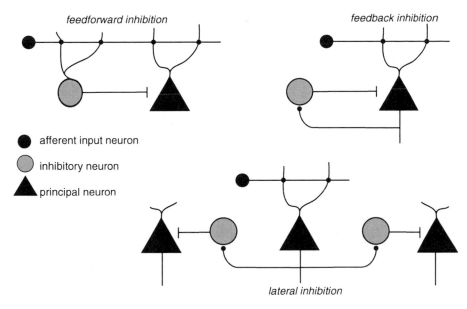

Fig. 7.2 Three circuit-level mechanisms of inhibition. In feed-forward inhibition, common afferent fibers terminate on both a principal cell and the inhibitory cell, ensuring generation of an EPSP-IPSP sequence. Feedback inhibition occurs when the inhibitory cell and principal contact one another. Here, the inhibition occurs only after the principal cell generates spikes. Lateral inhibition occurs when the principal cells excite inhibitory neurons connected to neighboring principal cells, thus spreading inhibition away from the center of the circuit

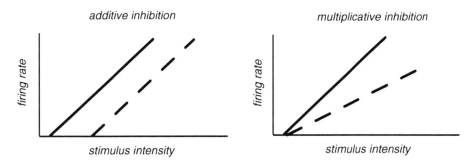

Fig. 7.3 Two forms of inhibitory computation. The normal input-output relation for neurons (*solid*) may be affected by ongoing inhibitory activity (*dashed*) by changing the slope of the relation (multiplicative) or the minimum level for excitation (additive)

excitatory input can be expected to increase the frequency of firing in the postsynaptic neuron. Inhibition may reduce the slope of this relationship (multiplicative inhibition) or shift the relation to the left (additive inhibition) depending on the temporal characteristics of the inputs (Semyanov et al. 2004). In summary, inhibition is not merely a switch for excitatory inputs, but has the capacity to play sophisticated roles in refining the output of a neural circuit.

2   Inhibitory Neurons Sculpt the Responses to Auditory Input

As more information is obtained about the connectivity and cellular properties of
inhibitory neurons of the auditory system, strong hypotheses have been advanced
about their function in hearing, functions with deep implications. For example, studies
of sound localization using cues from interaural time differences have classically been
based on the Jeffress model of excitation of coincidence detector neurons by converging bilateral sources. As described in Chap. 6 by MacLeod and Carr, newer models
have incorporated a prominent role for inhibition and as a result have cast some doubt
on some elements of the Jeffress model (Brand et al. 2002; Grothe 2003). Inhibition
has also been proposed to play a role in monoaural echo suppression (Wickesberg and
Oertel 1990). This proposal was consistent with the observation of feedback inhibition of principal cells of the ventral cochlear nucleus (VCN) by glycinergic tuberculoventral cells of the dorsal cochlear nucleus (DCN) (Wickesberg et al. 1991).

   This section examines in some detail two inhibitory projection neurons, which
serve as examples of the rich potential of inhibitory neurons to sculpt responses of
auditory brainstem neurons to acoustic stimuli. In particular, they illustrate that each
subtype of inhibitory neuron has highly specialized functions dependent on the
unique pattern of synaptic input, axonal projection, and cellular physiology.

2.1   D-Stellate Cells

D-stellate cells, also called radiate multipolar or Type II multipolar neurons, are large
glycine-containing cells of the VCN (Wenthold 1987; Alibardi 1998; Doucet et al.
1999). They have been studied extensively with respect to their branching pattern,
neurochemistry, electrophysiology, and ultrastructure. The distribution of their dendrites and axonal arbor in particular are a key to understanding their function. Dendrites
of D-stellate cells distribute broadly across the posterior ventral cochlear nucleus
(PVCN), receiving auditory nerve input encoding a broad range of sound frequencies
(Smith and Rhode 1989; Oertel et al. 1990). D-stellate cell axons extend through the
dorsal acoustic stria and terminate in the contralateral DCN and VCN (Cant and Gaston
1982; Schofield and Cant 1996; Doucet and Ryugo 2006). Thus, ipsi- and contralateral
cochlear nuclei are connected by this commissural inhibitory system (Fig. 7.4). An
additional commissural connection is made by an excitatory multipolar cell, the
T-stellate neuron (Cant and Gaston 1982; Schofield and Cant 1996; Doucet and Ryugo
2006). D-stellate cells terminate on VCN T-stellate cells both ipsi- and contralaterally
and on neurons of the ipsilateral DCN (Schofield and Cant 1996; Ferragamo et al.
1998; Alibardi 2006). The DCN targets of D-stellate cells are probably fusiform cells
(a major principal cell) and tuberculoventral cells. It may be that not all D-stellate cells
feature both ispi- and contralateral projections (Doucet et al. 2009). Thus, D-stellate
cells serve simultaneously as inhibitory local interneurons and projection neurons.

   The dendritic and axonal projections of D-stellate cells enable these cells to
respond to broad-band sound and thus control the output of both cochlear nuclei.

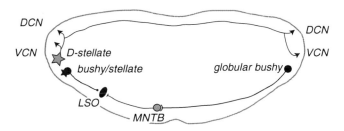

Fig. 7.4 Circuitry discussed in Sect. 7.3. D-stellate cells of the VCN make inputs to the ipsi- and contralateral VCN and DCN. MNTB principal cells receive excitation from contralateral globular bushy cells and forward inhibition to neurons in the LSO. LSO neurons receive excitatory input from bushy and possibly T-stellate cells

Further insight into their function is obtained by considering electrophysiological studies in vivo and in vitro. In vivo recordings indicate that D-stellate cells respond to best-frequency tones with a response profile termed onset chopper (O_C) or onset-late (O_L), indicating that they respond well at the onset of a sound and continue to fire but at a much lower rate through rest of the stimulus (Smith and Rhode 1989; Needham and Paolini 2007). Accordingly, intracellular recordings show that these cells can fire spikes throughout a long stimulus, but the individual spikes are not well timed from trial to trial, except at the onset (Smith and Rhode 1989; Oertel et al. 1990; Rodrigues and Oertel 2006). However, broad-band noise can elicit more spikes during a prolonged stimulus (Smith and Rhode 1989), indicating that these cells would respond well to spectrally complex stimuli.

Thus, D-stellate cells provide ongoing inhibition to multiple circuits across the cochlear nuclear complex. Two main areas of research, looking at inhibition in the VCN and the DCN respectively, indicate what may be the functions of this inhibition. Extracellular recordings from VCN showed that contralateral acoustic stimuli could enhance or suppress firing in cells of the VCN (Shore et al. 2003). Needham and Paolini (2003, 2007) have made intracellular recordings from D- and T-stellate cells in rat VCN, identified by the profile of spiking in response to sound stimuli. Both cell types showed evidence of a commissural projection in the form of an antidromic spike recorded on electrical stimulation of the contralateral cochlear nucleus. Recordings from the T-stellate cells also revealed large brief IPSPs elicited by either electrical stimulation to the contralateral cochlear nucleus or clicks delivered to the contralateral ear. These IPSPs followed well the timing of the stimulus and were interpreted to represent inputs from contralateral D-stellate cells. Responses of T-stellate cells to tone stimuli delivered to the ipsilateral ear were delayed slightly by contralateral stimulation and suggested that one function of the D-stellate cell is to introduce delays in onset of firing of T-stellate cells. Although the meaning of this effect of inhibition has yet to be clarified, inhibition-induced changes in spike latency are suggested to have significant effects on auditory processing in other brain regions (Park et al. 1996; Klug et al. 2000).

The projection of D-stellate cells into the DCN has been proposed to play a key role in shaping the DCN's response to sound. This inhibitory pathway has been reviewed in depth in a previous volume in this series, so only the outlines of the circuit are given here (Young and Davis 2002; Oertel and Young 2004). Fusiform cells (also called pyramidal cells) of the DCN are excitatory principal cells that project to the inferior colliculus. They are excited by auditory nerve fibers and narrowly tuned T-stellate cells. Response maps, the change in firing rate of the cells at different sound frequency and intensity, are remarkably complex in fusiform cells. These so-called Type-IV responses exhibit excitation at the cell's best frequency (BF) in response to weak sound and inhibition to sound just above or below BF. As sound intensity increases, the region of excitation disappears or shifts to higher sound frequency. This profile predicts that fusiform cells would be inhibited by loud, broad-band noise. In fact, the opposite is true (Nelken and Young 1994).

Young and colleagues have explored in detail the circuit elements that could explain this result, concluding that the strange shifts in response areas and excitation to noise can be accounted for by two types of inhibitory neuron, the D-stellate cell and the tuberculoventral cell (also called vertical cell) (Nelken and Young 1994; Davis and Young 2000). Tuberculoventral cells are glycinergic neurons of the deep layer of the DCN that, unlike the D-stellates, are narrowly tuned to sounds just below the BF of the fusiform cell they innervate (Voigt and Young 1988; Rhode 1999). Thus, when these cells are recruited by louder sound they provide inhibition that "encroaches" on the effects of the excitatory inputs to the fusiform cell. As sound intensity increases and a broader range of auditory nerve fibers are excited, D-stellate cells are recruited. These target both fusiform and tuberculoventral cells. According to the model of Young and colleagues, the effect of D-stellate cells in response to broad-band noise is greatest on the tuberculoventral cell, thus leading to a disinhibition of the fusiform cell (Davis and Young 2000; Young and Davis 2002). Thus, the actions of D-stellate cells, along with tuberculoventral cells and other inhibitory sources, create the possibility of complex nonlinear responses to sensory stimuli that cannot be replicated with a purely excitatory circuitry.

These results highlight that the effects of inhibitory circuitry must account for the pattern of innervation of the interneuron (that is, whether it is excited by a narrow or broad range of sound frequencies), the array of target cells contacted by the interneuron, and the relative strengths of those contacts. The circuitry just described is but a sample of the diversity of interneurons that determine the output of the DCN, which also includes Golgi, cartwheel, superficial stellate, and Purkinje-like cells. Moreover, inhibitory neurons connect the different regions of the cochlear nucleus, the ipsi- and contralateral cochlear nuclei, and even project to higher brain regions as well. Thus, inhibitory networks are a major feature of the earliest levels of central auditory processing.

2.2   The MNTB-LSO Circuit

A second example illustrates the extraordinary speed of inhibitory function in the auditory system. Glycinergic neurons of the medial nucleus of the trapezoid body

(MNTB) make synaptic contact on targets within the superior olivary complex and the lateral lemniscal nuclei. A major target of the MNTB is the lateral superior olive (LSO), and this projection is essential for sound localization based on interaural intensity cues. As well reviewed elsewhere (Yin 2002; Tollin 2003), principal cells of the LSO receive excitatory input from small spherical bushy cells in the ipsilateral cochlear nucleus. However, another possible source of excitatory input to LSO is multipolar (T-stellate) cells of the posteroventral cochlear nucleus (Cant and Casseday 1986; Thompson 1998; Doucet and Ryugo 2003); it remains unclear what the significance is of these multiple sources of excitation in terms of subtypes of target cell or functional outcomes in LSO. Contralateral globular bushy cells provide excitatory input to the MNTB. Figure 7.4 summarizes this circuitry. Thus, when sound is louder at the contralateral ear, the MNTB acts as a sign-inverting relay to oppose excitation in the ispilateral LSO neurons. A variety of specializations at anatomical and biophysical levels ensure that this circuit operates efficiently over the range of sound frequencies in which intensity differences are relevant (see Chap. 5 by Borst and Rusu). Among these are the best frequency of the population of bushy and MNTB cells that map into the LSO, the responsiveness of bushy and MNTB cells to changing acoustic stimuli, and the extraordinarily rapid conduction of signals from the contralateral VCN, which must overcome delays inherent in both the added axonal length and the synaptic delays in the MNTB (Tollin 2003).

An added feature required for this circuit to perform its function is that inhibitory signals must be rapid, that is, their onset and offset must be brief. Were this not the case, inhibitory signals in LSO would not converge effectively with ipsilateral excitation. Moreover, the circuit would have limited ability to keep up with the temporal variation inherent in natural sound sources. The responsiveness of inhibition can be explored in two ways. In the first, a time window is determined for how long contralateral signals are effective in suppressing excitation in the LSO. This has been determined in vivo (Wu and Kelly 1992a, b; Joris and Yin 1995) and in vitro (Sanes 1990; Irvine et al. 2001), with estimates between 1 and 2 ms. This range would then indicate the maximal length for inhibitory signals in the LSO.

A second way to look at the temporal aspects of inhibition is to use intracellular recording to examine the time course of the IPSP and the underlying synaptic conductance change, during either spontaneous activity or activity evoked by stimulating presynaptic fibers (Fig. 7.5a). The conductance change describes the time course of action of the neurotransmitter at its receptors; it is recorded using voltage clamp and measured as an inhibitory postsynaptic current (IPSC). As discussed in Sect. 7.1.2, inhibition may be mediated by the hyperpolarizing or the shunting aspects of IPSPs. IPSPs in the LSO are strongly hyperpolarizing, and this would be expected therefore to contribute to inhibition. However, IPSPs in the LSO may have decay times longer than 1–2 ms (Sanes 1990; Wu and Kelly 1992a). Because changes in synaptic conductance are typically briefer than synaptic potentials (Curtis and Eccles 1959; Araki and Terzuolo 1962), the shunting aspect of inhibition may play a dominant role in mediating the brief period of inhibition in LSO. What properties of IPSCs would be expected in LSO? Clearly they should be rapid (<2 ms in duration), for the reasons just outlined. However, the IPSCs must also be

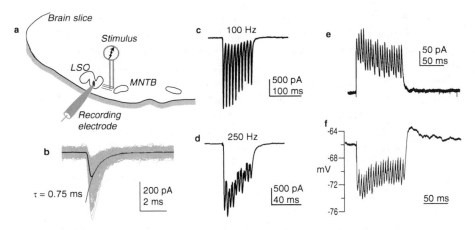

Fig. 7.5 Inhibitory inputs from the MNTB to the LSO. (**a**) cartoon of typical brain slice preparation shows the positions of recording and stimulus electrodes. (**b**) aligned spontaneous IPSCs in an LSO neuron at 34°C. *Gray* are 50 individual traces and *black* is their average. The decay of the average current was fitted with a single exponential curve with a 0.75 ms time constant. (**c** and **d**) show IPSCs recorded at two different stimulus frequencies using a recording electrode filled with an elevated Cl⁻, so that currents are inward at a holding potential of −70 mV. Note that currents return to the baseline immediately after termination of stimuli, indicated by rapid channel kinetics. (**e** and **f**) show response to 200-Hz stimuli from a different cell recorded in both voltage clamp and current clamp, respectively. The recording pipette contained a low Cl⁻ solution to ensure a hyperpolarizing IPSP. In current clamp there was a small transient depolarization, probably resulting from activation of the H-current during the IPSP train (Unpublished records from G. D. Price and L. O. Trussell)

large enough to shunt the effects of ipsilateral excitatory inputs, and they should attenuate little during an ongoing series of stimuli, that is, they should show little synaptic depression. There are relatively few published reports detailing these properties. While studies of glycinergic inhibition in LSO and neighboring nuclei show that the decay time of IPSCs accelerates with age (Awatramani et al. 2005; Magnusson et al. 2005), thus highlighting the importance of studying mature animals, gathering data from rodents older than 2 weeks after birth is technically difficult. Among published studies, glycinergic decay times for miniature inhibitory synaptic currents (mIPSCs) of LSO are about 3 ms at postnatal days 16–17 (P16–17) (Nabekura et al. 2004). In the medial superior olive (MSO), decay times of less than 2 ms have been recorded in P21 gerbils (Magnusson et al. 2005). Figure 7.5b illustrates patch-clamp recordings from P21 rats, showing spontaneous IPSCs with an average decay time of 0.75 ms. Thus, although studies from adult tissue are lacking, it is likely that in more mature animals the duration of inhibitory currents in LSO are sufficient to account for a narrow time window for inhibition.

Inhibition is not only fast, but quite powerful in the LSO. Developmental studies have shown that young LSO neurons receive large numbers of relatively weak inputs from MNTB; with maturation, the number of inputs drops dramatically, with a corresponding increase in the strength of the remaining fibers (Kim and Kandler

2003). In mature preparations, single action potentials in MNTB fibers can trigger an IPSP sufficient to cancel ipsilateral excitation (Sanes 1990; Wu and Kelly 1992a, 1994). Figure 7.5c and d show trains of IPSCs in one cell evoked by maximal stimulation of the MNTB. Although some synaptic depression is evident in the traces, all currents are large and decay rapidly after the termination of the stimulus. In Fig. 7.5e and f, trains of IPSCs and IPSPs are recorded from another neuron and illustrate that even the hyperpolarization evoked by MNTB stimulation shows rapid activation and deactivation. The developmental matching of excitatory and inhibitory inputs in LSO, which is critical for this sound localization circuit, has been the subject of intense research (Kotak and Sanes 2003), as addressed in Chap. 9 by Tzounopoulos and Leão.

It should be noted that the LSO is not the only target of MNTB fibers, and, like D-stellate cells, the function of the other targets, and therefore the "meaning" of the inhibition must vary. For example, during tones neurons of the superior periolivary nucleus (SPON) are tonically inhibited by inputs from the MNTB (Kulesza et al. 2003, 2007). Following cessation of the tone, firing of SPON neurons is transiently increased, a mechanism that may aid in detection of sound duration or gaps (Kadner et al. 2006; Kadner and Berrebi 2008). Terminals of MNTB axons in the MSO must also have functions distinct from those in the LSO, given that the two target nuclei encode interaural timing versus intensity cues and process sounds with different ranges of best frequencies (Tollin 2003; Joris and Yin 2007).

3 The IPSC in Auditory Brainstem

3.1 The Control of Intracellular Chloride Concentration

As discussed earlier, $GABA_A$ and glycine receptors are Cl^- channels, and therefore their effect on membrane potential depends on the concentration gradient of Cl^-. In young auditory neurons, within 1 week after birth, the concentration of Cl^- is sufficiently high that the transmitters GABA and glycine may cause a depolarization of the neurons, a phenomenon common throughout the central nervous system (Kakazu et al. 1999; Kullmann et al. 2002). In other brain regions, the age-dependent reduction in intracellular Cl^-, which is critical to the establishment of hyperpolarizing inhibition, is thought to be due to up-regulation of expression of the ion cotransporter KCC2 (K-Cl cotransporter 2), in opposition to the Cl-extruding pathway mediated by NKCC1, the Na-K-Cl cotransporter (Rivera et al. 1999). However, in LSO, expression of these two transporters is not consistent with such a mechanism, raising the question of what accounts for the switch in Cl^- concentration (Balakrishnan et al. 2003). Posttranslational modifications of KCC2 function might account for this difference between auditory and nonauditory regions. First, there is an age-dependent oligomerization of KCC2 protein that correlates with the onset of Cl^- extrusion in LSO (Blaesse et al. 2006). Second, an interacting protein termed CIP2 (for CCC-interacting protein 2) associates with KCC2 and enhances its function while reducing the function of the Cl⁻ importer NKCC (Na-K-Cl cotransporter) (Wenz et al. 2009). Since some forms of CIP2 are developmentally up-regulated, it may play a role in increased Cl⁻ extrusion.

Some auditory neurons do not lose their elevated Cl⁻ levels or are able to regulate Cl⁻ to shift quickly between depolarizing and hyperpolarizing. In the avian auditory brainstem, principal cells of the cochlear nuclei and the nucleus laminaris (the avian MSO; see Chap. 6 by MacLeod and Carr) maintain high [Cl⁻] into maturity, with Cl⁻ Nernst potentials of about −30 mV. Thus, GABA and glycine are strongly depolarizing and can even trigger action potentials (Fig. 7.1). In this setting, inhibition is mediated indirectly, as the depolarization caused sustained activation of K⁺ leak currents to shunt phasic excitatory signals and to inactivate Na⁺ channels (Lu and Trussell 2001; Monsivais and Rubel 2001; Burger et al. 2005). Such inhibition by depolarization block is characteristic of avian but not mammalian auditory brainstem. However, cartwheel cells, glycinergic interneurons in the mammalian DCN, exhibit a striking control of Cl⁻, which permits changes from depolarizing to hyperpolarizing inhibition. Cartwheel cells exert powerful hyperpolarizing inhibition of fusiform cells. Moreover, these interneurons also make synaptic contact onto one another, and this input can be depolarizing or hyperpolarizing (Golding and Oertel 1996). Hyperpolarizing responses correlated with cells that spontaneously generated Ca^{2+}- and Na⁺-dependent spikes called complex spikes. Kim and Trussell (2009) found that Ca^{2+} flux during the complex spike triggered a drop in intracellular pH, which led to a stimulation of the Na⁺-dependent Cl⁻- bicarbonate exchanger, thus extruding Cl⁻ and lowering Cl⁻ concentration. One functional interpretation of this phenomenon is that when the interneurons are overactive, glycine becomes more inhibitory, and the firing activity is tempered. Because this ion exchanger is widely expressed in the nervous system (Boron et al. 2009), it is possible that the Ca^{2+}- and pH-dependent mechanism is a common regulatory to control the magnitude of inhibition.

3.2 Biophysical Properties of Receptors

At fast-acting synapses in the brain, the neurotransmitter receptor site is part of a single receptor-channel complex, a structure that ensures rapid onset of transmission. Typically, such receptors are heteromultimeric complexes of four to five subunits in which binding of two or more transmitter molecules leads directly to a conformational change in the complex that results in the opening of the channel pore (Wollmuth and Sobolevsky 2004; Miller and Smart 2010). The length of time or number of times the pore opens (termed "gating kinetics") is a function of the unique biophysical properties of the protein and the transmitter. A major principle that emerged from biophysical studies of excitatory synapses in the auditory system is that the gating kinetics of the neurotransmitter receptors is a major determinant of the duration of excitatory synaptic transmission (Trussell 1999).

With some interesting exceptions discussed later (Sect. 7.3.4), it is likely that this principle applies also to inhibitory synapses in the auditory system. Most postsynaptic glycine receptors at auditory synapses are composed of the $\alpha 1$ and β subunits (Piechotta et al. 2001). Biophysical studies have been performed in which cDNA for these subunits were expressed in cell lines and the properties of the resulting channels were analyzed. When receptors are exposed very briefly to glycine, the ionic (chloride) current they generate decays with a time course that is very similar to the decay time of synaptic currents at many glycinergic synapses in the brain, about 4 ms at room temperature (Beato and Sivilotti 2007). Thus, one could expect that factors, either developmental or modulatory, that act on the properties of these channels could have profound effects on inhibitory transmission. This should be an area of active research in the future. In the following sections, some important departures from this principle are discussed, in which responses at auditory synapses are either faster or slower than predicted from the kinetic analysis of glycine channels.

3.3 Co-transmission

In addition to a developmental shift in the Cl^- gradient, there is a well-known shift from GABA-dominant inhibition to glycine dominant. This shift has been proposed to be mediated by a shift in the type of neurotransmitter that is produced by a given inhibitory neuron (as opposed to a shift in the expression of postsynaptic receptors), and indeed the MNTB-LSO connection is considered a classical example of this switch. However, in many auditory neurons, GABA and glycine remain co-expressed within single cells, even in adults (Wenthold et al. 1987; Ostapoff et al. 1997; Rubio and Juiz 2004). This co-expression is characteristic of specific cell types and is species-dependent, suggesting that it may subserve specific computational needs. Are the two transmitters not only co-expressed but also co-released from each synapse? The answer is almost certainly yes, as a single type of transporter, VGAT, moves both transmitters into synaptic vesicles and genetic knockout of VGAT nearly abolishes both kinds of inhibition (Wojcik et al. 2006). Moreover, synaptic currents mediated by release of both transmitters from single neurons have been recorded in LSO, MNTB, and DCN (Nabekura et al. 2004; Awatramani et al. 2005; Roberts et al. 2008). In these cases, GABA acts at $GABA_A$ receptors and glycine at glycine receptors, each with a distinct time course of action. However, it has also been shown that both transmitters may act at glycine receptors, thereby producing a response distinct from what glycine alone would produce. Specifically, it was found that synaptic responses mediated by co-release of GABA and glycine in the MNTB were faster than what would be expected for glycine alone, with synaptic decay times of 1–2 ms (Lu et al. 2008). Thus, cotransmssion may be a mechanism to ensure extra fast inhibition.

In the young LSO, synapses from the MNTB co-release not only GABA and glycine, but also the transmitter glutamate (Gillespie et al. 2005). This amazing result appears to be due to co-expression of both VGAT and VGLUT3 (vesicular glutamate transporter 3). Glutamate co-released acts on NMDA glutamate receptors and appears to have a significant developmental consequence for generation of tonotopic maps in the LSO. Mice in which VGLUT3 is genetically knocked out have a deficit in the reduction of the number of MNTB to LSO inputs and the corresponding strengthening of the synapses (Noh et al. 2010). Thus, at least in this case, cotransmission may serve not only to refine computations but to refine the circuit itself.

3.4   Mechanisms for Prolonging Synaptic Inhibition

Although the auditory brainstem features extremely rapid forms of synaptic inhibition, there are also multiple mechanisms that serve to lengthen the time course of inhibition. These act at the level of transmitter and receptor, the clearance of transmitter, and the time course of release of the transmitter.

GABAergic responses are often much slower than glycinergic responses in the auditory system, and this difference may be used in a pathway-specific manner. For example, whereas glycinergic IPSCs may have decay times between 1 and 4 ms, GABAergic IPSCs can decay far more slowly than glycinergic IPSCs (Jonas et al. 1998; Awatramani et al. 2005). In the MNTB, which receives both GABAergic and glycinergic inputs, GABAergic IPSCs decayed exponentially with a dominant fast phase of decay of about 10 ms (Awatramani et al. 2005). Similarly, in the chick auditory system, GABAergic IPSCs have average decay times of 8–30 ms (Kuo et al. 2009). In the dorsal nucleus of the lateral lemniscus (DNLL) of gerbils, these comparatively slow GABAergic decays may have a very specific physiological function. In vivo recordings show that neurons excited by contralateral sound can be inhibited by a simultaneous ipsilateral sound stimulus as a result of the excitatory and inhibitory inputs to the LSO. However, inhibition has a delayed offset lasting up to 20 ms (Pecka et al. 2007). This "persistent inhibition" is attributed to GABAergic crosstalk between ipsi- and contralateral DNLL via the commissure of Probst. Voltage-clamp recordings of DNLL neurons revealed that stimulation of the commissure of Probst resulted in a GABAergic IPSC with biexponential decay times of 13 and 40 ms. The IPSPs recorded in current clamp were sufficient to halt firing in DNLL for a period consistent with the "persistent inhibition" recorded in vivo. This example shows that the kinetics of inhibition at the transmitter receptor level may account for temporal aspects of auditory processing at the systems level.

The time course of inhibition is also determined by the period over which synaptic vesicles fuse after arrival at the nerve terminal of a presynaptic action potential. In many auditory connections that use glycine, an action potential arriving at the various boutons associated with a single axon triggers the release of multiple synaptic vesicles within a very narrow time window. This narrow window for exocytosis is required for postsynaptic responses to be both quick and well phase-locked to presynaptic signals. One measure of this window is the rise time of the IPSC and its jitter from trial to trial. In the DCN, cartwheel cell IPSCs arise 600 μs after triggering a presynaptic action potential, have rise times of about 0.5 ms and trial-to-trial jitter of nearly 40 μs (Roberts et al. 2008). This precision requires that the probability of release of vesicles rises very quickly and uniformly to a peak level after arrival of the action potential. However, at GABAergic synapses in the avian cochlear nucleus, exocytosis following trains of stimuli is initially very rapid but after a period of time switches into an "asynchronous" mode, in which exocytosis of GABA is poorly phase-locked to the timing of presynaptic stimuli (Lu and Trussell 2000). Asynchronous release, a process also seen at subtypes of cortical and hippocampal interneurons of mammals, results in tonic inhibition. Because the Nernst potential for Cl⁻ is quite depolarized in these neurons, the result of asynchronous release is a plateau depolarization that effectively blocks excitation, as discussed earlier (Yang et al. 1999; Monsivais et al. 2000; Lu and Trussell 2001).

Lastly, inhibition may be prolonged when transmitter clearance from the area of the synapse is delayed. This appears to be the case for glycinergic inhibition of granule cells of the DCN (Balakrishnan et al. 2009). Weak stimuli, just sufficient to activate a few presynaptic inputs, result in a glycinergic IPSC of about 4 ms, similar to that expected from the biophysical properties of glycine receptors. However, when strong stimuli or high-frequency trains are delivered, the decay is prolonged to 20–60 ms. This prolongation was attributed to pooling of the transmitter glycine surrounding the granule cell that occurred when a large number of synapses released their transmitter content in a narrow time window. It was suggested that this activity-dependent pooling could scale the duration of inhibition according to the intensity of excitation.

4 Summary

Inhibition is a major feature at all levels of central auditory processing. Perhaps because of the great complexity of these circuits, with multiple ascending and descending pathways, multiple mechanisms of inhibition are employed. Based on the discussion in this chapter, these mechanisms can be divided into three broad groups. Each area exemplifies fundamental principles of synaptic organization and function but also highlights important topics for further research. The first is anatomical. A given subtype of inhibitory neuron may act as both proper local interneurons and a projection neuron, targeting multiple regions. With respect to the kind of information being encoded, the target neurons may be of different functions. How a common inhibitory neuron can affect firing of target cells with different functions, and presumably different electrical properties, is a topic for future research. The second group of mechanisms is defined by the transmitter. While GABA or glycine is used in different cell types, many inhibitory neurons may use both transmitters simultaneously. The specification, function, and regulation of these transmitter phenotypes are not yet understood. Finally, inhibitory systems may use different modes of transmitter release and removal, as well as receptor activation, to sculpt postsynaptic responses. Thus, defining subtypes of inhibitory neurons and their modes of action constitutes a fascinating and rich area of research, essential to understanding function and dysfunction in auditory processing.

Acknowledgments I wish to thank Mr. Dan Yaeger and Dr. Donata Oertel for comments on the manuscript. Dr. Gareth Price provided data for Fig. 7.5. Support was provided by the NIH (grants NS028901 and DC004450).

References

Alibardi, L. (1998). Ultrastructural and immunocytochemical characterization of commissural neurons in the ventral cochlear nucleus of the rat. *Annals of Anatomy, 180*(5), 427–438.

Alibardi, L. (2006). Review: Cytological characteristics of commissural and tuberculo-ventral neurons in the rat dorsal cochlear nucleus. *Hearing Research, 216–217*, 73–80. doi: S0378-5955(06)00010-4[pii]10.1016/j.heares.2006.01.005.

Araki, T., & Terzuolo, C. A. (1962). Membrane currents in spinal motoneurons associated with the action potential and synaptic activity. *Journal of Neurophysiology, 25*, 772–789.

Awatramani, G. B., Turecek, R., & Trussell, L. O. (2005). Staggered development of GABAergic and glycinergic transmission in the MNTB. *Journal of Neurophysiology, 93*(2), 819–828. doi: 10.1152/jn.00798.200400798.2004[pii.]

Backoff, P. M., Palombi, P. S., & Caspary, D. M. (1997). Glycinergic and GABAergic inputs affect short-term suppression in the cochlear nucleus. *Hearing Research, 110*(1–2), 155–163.

Backoff, P. M., Shadduck Palombi, P., & Caspary, D. M. (1999). Gamma-aminobutyric acidergic and glycinergic inputs shape coding of amplitude modulation in the chinchilla cochlear nucleus. *Hearing Research, 134*(1–2), 77–88. doi: S0378-5955(99)00071-4 [pii].

Balakrishnan, V., Becker, M., Lohrke, S., Nothwang, H. G., Guresir, E., & Friauf, E. (2003). Expression and function of chloride transporters during development of inhibitory neurotransmission in the auditory brainstem. *Journal of Neuroscience, 23*(10), 4134–4145. doi: 23/10/4134[pii].

Balakrishnan, V., Kuo, S. P., Roberts, P. D., & Trussell, L. O. (2009). Slow glycinergic transmission mediated by transmitter pooling. *Nature Neuroscience, 12*(3), 286–294. doi: nn.2265[pii]10.1038/nn.2265.

Beato, M., & Sivilotti, L. G. (2007). Single-channel properties of glycine receptors of juvenile rat spinal motoneurones in vitro. *Journal of Physiology, 580*(Pt. 2), 497–506. doi: jphysiol.2006.1 25740[pii]10.1113/jphysiol.2006.125740.

Blaesse, P., Guillemin, I., Schindler, J., Schweizer, M., Delpire, E., Khiroug, L., Friauf, E., & Nothwang, H. G. (2006). Oligomerization of KCC2 correlates with development of inhibitory neurotransmission.*Journal of Neuroscience, 26*(41),10407–10419.doi:26/41/10407[pii]10.1523/ JNEUROSCI.3257-06.2006.

Boron, W. F., Chen, L., & Parker, M. D. (2009). Modular structure of sodium-coupled bicarbonate transporters.*Journal of Experimental Biology, 212*(Pt. 11),1697–1706.doi:212/11/1697[pii]10.1242/ jeb.028563.

Brand, A., Behrend, O., Marquardt, T., McAlpine, D., & Grothe, B. (2002). Precise inhibition is essential for microsecond interaural time difference coding. *Nature, 417*(6888), 543–547. doi: 10.1038/417543a417543a[pii].

Brawer, J. R., Morest, D. K., & Kane, E. C. (1974). The neuronal architecture of the cochlear nucleus of the cat. *Journal of Comparative Neurology, 155*(3), 251–300. doi: 10.1002/ cne.901550302.

Burger, R. M., Cramer, K. S., Pfeiffer, J. D., & Rubel, E. W. (2005). Avian superior olivary nucleus provides divergent inhibitory input to parallel auditory pathways. *Journal of Comparative Neurology, 481*(1), 6–18. doi: 10.1002/cne.20334.

Butt, S. J., Fuccillo, M., Nery, S., Noctor, S., Kriegstein, A., Corbin, J. G., & Fishell, G. (2005). The temporal and spatial origins of cortical interneurons predict their physiological subtype. *Neuron, 48*(4), 591–604. doi: S0896-6273(05)00934-7[pii]10.1016/j.neuron.2005.09.034.

Cant, N. B., & Casseday, J. H. (1986). Projections from the anteroventral cochlear nucleus to the lateral and medial superior olivary nuclei. *Journal of Comparative Neurology, 247*(4), 457–476. doi: 10.1002/cne.902470406.

Cant, N. B., & Gaston, K. C. (1982). Pathways connecting the right and left cochlear nuclei. *Journal of Comparative Neurology, 212*(3), 313–326. doi: 10.1002/cne.902120308.

Curtis, D. R., & Eccles, J. C. (1959). The time courses of excitatory and inhibitory synaptic actions. *Journal of Physiology, 145*(3), 529–546.

Davis, K. A., & Young, E. D. (2000). Pharmacological evidence of inhibitory and disinhibitory neuronal circuits in dorsal cochlear nucleus. *Journal of Neurophysiology, 83*(2), 926–940.

de la Rocha, J., Marchetti, C., Schiff, M., & Reyes, A. D. (2008). Linking the response properties of cells in auditory cortex with network architecture: Cotuning versus lateral inhibition. *Journal of Neuroscience, 28*(37), 9151–9163. doi: 28/37/9151[pii]10.1523/JNEUROSCI.1789-08.2008.

Doucet, J. R., & Ryugo, D. K. (2003). Axonal pathways to the lateral superior olive labeled with biotinylated dextran amine injections in the dorsal cochlear nucleus of rats. *Journal of Comparative Neurology, 461*(4), 452–465. doi: 10.1002/cne.10722.

Doucet, J. R., & Ryugo, D. K. (2006). Structural and functional classes of multipolar cells in the ventral cochlear nucleus. *Anatomical Record Part A: Discoveries in Molecular, Cellular, and Evolutionary Biology, 288*(4), 331–344. doi: 10.1002/ar.a.20294.

Doucet, J. R., Ross, A. T., Gillespie, M. B., & Ryugo, D. K. (1999). Glycine immunoreactivity of multipolar neurons in the ventral cochlear nucleus which project to the dorsal cochlear nucleus. *Journal of Comparative Neurology, 408*(4), 515–531. doi: 10.1002/(SICI)1096-9861 (19990614)408:4<515::AID-CNE6>3.0.CO;2-O[pii].

Doucet, J. R., Lenihan, N. M., & May, B. J. (2009). Commissural neurons in the rat ventral cochlear nucleus. *Journal of the Association for Research in Otolaryngology, 10*(2), 269–280. doi: 10.1007/s10162-008-0155-6.

Farrant, M., & Nusser, Z. (2005). Variations on an inhibitory theme: Phasic and tonic activation of GABA(A) receptors. *Nature Reviews Neuroscience, 6*(3), 215–229. doi: nrn1625[pii]10.1038/ nrn1625.

Fatt, P., & Katz, B. (1953). The effect of inhibitory nerve impulses on a crustacean muscle fibre. *Journal of Physiology, 121*(2), 374–389.

Ferragamo, M. J., Golding, N. L., & Oertel, D. (1998). Synaptic inputs to stellate cells in the ventral cochlear nucleus. *Journal of Neurophysiology, 79*(1), 51–63.

Gillespie, D. C., Kim, G., & Kandler, K. (2005). Inhibitory synapses in the developing auditory system are glutamatergic. *Nature Neuroscience, 8*(3), 332–338. doi: nn1397[pii]10.1038/nn1397.

Golding, N. L., & Oertel, D. (1996). Context-dependent synaptic action of glycinergic and GABAergic inputs in the dorsal cochlear nucleus. *Journal of Neuroscience, 16*(7), 2208–2219.

Grothe, B. (2003). New roles for synaptic inhibition in sound localization. *Nature Reviews Neuroscience, 4*(7), 540–550. doi: 10.1038/nrn1136nrn1136[pii].

Gulledge, A. T., & Stuart, G. J. (2003). Excitatory actions of GABA in the cortex. *Neuron, 37*(2), 299–309. doi: S0896627302011467[pii].

Irvine, D. R., Park, V. N., & McCormick, L. (2001). Mechanisms underlying the sensitivity of neurons in the lateral superior olive to interaural intensity differences. *Journal of Neurophysiology, 86*(6), 2647–2666.

Jonas, P., Bischofberger, J., & Sandkuhler, J. (1998). Corelease of two fast neurotransmitters at a central synapse. *Science, 281*(5375), 419–424.

Joris, P. X., & Yin, T. C. (1995). Envelope coding in the lateral superior olive. I. Sensitivity to interaural time differences. *Journal of Neurophysiology, 73*(3), 1043–1062.

Joris, P., & Yin, T. C. (2007). A matter of time: Internal delays in binaural processing. *Trends in Neurosciences, 30*(2), 70–78. doi: S0166-2236(06)00275-X[pii]10.1016/j.tins.2006.12.004.

Kadner, A., & Berrebi, A. S. (2008). Encoding of temporal features of auditory stimuli in the medial nucleus of the trapezoid body and superior paraolivary nucleus of the rat. *Neuroscience, 151*(3), 868–887. doi: S0306-4522(07)01408-X[pii]10.1016/j.neuroscience.2007.11.008.

Kadner, A., Kulesza, R. J. Jr., & Berrebi, A. S. (2006). Neurons in the medial nucleus of the trapezoid body and superior paraolivary nucleus of the rat may play a role in sound duration coding. *Journal of Neurophysiology, 95*(3), 1499–1508. doi: 00902.2005[pii]10.1152/jn.00902.2005.

Kakazu, Y., Akaike, N., Komiyama, S., & Nabekura, J. (1999). Regulation of intracellular chloride by cotransporters in developing lateral superior olive neurons. *Journal of Neuroscience, 19*(8), 2843–2851.

Kim, G., & Kandler, K. (2003). Elimination and strengthening of glycinergic/GABAergic connections during tonotopic map formation. *Nature Neuroscience, 6*(3), 282–290. doi: 10.1038/nn1015nn1015[pii].

Kim, Y., & Trussell, L. O. (2009). Negative shift in the glycine reversal potential mediated by a $Ca2+-$ and pH-dependent mechanism in interneurons. *Journal of Neuroscience, 29*(37), 11495–11510. doi: 29/37/11495[pii]10.1523/JNEUROSCI.1086-09.2009.

Klug, A., Khan, A., Burger, R. M., Bauer, E. E., Hurley, L. M., Yang, L., Grothe, B., Halvorsen, M. B., & Park, T. J. (2000). Latency as a function of intensity in auditory neurons: Influences of central processing. *Hearing Research, 148*(1–2), 107–123. doi: S0378-5955(00)00146-5[pii].

Kolston, J., Osen, K. K., Hackney, C. M., Ottersen, O. P., & Storm-Mathisen, J. (1992). An atlas of glycine- and GABA-like immunoreactivity and colocalization in the cochlear nuclear complex of the guinea pig. *Anatomy and Embryology, 186*(5), 443–465.

Kotak, V. C., & Sanes, D. H. (2003). Gain adjustment of inhibitory synapses in the auditory system. *Biological Cybernetics, 89*(5), 363–370. doi: 10.1007/s00422-003-0441-7.

Kulesza, R. J. Jr., Spirou, G. A., & Berrebi, A. S. (2003). Physiological response properties of neurons in the superior paraolivary nucleus of the rat. *Journal of Neurophysiology, 89*(4), 2299–2312. doi: 10.1152/jn.00547.200200547.2002[pii].

Kulesza, R. J. Jr., Kadner, A., & Berrebi, A. S. (2007). Distinct roles for glycine and GABA in shaping the response properties of neurons in the superior paraolivary nucleus of the rat. *Journal of Neurophysiology, 97*(2), 1610–1620. doi: 00613.2006[pii]10.1152/jn.00613.2006.

Kullmann, P. H., Ene, F. A., & Kandler, K. (2002). Glycinergic and GABAergic calcium responses in the developing lateral superior olive. *European Journal of Neuroscience, 15*(7), 1093–1104. doi: 1946[pii].

Kuo, S. P., Bradley, L. A., & Trussell, L. O. (2009). Heterogeneous kinetics and pharmacology of synaptic inhibition in the chick auditory brainstem. *Journal of Neuroscience, 29*(30), 9625–9634. doi: 29/30/9625[pii]10.1523/JNEUROSCI.0103-09.2009.

Lorente de No, V. (1981). *The Primary Acoustic Nuclei.* New York: Raven Press.

Lu, T., & Trussell, L. O. (2000). Inhibitory transmission mediated by asynchronous transmitter release. *Neuron, 26*(3), 683–694. doi: S0896-6273(00)81204-0[pii].

Lu, T., & Trussell, L. O. (2001). Mixed excitatory and inhibitory GABA-mediated transmission in chick cochlear nucleus. *Journal of Physiology, 535*(Pt. 1), 125–131. doi: PHY_12754 [pii].

Lu, T., Rubio, M. E., & Trussell, L. O. (2008). Glycinergic transmission shaped by the corelease of GABA in a mammalian auditory synapse. *Neuron, 57*(4), 524–535. doi: S0896-6273(07)01010-0[pii]10.1016/j.neuron.2007.12.010.

Magnusson, A. K., Kapfer, C., Grothe, B., & Koch, U. (2005). Maturation of glycinergic inhibition in the gerbil medial superior olive after hearing onset. *Journal of Physiology, 568*(Pt. 2), 497–512. doi: jphysiol.2005.094763[pii]10.1113/jphysiol.2005.094763.

McBain, C. J., & Fisahn, A. (2001). Interneurons unbound. *Nature Reviews Neuroscience, 2*(1), 11–23. doi: 10.1038/35049047.

Miller, P. S., & Smart, T. G. (2010). Binding, activation and modulation of Cys-loop receptors. *Trends in Pharmacological Sciences, 31*(4), 161–174. doi: S0165-6147(09)00211-9[pii]10.1016/j.tips.2009.12.005.

Monsivais, P., & Rubel, E. W. (2001). Accommodation enhances depolarizing inhibition in central neurons. *Journal of Neuroscience, 21*(19), 7823–7830. doi: 21/19/7823[pii].

Monsivais, P., Yang, L., & Rubel, E. W. (2000). GABAergic inhibition in nucleus magnocellularis: Implications for phase locking in the avian auditory brainstem. *Journal of Neuroscience, 20*(8), 2954–2963.

Moore, J. K., Osen, K. K., Storm-Mathisen, J., & Ottersen, O. P. (1996). Gamma-aminobutyric acid and glycine in the baboon cochlear nuclei: An immunocytochemical colocalization study with reference to interspecies differences in inhibitory systems. *Journal of Comparative Neurology, 369*(4), 497–519. doi: 10.1002/(SICI)1096-9861(19960610)369:4<497::AID-CNE2>3.0.CO;2-#[pii].

Mugnaini, E. (1985). GABA neurons in the superficial layers of the rat dorsal cochlear nucleus: Light and electron microscopic immunocytochemistry. *Journal of Comparative Neurology, 235*(1), 61–81. doi: 10.1002/cne.902350106.

Nabekura, J., Katsurabayashi, S., Kakazu, Y., Shibata, S., Matsubara, A., Jinno, S., Mizoguchi, Y., Sasaki, A., & Ishibashi, H. (2004). Developmental switch from GABA to glycine release in single central synaptic terminals. *Nature Neuroscience, 7*(1), 17–23. doi: 10.1038/nn1170nn1170[pii].

Needham, K., & Paolini, A. G. (2003). Fast inhibition underlies the transmission of auditory information between cochlear nuclei. *Journal of Neuroscience, 23*(15), 6357–6361. doi: 23/15/6357[pii].

Needham, K., & Paolini, A. G. (2007). The commissural pathway and cochlear nucleus bushy neurons: An in vivo intracellular investigation. *Brain Research, 1134*(1), 113–121. doi: S0006-8993(06)03459-7[pii]10.1016/j.brainres.2006.11.058.

Nelken, I., & Young, E. D. (1994). Two separate inhibitory mechanisms shape the responses of dorsal cochlear nucleus type IV units to narrowband and wideband stimuli. *Journal of Neurophysiology, 71*(6), 2446–2462.

Noh, J., Seal, R. P., Garver, J. A., Edwards, R. H., & Kandler, K. (2010). Glutamate co-release at GABA/glycinergic synapses is crucial for the refinement of an inhibitory map. *Nature Neuroscience, 13*(2), 232–238. doi: nn.2478[pii]10.1038/nn.2478.

Oertel, D., & Young, E. D. (2004). What's a cerebellar circuit doing in the auditory system? *Trends in Neurosciences, 27*(2), 104–110. doi: 10.1016/j.tins.2003.12.001S0166223603003862[pii].

Oertel, D., Wu, S. H., Garb, M. W., & Dizack, C. (1990). Morphology and physiology of cells in slice preparations of the posteroventral cochlear nucleus of mice. *Journal of Comparative Neurology, 295*(1), 136–154. doi: 10.1002/cne.902950112.

Osen, K. K. (1969). The intrinsic organization of the cochlear nuclei. *Acta Otolaryngologica, 67*(2), 352–359.

Ostapoff, E. M., Benson, C. G., & Saint Marie, R. L. (1997). GABA- and glycine-immunoreactive projections from the superior olivary complex to the cochlear nucleus in guinea pig. *Journal of Comparative Neurology, 381*(4), 500–512. doi: 10.1002/(SICI)1096-9861(19970519)381:4<500::AID-CNE9>3.0.CO;2-6 [pii].

Palombi, P. S., & Caspary, D. M. (1992). GABAA receptor antagonist bicuculline alters response properties of posteroventral cochlear nucleus neurons. *Journal of Neurophysiology, 67*(3), 738–746.

Park, T. J., Grothe, B., Pollak, G. D., Schuller, G., & Koch, U. (1996). Neural delays shape selectivity to interaural intensity differences in the lateral superior olive. *Journal of Neuroscience, 16*(20), 6554–6566.

Pecka, M., Zahn, T. P., Saunier-Rebori, B., Siveke, I., Felmy, F., Wiegrebe, L., Klug, A., Pollak, G. D., & Grothe, B. (2007). Inhibiting the inhibition: A neuronal network for sound localization in reverberant environments. *Journal of Neuroscience, 27*(7), 1782–1790. doi: 27/7/1782[pii]10.1523/JNEUROSCI.5335-06.2007.

Piechotta, K., Weth, F., Harvey, R. J., & Friauf, E. (2001). Localization of rat glycine receptor alpha1 and alpha2 subunit transcripts in the developing auditory brainstem. *Journal of Comparative Neurology, 438*(3), 336–352. doi: 10.1002/cne.1319[pii].

Rhode, W. S. (1999). Vertical cell responses to sound in cat dorsal cochlear nucleus. *Journal of Neurophysiology, 82*(2), 1019–1032.

Rhode, W. S., Oertel, D., & Smith, P. H. (1983a). Physiological response properties of cells labeled intracellularly with horseradish peroxidase in cat ventral cochlear nucleus. *Journal of Comparative Neurology, 213*(4), 448–463. doi: 10.1002/cne.902130408.

Rhode, W. S., Smith, P. H., & Oertel, D. (1983b). Physiological response properties of cells labeled intracellularly with horseradish peroxidase in cat dorsal cochlear nucleus. *Journal of Comparative Neurology, 213*(4), 426–447. doi: 10.1002/cne.902130407.

Rivera, C., Voipio, J., Payne, J. A., Ruusuvuori, E., Lahtinen, H., Lamsa, K., Pirvola, U., Saarma, M., & Kaila, K. (1999). The K+/Cl- co-transporter KCC2 renders GABA hyperpolarizing during neuronal maturation. *Nature, 397*(6716), 251–255. doi: 10.1038/16697.

Roberts, M. T., & Trussell, L. O. (2010). Molecular layer inhibitory interneurons provide feedforward and lateral inhibition in the dorsal cochlear nucleus. *Journal of Neurophysiology.* doi: jn.00312.2010[pii]10.1152/jn.00312.2010.

Roberts, M. T., Bender, K. J., & Trussell, L. O. (2008). Fidelity of complex spike-mediated synaptic transmission between inhibitory interneurons. *Journal of Neuroscience, 28*(38), 9440–9450. doi: 28/38/9440[pii]10.1523/JNEUROSCI.2226-08.2008.

Rodrigues, A. R., & Oertel, D. (2006). Hyperpolarization-activated currents regulate excitability in stellate cells of the mammalian ventral cochlear nucleus. *Journal of Neurophysiology, 95*(1), 76–87. doi: 00624.2005[pii]10.1152/jn.00624.2005.

Rubio, M. E., & Juiz, J. M. (2004). Differential distribution of synaptic endings containing glutamate, glycine, and GABA in the rat dorsal cochlear nucleus. *Journal of Comparative Neurology, 477*(3), 253–272. doi: 10.1002/cne.20248.

Sanes, D. H. (1990). An in vitro analysis of sound localization mechanisms in the gerbil lateral superior olive. *Journal of Neuroscience, 10*(11), 3494–3506.

Schofield, B. R., & Cant, N. B. (1996). Origins and targets of commissural connections between the cochlear nuclei in guinea pigs. *Journal of Comparative Neurology, 375*(1), 128–146. doi: 10.1002/(SICI)1096-9861(19961104)375:1<128::AID-CNE8>3.0.CO;2-5[pii].

Semyanov, A., Walker, M. C., Kullmann, D. M., & Silver, R. A. (2004). Tonically active GABA A receptors: Modulating gain and maintaining the tone. *Trends in Neurosciences, 27*(5), 262–269. doi: 10.1016/j.tins.2004.03.005S0166223604000906[pii].

Shore, S. E., Sumner, C. J., Bledsoe, S. C., & Lu, J. (2003). Effects of contralateral sound stimulation on unit activity of ventral cochlear nucleus neurons. *Experimental Brain Research, 153*(4), 427–435. doi: 10.1007/s00221-003-1610-6.

Simat, M., Parpan, F., & Fritschy, J. M. (2007). Heterogeneity of glycinergic and gabaergic interneurons in the granule cell layer of mouse cerebellum. *Journal of Comparative Neurology, 500*(1), 71–83. doi: 10.1002/cne.21142.

Smith, P. H., & Rhode, W. S. (1989). Structural and functional properties distinguish two types of multipolar cells in the ventral cochlear nucleus. *Journal of Comparative Neurology, 282*(4), 595–616. doi: 10.1002/cne.902820410.

Street, S. E., & Manis, P. B. (2007). Action potential timing precision in dorsal cochlear nucleus pyramidal cells. *Journal of Neurophysiology, 97*(6), 4162–4172. doi: 00469.2006[pii]10.1152/jn.00469.2006.

Thompson, A. M. (1998). Heterogeneous projections of the cat posteroventral cochlear nucleus. *Journal of Comparative Neurology, 390*(3), 439–453. doi: 10.1002/(SICI)1096-9861(19980119)390:3<439::AID-CNE10>3.0.CO;2-J[pii].

Tollin, D. J. (2003). The lateral superior olive: A functional role in sound source localization. *Neuroscientist, 9*(2), 127–143.

Trussell, L. O. (1999). Synaptic mechanisms for coding timing in auditory neurons. *Annual Review of Physiology, 61*, 477–496. doi: 10.1146/annurev.physiol.61.1.477.

Voigt, H. F., & Young, E. D. (1988). Neural correlations in the dorsal cochlear nucleus: Pairs of units with similar response properties. *Journal of Neurophysiology, 59*(3), 1014–1032.

Wenthold, R. J. (1987). Evidence for a glycinergic pathway connecting the two cochlear nuclei: An immunocytochemical and retrograde transport study. *Brain Research, 415*(1), 183–187. doi: 0006-8993(87)90285-X[pii].

Wenthold, R. J., Huie, D., Altschuler, R. A., & Reeks, K. A. (1987). Glycine immunoreactivity localized in the cochlear nucleus and superior olivary complex. *Neuroscience, 22*(3), 897–912. doi: 0306-4522(87)92968-X[pii].

Wenz, M., Hartmann, A. M., Friauf, E., & Nothwang, H. G. (2009). CIP1 is an activator of the K+−Cl- cotransporter KCC2. *Biochemical and Biophysical Research Communications, 381*(3), 388–392. doi: S0006-291X(09)00311-8[pii]10.1016/j.bbrc.2009.02.057.

Wickesberg, R. E., & Oertel, D. (1990). Delayed, frequency-specific inhibition in the cochlear nuclei of mice: A mechanism for monaural echo suppression. *Journal of Neuroscience, 10*(6), 1762–1768.

Wickesberg, R. E., Whitlon, D., & Oertel, D. (1991). Tuberculoventral neurons project to the multipolar cell area but not to the octopus cell area of the posteroventral cochlear nucleus. *Journal of Comparative Neurology, 313*(3), 457–468. doi: 10.1002/cne.903130306.

Wojcik, S. M., Katsurabayashi, S., Guillemin, I., Friauf, E., Rosenmund, C., Brose, N., & Rhee, J. S. (2006). A shared vesicular carrier allows synaptic corelease of GABA and glycine. *Neuron, 50*(4), 575–587. doi: S0896-6273(06)00307-2[pii]10.1016/j.neuron.2006.04.016.

Wollmuth, L. P., & Sobolevsky, A. I. (2004). Structure and gating of the glutamate receptor ion channel. *Trends in Neurosciences, 27*(6), 321–328. doi: 10.1016/j.tins.2004.04.005S0166223604001250[pii].

Wouterlood, F. G., Mugnaini, E., Osen, K. K., & Dahl, A. L. (1984). Stellate neurons in rat dorsal cochlear nucleus studies with combined Golgi impregnation and electron microscopy: Synaptic connections and mutual coupling by gap junctions. *Journal of Neurocytology, 13*(4), 639–664.

Wu, S. H., & Kelly, J. B. (1992a). Binaural interaction in the lateral superior olive: Time difference sensitivity studied in mouse brain slice. *Journal of Neurophysiology, 68*(4), 1151–1159.

Wu, S. H., & Kelly, J. B. (1992b). Synaptic pharmacology of the superior olivary complex studied in mouse brain slice. *Journal of Neuroscience, 12*(8), 3084–3097.

Wu, S. H., & Kelly, J. B. (1994). Physiological evidence for ipsilateral inhibition in the lateral superior olive: Synaptic responses in mouse brain slice. *Hearing Research, 73*(1), 57–64.

Yang, L., Monsivais, P., & Rubel, E. W. (1999). The superior olivary nucleus and its influence on nucleus laminaris: A source of inhibitory feedback for coincidence detection in the avian auditory brainstem. *Journal of Neuroscience, 19*(6), 2313–2325.

Yin, T. C. T. (2002). Neural mechanisms of encoding binaural localization cue in the auditory brainstem. In D. Oertel, R. R. Fay, & A. N. Popper (Eds.), *Integrative Functions in the Mammalian Auditory Pathway* (pp. 99–159). New York: Springer.

Young, E. D., & Davis, K. A. (2002). Circuitry and function of the dorsal cochlear nucleus. In D. Oertel, R. R. Fay, & A. N. Popper (Eds.), *Integrative Functions in the Mammalian Auditory Pathway* (pp. 160–206). New York: Springer.

Zhang, S., & Oertel, D. (1993). Cartwheel and superficial stellate cells of the dorsal cochlear nucleus of mice: Intracellular recordings in slices. *Journal of Neurophysiology, 69*(5), 1384–1397.

Zhang, S., & Oertel, D. (1994). Neuronal circuits associated with the output of the dorsal cochlear nucleus through fusiform cells. *Journal of Neurophysiology, 71*(3), 914–930.

Chapter 8
Modulatory Mechanisms Controlling Auditory Processing

Raju Metherate

1 Introduction

> The trick, too, is setting the right tone: too hot and people recoil, too cool and they ignore you, too boring and they nod off. Message modulation has tripped Mr. Gore up before.
>
> *New York Times* editorial on the politics of climate change, July 2008

The importance of neuromodulation for auditory processing is, essentially, that "tone" matters. The tone of a message can carry as much information as the message itself, and clearly there are parallels in everyday experience as illustrated by the above quote. In the brain, the tone of auditory processing is determined by neural systems that also contribute to changes in behavioral state, such as arousal and attention. The extent to which these changes involve neuromodulation depends, at least in part, on how one defines the term. Yet the end result – setting the tone of a message – is fundamental to auditory function. The goals of this chapter is to explain what neuromodulation is (hint: it depends on who you're talking to) and to describe some of its mechanisms and how they affect neural processing. To illustrate the main points, examples of neural processing focus on the auditory system where possible, and examples of neuromodulation focus on cholinergic modulation of auditory processing.

R. Metherate (✉)
Department of Neurobiology and Behavior and Center for Hearing Research,
University of California, 2205 McGaugh Hall, Irvine, CA 92697-4550, USA
e-mail: rmethera@uci.edu

L.O. Trussell et al. (eds.), *Synaptic Mechanisms in the Auditory System*,
Springer Handbook of Auditory Research 41, DOI 10.1007/978-1-4419-9517-9_8,
© Springer Science+Business Media, LLC 2012

1.1 What Is Neuromodulation?

The term *neuromodulation* can prove difficult to define, as it means different things to different people. In terms of cellular physiology it most often refers to biochemical processes induced by synaptic stimulation that alter neuronal excitability in a variety of ways (Kaczmarek and Levitan 1987). A strict definition is harder to come by at the sensory-system and behavioral levels, but the various usages are linked by a common thread: neuromodulation refers to regulation of the brain's response to afferent sensory input, rather than mechanisms of the sensory response, per se. This can be conceptualized by contrasting "mediating" and "modulating" mechanisms, that is, how the nervous system encodes the physical characteristics of an acoustic stimulus (mediating) versus regulating the processing of that encoded input (modulating).

Because the most explicit definition of neuromodulation (at the cellular level) is in terms of intracellular biochemical processes, for many systems-level investigators the mere demonstration of the involvement of metabotropic, or G-protein-coupled, receptors (discussed later in this chapter) is sufficient to identify a synaptic action as neuromodulatory. Although at first glance this may seem to be a reasonable criterion, as a general definition it has limits: it excludes phenomena that do not involve metabotropic receptors but nonetheless appear to be functionally neuromodulatory, such as postsynaptic consequences of activating ion channels that permit Ca^{2+} influx (e.g., NMDA receptors), and it includes other phenomena that mediate, rather than modulate, sensory inputs (e.g., transmission at photoreceptor synapses in the retina). Thus, neuromodulation is not well defined at the systems or behavioral level. However, its usage is consistent with the contrasting notions of modulation versus mediation. For example, behavioral arousal is associated with more robust processing of sensory signals due to neurotransmitter actions that regulate, but do not mediate, the sensory signal itself (see later and Fig. 8.1). Similarly, neuromodulation is thought to enhance or reduce the strength of a memory rather than encode the memory itself (McGaugh 2000). Typically, the altered processing is either presumed or explicitly shown to be linked to metabotropic receptors or lasting biochemical changes.

At all levels, molecular to behavioral, studies that refer to so-called neuromodulators like acetylcholine and norepinephrine generally imply the effects of neuromodulation because of the presumed involvement of G-protein-coupled receptors. Though widely used, the term "neuromodulator" is inaccurate (as opposed to the term "neuromodulation," which is merely ambiguous), because many neurotransmitters act at a variety of receptors, both ionotropic and metabotropic. Strictly speaking, the term should not be used to describe the neurotransmitter itself; rather one should talk about the transmitter's neuromodulatory actions (typically mediated via its subset of metabotropic receptors).

This chapter expands on each of these points, focusing on describing neuromodulatory mechanisms and what they do for auditory processing at the synaptic and systems levels, motivated by the kinds of changes that accompany changes in behavior. The chapter is not about the cell biology of neuromodulation, which has

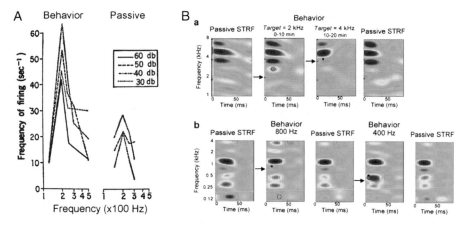

Fig. 8.1 Behavioral-state-dependent changes of receptive fields in the auditory cortex. (**A**) Neurons in the auditory cortex of a monkey performing an auditory discrimination task ("Behavior") exhibit stronger responses (greater spike discharge) to tone stimuli than when the animal is not performing ("Passive") (Reprinted from Miller et al. (1972) with permission of AAAS). (**B**) Spectrotemporal receptive fields (STRFs) of neurons in auditory cortex when an animal is expecting a target tone ("Behavior," target frequency indicated by arrow) versus when the animal is idle ("Passive STRF"). Warmer colors indicate stronger responses; cooler colors indicate weaker responses. Data in (**Ba**) and (**Bb**) are from two different animals, each undergoing repeated sequential determination of receptive fields under passive and behavior conditions. During behavior, receptive fields show enhanced excitation and/or reduced inhibition near the target frequency (Reprinted from Fritz et al. (2003) with permission from Macmillan Publishers Ltd, *Nature Neuroscience*, copyright 2003)

been extensively studied and reviewed (Kaczmarek and Levitan 1987; Weston et al. 2002; Gill et al. 2006). Finally, it touches on behaviorally induced plasticity of auditory processing, a relatively recent, still-unfolding, and exciting area of auditory research.

2 Neuromodulation and Behavior: State-Dependent Auditory Processing

Neuromodulation in the service of behavior is exemplified by auditory processing at different levels of arousal, for example, during sleep versus wakefulness. As an animal's state shifts from deep sleep to wakefulness, acoustic responsiveness in the auditory forebrain increases in general, although specific changes in receptive fields can be more complex (Edeline et al. 2000; Edeline 2003). The effect of arousal on sensory processing was recognized decades ago and was formalized in the concept of the ascending reticular activating system (ARAS) (Moruzzi and Magoun 1949). In that conceptualization, sensory pathways were separate from, but regulated by, the ARAS whose cells of origin were located in the rostral reticular core of the brainstem. Electrophysiologically, the ARAS was defined as that region of the brainstem within which brief electrical stimulation would produce desynchronization (activation) of the cortical electroencephalogram (EEG) in a sleeping or anesthetized animal, mimicking the EEG change seen on behavioral arousal. This general concept still holds (Steriade 1996), although the term ARAS disappeared from the literature as chemical neuroanatomy methods were developed that revealed its true nature. The role of the ARAS is now attributed to diffuse modulatory systems (see Sect. 8.3).

State-dependent changes in auditory processing in an awake animal occur when the animal attends to a behaviorally relevant acoustic stimulus. The change in response to the stimulus can be sustained for the duration of attention, and sometimes longer (e.g., as a result of experience-dependent plasticity, see later). One of the earliest auditory studies found suppression of click-evoked neural responses in the cochlear nucleus when the animal, a cat with chronically implanted electrodes, viewed an engaging visual stimulus – a mouse placed out of reach inside a glass beaker (Hernandez-Peon et al. 1956). Control (nonsuppressed) responses were obtained when the animal was resting quietly and not overtly attending. Although this early example was striking, the lack of controls for stimulus constancy in the attended versus unattended conditions (the animal was freely moving) and the lack of objective measures of attention made interpretation difficult. In a later study (Miller et al. 1972) neurons in the auditory cortex of a monkey performing an auditory discrimination task exhibited stronger responses to tone stimuli ("Behavior" in Fig. 8.1a) relative to responses when the animal was not performing ("Passive"). Head fixation ensured stimulus constancy across conditions, and correct behavioral responses verified that the animal was attending to the stimulus in the "behavior" condition. Thus, enhanced responses during behavior can be attributed to state-dependent processes such as those underlying attention to the stimulus.

More recent studies have gone beyond demonstrating generally increased responsiveness during attentive behavior to demonstrating that changes in responsiveness can be highly selective for the stimulus being attended. Early demonstrations used a paradigm whereby frequency receptive fields for neurons in the auditory cortex were determined before and after classical conditioning using a single-frequency tone paired with electrical shock (Weinberger et al. 1984; Diamond and Weinberger 1989; Weinberger 2004). In addition to behavioral measures that confirmed the effectiveness of the conditioning (i.e., learning did occur), posttraining receptive fields showed increased responsiveness that was restricted to frequencies at or near that used for conditioning. Moreover, in some cases, responses to frequencies adjacent to the conditioning frequency were reduced, further highlighting enhancement at the paired frequency. Because the changes in receptive fields outlasted the period of attention to the stimulus, they may reflect information acquired about the stimulus, that is, they may reflect sensory learning. Eventually, with repeated behavior, similar changes in receptive fields may occur in many neurons and produce global changes in tonotopic representations, or "maps" (Recanzone et al. 1993; Weinberger 2004).

Finally, in other recent studies, the use of experimental protocols that embed a to-be-attended frequency within a sequence of spectrally complex stimuli has allowed determination of frequency-receptive fields *during* attention to a pure-tone stimulus (Fritz et al. 2003). These studies have shown that changes in receptive fields can be rapid, selective – that is, increased and/or decreased responses at multiple frequencies that serve to highlight the attended frequency – and may or may not outlast the task duration (Fig. 8.1b). These and other studies of behaviorally induced receptive field plasticity show that a full understanding of state-dependent processing must account for complex, rapid, and rapidly reversible changes in receptive fields. A complete solution to this problem is difficult to envisage but remains the goal in an intensely investigated area of research.

Because changes in behavioral state have long been thought to involve neuromodulation (and initially, the ARAS), it seems likely that neuromodulation underlies, or at least contributes to, state-dependent changes in auditory processing. Exact mechanisms remain unknown, especially for highly selective and rapidly reversible receptive field plasticity (Fig. 8.1b), but the eventual understanding, along with an understanding of general processing changes (Fig. 8.1a), will likely involve mechanisms discussed in this chapter. It is almost a foregone conclusion that following the presentation of a behaviorally relevant auditory stimulus, neuromodulation will alter the processing of that stimulus, thereby altering the perception, storage, and recall of the original auditory experience.

3   The Neuroanatomy of Neuromodulation: Diffuse Modulatory Systems

As mentioned earlier, the notion of the ARAS (Moruzzi and Magoun 1949) faded following the introduction in the 1960s of histological, and later immunocytochemical, methods to reveal chemical neuroanatomy (Shute and Lewis 1967; Steriade et al. 1990; Wainer and Mesulam 1990). The "activating system" that initially was thought to occupy a large portion of the rostral brainstem became a well-defined set of small nuclei located mostly, but not exclusively, in the brainstem (Fig. 8.2a). Each nucleus was distinguished by dependence on a single neurotransmitter that has since been implicated in regulation of brain state (Table 8.1). In terms of anatomical projection patterns, these systems differ qualitatively from typical sensory or motor systems that project from one relay nucleus to another in a topographic arrangement and instead project widely across many brain regions. As a result, they sometimes are grouped together as "diffuse modulatory systems" (Bear et al. 2007). Each of these nuclei encompasses a small number of neurons that project widely throughout the brain. In some cases, their axons give off *en passant* ("in passing") synapses, with no obvious postsynaptic specialization, that release transmitter over large brain regions and influence large numbers of neurons simultaneously (Descarries and Mechawar 2000). In this way, these small specialized groups of neurons can regulate global brain state.

As one might expect given the anatomy of diffuse modulatory systems (and consistent with the earlier ARAS concept), stimulation of one of these small regions can produce dramatic effects over large brain areas. As shown in Fig. 8.2b, brief

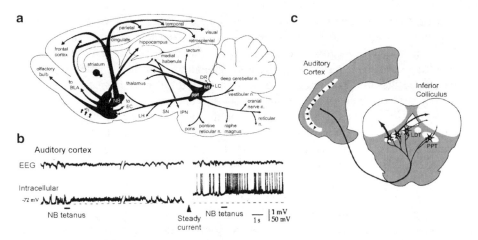

Fig. 8.2 Anatomy and physiology of a diffuse modulatory system. (**a**) The central cholinergic system arising from a few nuclei in the basal forebrain and rostral brainstem projects widely throughout the brain. (*NB* nucleus basalis, *ppt* pedunculopontine tegmental nucleus, *ldt* laterodorsal tegmental nucleus.) (Reprinted from Woolf (1991) with permission from Elsevier). (**b**) Brief tetanic stimulation of the nucleus basalis activates the EEG in auditory cortex (*top trace*), while a simultaneous intracellular recording (*bottom*) from a neuron in the auditory cortex reveals a parallel change in membrane potential fluctuations. After recovery of the EEG and intracellular fluctuations, a second nucleus basalis tetanus at a more depolarized membrane potential induced by steady current injection shows the effect of stimulation on action potential discharge (Reprinted from Metherate et al. (1992) with permission from the Society for Neuroscience). (**c**) Auditory cortex output neurons project to cholinergic nuclei in the brainstem that in turn project to the inferior colliculus (Reprinted from Schofield (2010) with permission from Elsevier)

Table 8.1 Diffuse modulatory systems

Neurotransmitter	Major nuclei of origin	Region of origin
Norepinephrine	Locus coeruleus	Pons
	Lateral tegmental area	Pons, medulla
Acetylcholine	Nucleus basalis	Basal forebrain
	Pedunculopontine nucleus	Pons
	Laterodorsal tegmental nucleus	Pons
Serotonin	Raphe nuclei	Midbrain, pons, medulla
Dopamine	Substantia nigra	Midbrain
	Ventral tegmental area	Midbrain

stimulation of the nucleus basalis region of the basal forebrain can activate the EEG as recorded in auditory cortex, indicating modified activity of hundreds or thousands of neurons in the auditory cortex and surrounding regions (Metherate et al. 1992). The simultaneous intracellular recording from one neuron in the auditory cortex reveals a remarkable similarity between intracellular and EEG fluctuations, indicating that similar membrane potential fluctuations are happening simultaneously in neurons across the cortex. The effect of basal forebrain stimulation on EEG fluctuations is blocked by muscarinic receptor antagonist delivered directly into auditory cortex (Metherate et al. 1992), indicating the involvement of cholinergic nucleus basalis neurons. Functional implications of the nucleus basalis–mediated effect for auditory processing are described in Sect. 8.5.

While the physiologic effects of activating diffuse modulatory systems are consistent with the nature of their anatomical projections, the term "diffuse" should not imply imprecise. There are major differences in projection patterns among the various nuclei, and even within nuclei there is topography of inputs and outputs, though not nearly as organized and restricted as those of lemniscal sensory or motor systems. For example, the cholinergic nuclei in the brainstem provide acetylcholine to most of the diencephalon, whereas the nucleus basalis provides acetylcholine to the entire cerebral cortex (Fig. 8.2a), and within the nucleus basalis, the caudal, but not rostral, portion projects to auditory and other temporal cortical areas (Wainer and Mesulam 1990; Baskerville et al. 1993). Moreover, within a target region there can be clear differential distribution even of *en passant* synapses, for example, with some cortical layers receiving dense input and others significantly less (Mechawar et al. 2000). Finally, afferent input to diffuse modulatory systems itself can be precise: Fig. 8.2c shows schematically that auditory cortex output neurons drive cholinergic neurons in the brainstem that then regulate auditory processing in the inferior colliculus (Schofield 2010). There remains much to be understood not only about how modulatory systems regulate sensory and motor systems, but also about how they themselves are regulated and interact with each other.

4 Synaptic Mechanisms of Modulation (or, "There's No Such Thing as a Neuromodulator")

The previous example of acetylcholine exerting a modulating action via muscarinic acetylcholine receptors serves to illustrate an important point: receptors determine the actions of neurotransmitters. While this may seem obvious, it is common to hear neurotransmitters such as acetylcholine or norepinephrine referred to as neuromodulators although they do not produce modulatory effects in every instance. A case in point is that whereas acetylcholine often exerts neuromodulatory effects via activation of muscarinic (and sometimes nicotinic, discussed later) receptors, at the neuromuscular junction it produces rapid activation of nicotinic acetylcholine receptors and subsequent muscle contraction – clearly a mediating, rather than modulating, effect. Strictly speaking, and despite common usage, there is no such thing as a neuromodulator, there are only neuromodulatory actions of neurotransmitters.

4.1 Ionotropic and Metabotropic Receptors Mediate Fast and Slow Synaptic Potentials

Many neurotransmitters can exert multiple actions, depending on which receptors are present at a particular synapse. There are two broad classes of receptors – ionotropic and metabotropic – and the receptors within each class bear more structural similarity

Table 8.2 Some neurotransmitters with ionotropic and metabotropic receptors

Neurotransmitter	Ionotropic receptors	Metabotropic receptors
Acetylcholine	Nicotinic	Muscarinic (M1-M5)
Serotonin	5-HT3	5-HT1, 5-HT2, 5-HT4
Glutamate	NMDA	mGluR
	Non-NMDA (e.g., AMPA)	
GABA	GABA-A	GABA-B

5-HT 5-hydroxy-triptamine, *NMDA* N-methyl-D-aspartate, *AMPA* α-amino-3-hydroxyl-5-methyl-4-isoxazole-propionate, *GABA* gamma aminobutyric acid

to each other than to those members of the other class that bind the same neurotransmitter (Table 8.2). Ionotropic receptors are ligand-gated ion channels, with each receptor having four or five protein subunits that combine to form a membrane-spanning ion channel (Barrera and Edwardson 2008). Metabotropic receptors are G-protein-coupled receptors that do not directly alter ion channel function but may do so indirectly (via second messengers) and also may affect a host of cellular functions including gene transcription and translation (Rosenbaum et al. 2009). Thus, for example, GABA-A and nicotinic acetylcholine receptors are structurally related, pentameric, ligand-binding ion channels, whereas GABA-B and muscarinic acetylcholine receptors are both coupled to GTP-binding proteins (G-proteins) that indirectly regulate K^+ channels. Some common examples of each type of receptor are listed in Table 8.2, which also makes the point that a single neurotransmitter can act at both types of receptor.

The synaptic consequences of binding to ionotropic or metabotropic receptors often involve fast or slow synaptic potentials, respectively. Changes in neural excitability due to activation of metabotropic receptors, while not as rapid as those due to activation of ionotropic receptors, can be subtle and long-lasting due to the biochemical nature of the underlying mechanism. Such changes are typical of neuromodulatory effects. Fast excitatory postsynaptic potentials (EPSPs) result from activation of ionotropic receptors that rapidly increase cation permeability and typically mediate a net influx of Na^+. In contrast, slow EPSPs resulting from activation of metabotropic receptors depend on a very different mechanism, *decreased* K^+ permeability. To understand how decreased permeability could produce an EPSP, consider that K^+ channels that are active (open) at the resting membrane potential exhibit a slow but steady efflux of K^+. Closing them, therefore, will reduce the steady efflux, producing a slow small-amplitude depolarization, that is, a slow EPSP, that can be long-lasting because of the sustained biochemical signals underlying it. In addition, some K^+ channels are activated only on depolarization (voltage-dependent K^+ channels); thus, the effects of closing them will be evident only during membrane depolarization in response to other inputs, because a greater depolarization will result when the normally opposing K^+ flux does not occur. Some common features of fast and slow EPSPs are listed in Table 8.3.

The fact that a neurotransmitter can exert multiple actions at a synapse, depending on which receptors are present, is illustrated by the example in Fig. 8.3a. At this ganglionic synapse, acetylcholine released from the presynaptic terminal activates

Table 8.3 Typical features of fast and slow EPSPs

	Fast EPSP	Slow EPSP
Receptor type	Ionotropic	Metabotropic
Ion species involved	Na^+, K^+ (Ca^{2+})	K^+
Conductance change	Increase	Decrease
Time course	Milliseconds	Can last minutes
Intracellular messenger	None	G-proteins and second messengers
Function	Mediating	Modulating

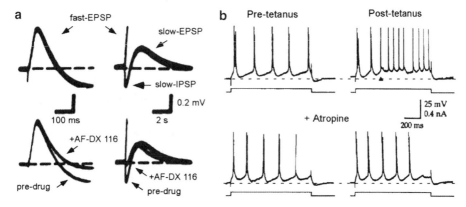

Fig. 8.3 Multiple effects of one neurotransmitter acting via ionotropic and metabotropic receptors. (a) Acetylcholine release in sympathetic ganglia elicits a fast EPSP mediated by nicotinic acetylcholine receptors, a slow IPSP mediated by muscarinic M2 receptors, and a slow EPSP mediated by muscarinic M1 receptors. *Top* traces are control, *bottom* traces show effect of AF-DX 116, a selective M2 receptor antagonist, to block the slow IPSP (predrug traces are also shown at *bottom*). Traces on the *left* and *right* are the same records at different time scales (Reprinted from Yarosh et al. (1988) with permission of the American Society of Pharmacology and Experimental Therapeutics). (b) Electrical stimulation (tetanus) of cholinergic fibers afferent to the auditory cortex in vitro produces little membrane depolarization (approximately 2–3 mV; bath contains glutamate and GABA receptor antagonists) but does change the discharge pattern in response to intracellular current pulse (the current record is under membrane potential records). The firing pattern changes from rhythmic bursting to largely single-spike firing (the transition is at the arrowhead). The effect of tetanus was blocked by atropine, indicating the involvement of muscarinic receptors (Reprinted from Metherate et al. (1992) with permission from the Society for Neuroscience)

one ionotropic and two metabotropic receptors to produce: (1) a fast EPSP mediated by nicotinic acetylcholine receptors that increase cation permeability, (2) a slow IPSP mediated by muscarinic M2 receptors that increase K^+ permeability, and (3) a slow EPSP mediated by muscarinic M1 receptors that decrease K^+ permeability (Yarosh et al. 1988). The durations of these synaptic responses range from tens of milliseconds to many seconds (note the different time scales in Fig. 8.3a). The same postsynaptic neuron also is regulated by a peptide neurotransmitter that decreases K^+ permeability to produce a late slow EPSP lasting minutes (not shown). Clearly, the types of synaptic responses that underlie neuromodulation (e.g., slow EPSPs) depend not only on the neurotransmitters released but also on which receptors are present.

4.2   Slow EPSPs and Synaptic Integration

A feature of slow EPSPs that is apparent from the previous example is that in most cases they produce little depolarization (<1 mV in Fig. 8.3a). While this may make sense given that slow EPSPs result from the reduction of an already-small resting K⁺ efflux, it also suggests that their major effects, unlike those of fast EPSPs, have little to do with depolarizing the membrane potential closer to spike threshold (although small effects multiplied over the thousands of neurons influenced by a diffuse modulatory system may become substantial). One likely possibility, in keeping with a potential neuromodulatory function, is that the increased membrane resistance produced by decreased K⁺ permeability will increase the postsynaptic neuron's response to other inputs. (Remember Ohm's law: voltage = current × resistance; or, for synapses, EPSP = synaptic current × membrane resistance.) Similarly, increased membrane resistance will enhance the temporal and spatial extent of synaptic integration; per the cable properties of neurons, both membrane time and length constants increase with increased membrane resistance.

Thus, an important consequence of slow EPSPs is to enhance neurons' responses to other, "mediating" afferent inputs. Because there are dozens of K⁺ channels that regulate neuronal excitability, the actual effect of neuromodulation via reduced K⁺ permeability will depend on which K⁺ channels are affected. Reduction of permeability for K⁺ channels that are active at rest will produce a slow EPSP, as described earlier. On the other hand, reduction of voltage- or Ca^{2+}-dependent K⁺ channels that regulate action potential duration or subsequent activity (e.g., afterhyperpolarization), would not necessarily produce a slow depolarization but would alter patterns of discharge in response to other inputs. An example is shown in Fig. 8.3b, where stimulation of cholinergic fibers afferent to auditory cortex results in little membrane depolarization but a dramatic change in firing pattern (from rhythmic bursting to single-spike firing). This effect was blocked by atropine, indicating the involvement of muscarinic acetylcholine receptors.

4.3   Neuromodulation via Ionotropic Receptors

At the synaptic level, it is increasingly clear that contrary to common thinking (and some definitions), even ionotropic receptors at times can produce neuromodulatory effects. A common mechanism may be the involvement of receptor-mediated Ca^{2+} influx. Thus, activation of NMDA receptors can produce EPSPs that are small-amplitude and long-lasting (in comparison to glutamatergic EPSPs mediated by AMPA receptors) and, due to activation of signal transduction pathways, can produce even longer-lasting changes (Yuste and Sur 1999; Cull-Candy et al. 2001). Similarly, nicotinic acetylcholine receptors that gate Ca^{2+} influx, especially those containing α7 subunits that confer Ca^{2+} permeability as great or greater than that of NMDA receptors (Seguela et al. 1993), can produce a variety of effects via activation of signal transduction pathways (Broide and Leslie 1999; Berg et al. 2006).

Because the range of ionotropic receptor-mediated effects resembles that mediated by metabotropic receptors, it seems appropriate that a definition of neuromodulation should include them.

5 Functional Consequences of Modulation for Systems-Level and Sensory Processing

An important goal for future studies is to understand how the ever-increasing variety of neuromodulatory effects documented at the cellular and synaptic levels contribute to regulation of systems-level and sensory function, and ultimately, behavior. A well-studied example, related to the changes in sensory processing that occur during behavioral arousal, serves to demonstrate how a variety of neuromodulatory cellular actions can integrate to produce qualitative changes in systems-level function in the sensory forebrain (Steriade et al. 1993). As mentioned earlier, animals in deep sleep or under anesthesia exhibit a characteristic forebrain EEG pattern consisting of large-amplitude slow-wave oscillations (Fig. 8.2). Recordings from thalamic relay neurons under similar conditions reveal relatively hyperpolarized membrane potentials and characteristic "burst mode" responses to excitatory inputs (Fig. 8.4). The burst mode of thalamic neurons and sleep-like EEG pattern both are associated with poor relay of sensory information to the cortex (Livingstone and Hubel 1981; McCormick and Feeser 1990; Edeline et al. 2000). The awake animal, in contrast, exhibits an activated or desynchronized EEG and thalamic neurons that are depolarized into a

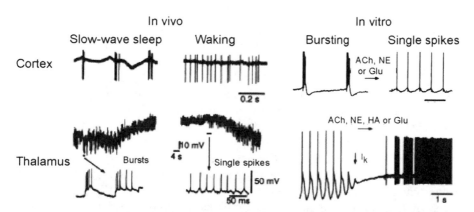

Fig. 8.4 Neuromodulatory actions mimic some effects of behavioral arousal. Recordings from cortical (*top*) and thalamic (*bottom*) neurons in vivo (*left*) and in vitro (*right*) show changes in neuronal function on behavioral arousal (in vivo) or activation of metabotropic receptors (in vitro). Waking produces a depolarization of thalamic relay neurons and a shift from single-spike to bursting mode for both thalamic and cortical neurons. Similar changes are seen in vitro on metabotropic receptor stimulation. (*ACh* acetylcholine, *NE* norepinephrine, *Glu* glutamate, *HA* histamine, I_K K$^+$ leak current.) (Reprinted from Steriade et al. (1993) with permission from AAAS)

"single-spike" firing mode that is more conducive to relaying sensory inputs to the cortex (Livingstone and Hubel 1981; Steriade et al. 1993; Edeline et al. 2000). A variety of studies using in vivo (anesthetized animals) and in vitro (brain slice) preparations have shown that activation of diffuse modulatory systems (in vivo) or metabotropic receptors (in vitro) can produce changes in neural activity and mode similar to those produced by behavioral arousal (Fig. 8.4) (McCormick and Prince 1986; Metherate et al. 1992; Steriade et al. 1993). These changes include:

1. Depolarization of thalamic relay neurons and a shift from single-spike to bursting mode; several neurotransmitters acting at metabotropic receptors produce effects that are consistent with this overall effect.
2. Muscarinic inhibition of thalamic reticularis neurons that themselves inhibit thalamic relay neurons, thereby disinhibiting the thalamocortical relay of information.
3. Increased responsiveness of cortical neurons produced by decreased spike frequency adaptation and, for some cell types, a shift from burst to single-spike mode; again, several neuromodulatory actions mimic this effect.

The overall effect is to increase thalamocortical relay of afferent sensory information and to enhance cortical responses to the same inputs. Thus, a cellular understanding of neuromodulatory actions combined with an understanding of the cellular basis of systems-level function can provide insight into the functional consequences of neuromodulation for sensory function and behavior.

5.1   Neuromodulation and Auditory Synaptic Plasticity

A large body of work points to the involvement of neuromodulation in enabling, or gating, plasticity in sensory systems. This has been studied extensively in the auditory cortex and subcortical auditory centers and has been extensively reviewed, including elsewhere in this volume (Chap. 9, Tzounopoulos and Leão). A single example here can serve to link that literature with this chapter on modulation. A number of studies have shown that stimulation of the cholinergic nucleus basalis can enhance cortical responses to acoustic inputs (Edeline et al. 1994; Yan and Zhang 2005). Moreover, repeated pairing of nucleus basalis stimulation with sensory stimuli can produce prolonged changes in responsiveness to subsequent stimuli, and this effect has been hypothesized to be involved in behaviorally induced plasticity of receptive fields and tonotopic maps (Bakin and Weinberger 1996; Kilgard and Merzenich 1998). That is, as mentioned earlier, learning the behavioral significance of an acoustic stimulus can produce enduring changes in frequency-receptive fields that serve to highlight the significant stimulus and a subsequent change in tonotopic maps in the cortex; this plasticity is thought to involve cholinergic inputs from the nucleus basalis that are paired with repeated acoustic stimulation. Consistent with these proposals, nucleus basalis stimulation enhances synaptic responses in the auditory cortex elicited by electrical stimulation of the auditory thalamus (Fig. 8.5a; note that the nucleus basalis stimulation

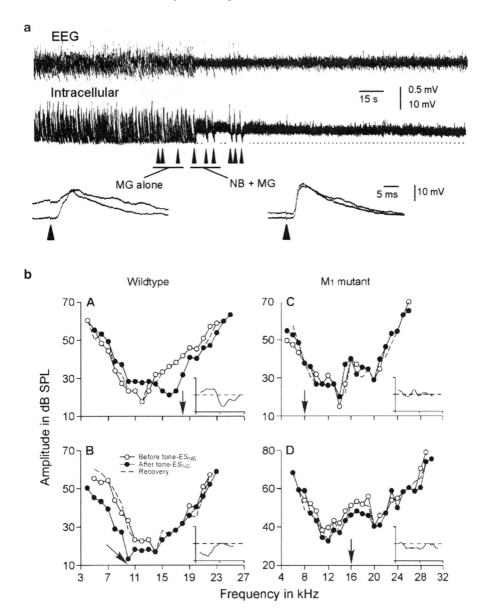

Fig. 8.5 Neuromodulation and synaptic plasticity in auditory cortex. (**a**) Nucleus basalis stimulation activates EEG in auditory cortex (*top trace*), depolarizes a cortical neuron recorded simultaneously (*middle*), and enhances the synaptic response in the auditory cortex to electrical stimulation of the auditory thalamus (*bottom*). (*NB* nucleus basalis, *MG* medial geniculate.) Arrowheads indicate time of MG or NB + MG stimulation (Reprinted from Metherate and Ashe (1993) with permission from John Wiley and Sons). (**b**) Frequency-tuning curves from two control animals (A, B) and two M1 receptor knock-out mice (C, D) before and after repeated pairing of a single-frequency (*arrow*) tone with electrical stimulation of the nucleus basalis. Insets show before and after tuning curve difference (*x* axes are the same as main axes; *y* axes are 30 to −30 dB). Pairing with nucleus basalis stimulation shifts tuning functions toward the paired frequency in control, but not mutant, mice (Reprinted from Zhang et al. (2006) with permission from Oxford University Press)

also simultaneously activates the cortical EEG) (Metherate and Ashe 1993). Similar effects following pairing of nucleus basalis stimulation and tone stimulation result in a reduced acoustic threshold at the paired tone frequency (Fig. 8.5b) (Zhang et al. 2006; Chen and Yan 2007). A detailed examination of the effects of nucleus basalis stimulation on tone-evoked excitatory and inhibitory synaptic responses in the auditory cortex show that nucleus basalis stimulation can simultaneously enhance the magnitude of EPSPs and reduce the magnitude of inhibitory potentials (Froemke et al. 2007). With a few pairings the enhanced response to paired frequencies is transient (Fig. 8.5), but it is thought that with additional pairings, or possibly additional conditions, long-lasting and frequency-specific changes in receptive fields and tonotopic maps result (Weinberger 2004). Thus, the mechanisms for prolonged changes in receptive fields may share features with synaptic long-term potentiation (see Chap. 9, Tzounopoulos and Leão).

6 Summary

As several decades of research have made abundantly clear, auditory processing depends not only on the physical characteristics of acoustic stimuli, but also on brain state and behavioral context. Changes in behavioral relevance can produce rapid and rapidly reversible changes in processing (e.g., during attention to, or away from, an auditory object). Other changes can endure, leading to plasticity of receptive fields and tonotopic representations during auditory learning. Neuromodulation is central to all of these processes, and while the present treatment has focused on cholinergic modulation, the unique contributions of each neurotransmitter must be delineated. There is much known about cellular neuromodulation, that is, synaptically induced biochemical processes that alter neuronal excitability. A major challenge will be to relate these cellular mechanisms to an understanding of modulation at the systems and behavioral levels. However, at those levels even a definition of neuromodulation is unclear. Current criteria often are based on: (1) the neurotransmitters involved ("neuromodulators"), (2) the involvement of metabotropic receptors, or (3) the speed and duration of effects. None of these criteria is ideal, because so-called neuromodulators can produce actions that are not modulatory; some modulatory effects depend on ionotropic receptors, and the time course of modulatory effects can be highly variable. A useful definition should be based on mechanism and should be clearer from functional and conceptual viewpoints, for example, intracellular biochemical processes that regulate the brain's response to afferent inputs encoding the physical characteristics of sensory stimuli. The goal is a full understanding of neuromodulation in auditory processing that spans molecules to behavior, that is, that shows how cellular modulation alters systems-level processing to reflect the behavioral relevance of the stimulus to the animal.

Acknowledgments The author's research cited here was supported by the National Institutes of Health (NIDCD DC02967 and NIDA DA12929) and the National Science Foundation (IBN 9510904).

References

Bakin, J. S., & Weinberger, N. M. (1996). Induction of a physiological memory in the cerebral cortex by stimulation of the nucleus basalis. *Proceedings of the National Academy of Sciences of the United States of America, 93*, 11219–11224.

Barrera, N. P., & Edwardson, J. M. (2008). The subunit arrangement and assembly of ionotropic receptors. *Trends in Neurosciences, 31*(11), 569–576.

Baskerville, K. A., Chang, H. T., & Herron, P. (1993). Topography of cholinergic afferents from the nucleus basalis of Meynert to representational areas of sensorimotor cortices in the rat. *Journal of Comparative Neurology, 335*, 552–562.

Bear, M. F., Connors, B. W., & Paradiso, M. A. (2007). *Neuroscience: Exploring the brain* (3 rd ed.). Philadelphia, PA: Lippincott Williams & Wilkins.

Berg, D. K., Conroy, W. G., Liu, Z., & Zago, W. M. (2006). Nicotinic signal transduction machinery. *Journal of Molecular Neuroscience, 30*(1–2), 149–152.

Broide, R. S., & Leslie, F. M. (1999). The alpha7 nicotinic acetylcholine receptor in neuronal plasticity. *Molecular Neurobiology, 20*(1), 1–16.

Chen, G., & Yan, J. (2007). Cholinergic modulation incorporated with a tone presentation induces frequency-specific threshold decreases in the auditory cortex of the mouse. *European Journal of Neuroscience, 25*(6), 1793–1803.

Cull-Candy, S., Brickley, S., & Farrant, M. (2001). NMDA receptor subunits: Diversity, development and disease. *Current Opinion in Neurobiology, 11*(3), 327–335.

Descarries, L., & Mechawar, N. (2000). Ultrastructural evidence for diffuse transmission by monoamine and acetylcholine neurons of the central nervous system. *Progress in Brain Research, 125*, 27–47.

Diamond, D. M., & Weinberger, N. M. (1989). Role of context in the expression of learning-induced plasticity of single neurons in auditory cortex. *Behavioral Neuroscience, 103*, 471–494.

Edeline, J.-M. (2003). The thalamo-cortical auditory receptive fields: Regulation by the states of vigilance, learning and the neuromodulatory systems. *Experimental Brain Research, 153*(4), 554–572.

Edeline, J.-M., Hars, B., Maho, C., & Hennevin, E. (1994). Transient and prolonged facilitation of tone-evoked responses induced by basal forebrain stimulations in the rat auditory cortex. *Experimental Brain Research, 97*, 373–386.

Edeline, J.-M., Manunta, Y., & Hennevin, E. (2000). Auditory thalamus neurons during sleep: Changes in frequency selectivity, threshold, and receptive field size. *Journal of Neurophysiology, 84*(2), 934–952.

Fritz, J., Shamma, S., Elhilali, M., & Klein, D. (2003). Rapid task-related plasticity of spectrotemporal receptive fields in primary auditory cortex. *Nature Neuroscience, 6*(11), 1216–1223.

Froemke, R. C., Merzenich, M. M., & Schreiner, C. E. (2007). A synaptic memory trace for cortical receptive field plasticity. *Nature, 450*(7168), 425–429.

Gill, D. L., Spassova, M. A., & Soboloff, J. (2006). Signal transduction: Calcium entry signals – trickles and torrents. *Science, 313*(5784), 183–184.

Hernandez-Peon, R., Scherrer, H., & Jouvet, M. (1956). Modification of electric activity in cochlear nucleus during attention in unanesthetized cats. *Science, 123*(3191), 331–332.

Kaczmarek, L. K., & Levitan, I. B. (Eds.). (1987). *Neuromodulation.* New York: Oxford University Press.

Kilgard, M. P., & Merzenich, M. M. (1998). Cortical map reorganization enabled by nucleus basalis activity. *Science, 279*, 1714–1718.

Livingstone, M. S., & Hubel, D. H. (1981). Effects of sleep and arousal on the processing of visual information in the cat. *Nature, 291*, 554–561.

McCormick, D. A., & Feeser, H. R. (1990). Functional implications of burst firing and single spike activity in lateral geniculate relay neurons. *Neuroscience, 39*, 103–113.

McCormick, D. A., & Prince, D. A. (1986). Acetylcholine induces burst firing in thalamic reticular neurones by activating a potassium conductance. *Nature, 319*, 402–405.

McGaugh, J. L. (2000). Memory – a century of consolidation. *Science, 287*(5451), 248–251.

Mechawar, N., Cozzari, C., & Descarries, L. (2000). Cholinergic innervation in adult rat cerebral cortex: A quantitative immunocytochemical description. *Journal of Comparative Neurology, 428*(2), 305–318.

Metherate, R., & Ashe, J. H. (1993). Nucleus basalis stimulation facilitates thalamocortical synaptic transmission in rat auditory cortex. *Synapse, 14*, 132–143.

Metherate, R., Cox, C. L., & Ashe, J. H. (1992). Cellular bases of neocortical activation: Modulation of neural oscillations by the nucleus basalis and endogenous acetylcholine. *Journal of Neuroscience, 12*, 4701–4711.

Miller, J. M., Sutton, D., Pfingst, B., Ryan, A., Beaton, R., & Gourevitch, G. (1972). Single cell activity in the auditory cortex of Rhesus monkeys: Behavioral dependency. *Science, 177*(47), 449–451.

Moruzzi, G., & Magoun, H. W. (1949). Brain stem reticular formation and activation of the EEG. *Electroencephalography and Clinical Neurophysiology, 1*(4), 455–473.

Recanzone, G. H., Schreiner, C. E., & Merzenich, M. M. (1993). Plasticity in the frequency representation of primary auditory cortex following discrimination training in adult owl monkeys. *Journal of Neuroscience, 13*(1), 87–103.

Rosenbaum, D. M., Rasmussen, S. G., & Kobilka, B. K. (2009). The structure and function of G-protein-coupled receptors. *Nature, 459*(7245), 356–363.

Schofield, B. R. (2010). Projections from auditory cortex to midbrain cholinergic neurons that project to the inferior colliculus. *Neuroscience, 166*(1), 231–240.

Seguela, P., Wadiche, J., Dineley-Miller, K., Dani, J. A., & Patrick, J. W. (1993). Molecular cloning, functional properties, and distribution of rat brain alpha7: A nicotinic cation channel highly permeable to calcium. *Journal of Neuroscience, 13*, 596–604.

Shute, C. C., & Lewis, P. R. (1967). The ascending cholinergic reticular system: Neocortical, olfactory and subcortical projections. *Brain, 90*(3), 497–520.

Steriade, M. (1996). Arousal: Revisiting the reticular activating system. *Science, 272*(5259), 225–226.

Steriade, M., Gloor, P., Llinas, R. R., Silva, F. H. L. d., & Mesulam, M.-M. (1990). Basic mechanisms of cerebral rhythmic activities. *Electroencephalography and Clinical Neurophysiology, 76*, 481–508.

Steriade, M., McCormick, D. A., & Sejnowski, T. J. (1993). Thalamocortical oscillations in the sleeping and aroused brain. *Science, 262*(5134), 679–685.

Wainer, B. H., & Mesulam, M.-M. (1990). Ascending cholinergic pathways in the rat brain. In M. Steriade & D. Biesold (Eds.), *Brain Cholinergic Systems* (pp. 65–119). Oxford: Oxford University Press.

Weinberger, N. M. (2004). Specific long-term memory traces in primary auditory cortex. *Nature Reviews Neuroscience, 5*(4), 279–290.

Weinberger, N. M., Hopkins, W., & Diamond, D. M. (1984). Physiological plasticity of single neurons in auditory cortex of the cat during acquisition of the pupillary conditioned response: I. Primary field (AI). *Behavioral Neuroscience, 98*(2), 171–188.

Weston, C. R., Lambright, D. G., & Davis, R. J. (2002). Signal transduction: MAP kinase signaling specificity. *Science, 296*(5577), 2345–2347.

Woolf, N. J. (1991). Cholinergic systems in mammalian brain and spinal cord. *Progress in Neurobiology, 37*(6), 475–524.

Yan, J., & Zhang, Y. (2005). Sound-guided shaping of the receptive field in the mouse auditory cortex by basal forebrain activation. *European Journal of Neuroscience, 21*(2), 563–576.

Yarosh, C. A., Olito, A. C., & Ashe, J. H. (1988). AF-DX 116: A selective antagonist of the slow inhibitory postsynaptic potential and methacholine-induced hyperpolarization in superior cervical ganglion of the rabbit. *Journal of Pharmacology and Experimental Therapeutics, 245*(2), 419–425.

Yuste, R., & Sur, M. (1999). Development and plasticity of the cerebral cortex: From molecules to maps. *Journal of Neurobiology, 41*(1), 1–6.

Zhang, Y., Hamilton, S. E., Nathanson, N. M., & Yan, J. (2006). Decreased input-specific plasticity of the auditory cortex in mice lacking M1 muscarinic acetylcholine receptors. *Cerebral Cortex, 16*(9), 1258–1265.

Chapter 9
Mechanisms of Memory and Learning in the Auditory System

Thanos Tzounopoulos and Ricardo M. Leão

1 Introduction

What are the cellular mechanisms underlying memory and learning? This question has puzzled scientists and philosophers from Aristotelis, who proposed the hypothesis that the heart is the site of learning (Aristotle 350 B.C.E), to John Locke and his wax tablet analogy (Locke 1689), to contemporary ideas regarding synaptic plasticity (Bliss and Lomo 1973). Indeed, long-term synaptic plasticity of synaptic transmission in the hippocampus is the leading experimental model for the synaptic changes that may underlie learning and memory. Activity-dependent long-lasting enhancement in synaptic strength (long-term potentiation, LTP) requires co-activation of a certain number of inputs ("cooperativity"). In addition, LTP exhibits "associativity," meaning that when weak stimulation of one input is insufficient for the induction of LTP, simultaneous (associative) strong stimulation of another input will induce LTP in both inputs. Persistence, cooperativity, and associativity have rendered LTP a candidate mechanism supporting learning behaviors, beginning with studies in the 1970s.

The first detailed description of long-term potentiation appeared in 1973 (Bliss and Gardner-Medwin 1973; Bliss and Lomo 1973). Terje Lomo and Tim Bliss recorded electrical responses from neurons in the hippocampus of anaesthetized rabbits. Brief stimulation of fibers connecting to the cell layers of the hippocampus led to an enhanced response by a single pulse to the same fibers

T. Tzounopoulos (✉)
Department of Otolaryngology, University of Pittsburgh, 3501 Fifth Avenue, BST3 10021, Pittsburgh, PA 15261, USA
e-mail: thanos@pitt.edu

L.O. Trussell et al. (eds.), *Synaptic Mechanisms in the Auditory System*, Springer Handbook of Auditory Research 41, DOI 10.1007/978-1-4419-9517-9_9, © Springer Science+Business Media, LLC 2012

(Bliss and Lomo 1973). Today, LTP still represents the leading model for cellular mechanisms underlying memory and learning. LTP has been found in almost every brain region. Although the properties and molecular mechanisms of synaptic plasticity have been extensively studied in a number of brain regions, much less is known about the significance of synaptic plasticity for neural computation in the context of particular neural circuits. Sensory systems and their known functions in sensory processing provide a good site for studying the role of synaptic plasticity.

A consistent finding has been that as sensory scenes are changing and new information is being processed, cells gain responsiveness to some sensory stimuli and lose responsiveness to other stimuli. In other words, neuronal receptive fields are modified by experience. A good example of receptive field plasticity can be found in the visual cortex. When a visual cortical neuron receives information from the two eyes that is correlated, as is often the case in normal binocular vision, the cell becomes responsive to both eyes. When the correlation breaks down, as occurs during a period of monocular deprivation or strabismus, the cell becomes monocularly responsive. Thus, input patterns can associate or compete depending on how well they are correlated (Wiesel 1982). Therefore, synaptic plasticity and its associative properties have been enthusiastically embraced as a likely basis for receptive field plasticity in the visual cortex.

In the auditory system, reports of learning-induced changes in the primary auditory cortex began in the 1950s with the observation that in adult cats, classical conditioning was associated with evoked potentials to the conditioned stimulus (Galambos et al. 1956). Subsequent studies related to long-term plasticity and learning-related phenomena have focused on the auditory cortex (Fritz et al. 2007a; Schreiner and Winer 2007). However, recent studies of plasticity in the auditory system revealed some surprising findings. Traditional views of sensory processing have maintained that early sensory processing is largely hard-wired. Therefore, the auditory brainstem has been considered as a nonadaptive site dedicated to the preservation of timing information and to the minimization of transmission delays. However, recent studies have revealed remarkable evidence for learning, adaptive sensory processing, and cellular mechanisms for memory and learning in the auditory brainstem (Tzounopoulos and Kraus 2009).

For the purposes of this chapter, synaptic plasticity refers to changes in synaptic strength that are activity-dependent and that can last for at least 30 min (short-term synaptic plasticity is not considered in this chapter). Intrinsic plasticity refers to the persistent modification of a neuron's intrinsic electrical properties by neuronal or synaptic activity. Intrinsic plasticity is mediated by changes in the expression level or biophysical properties of ion channels in the membrane. In this chapter the main types, mechanisms, and roles of synaptic and intrinsic plasticity that have been observed in the auditory system are reviewed.

2 Plasticity in the Auditory Brainstem

2.1 Evoked Auditory Brainstem Responses in Humans: Evidence for Plastic Auditory Brainstem

Auditory evoked potentials in humans represent synchronized neural activity. The auditory brainstem response (ABR) is a noninvasive measure of far-field representation of stimulus-locked synchronous electrical events reflecting the orderly pattern of responses to simple (brief nonspeech) stimuli. In response to auditory stimuli, a series of potential fluctuations measured at the scalp provide information about the functional integrity of brainstem nuclei along the ascending auditory pathway, making it a widely used clinical measure of auditory function. A component of the ABR that occurs in response to a periodic stimulus, the frequency following response (FFR), is well suited for examining how speech elements are encoded subcortically (Fig. 9.1). In the neuronal representation of speech, neural phase locking via the FFR reflects the period of the fundamental frequency (F_0) and its harmonics (Fig. 9.1, top, middle right, and bottom panels).

Evoked ABRs reflect the temporal and spectral characteristics of complex stimuli with remarkable precision (Akhoun et al. 2008; Kraus and Nicol 2005; Krishnan et al. 2005). The temporal fidelity of the evoked ABR has proven a useful tool for a wide array of studies and clinical applications. Because the major morphologic features of the response are stable over time within an individual (Russo et al. 2005) and the peaks are highly replicable between individuals (Akhoun et al. 2008), evoked ABRs are a very sensitive tool for measuring deviations from the normal range of temporal properties (Fig. 9.1, middle left).

The temporal fidelity of subcortical encoding of speech sounds measured in evoked ABRs has been thought to reflect a hard-wired automatic detection of the acoustic features of sound that is not affected by plastic changes usually associated with higher processing structures, such as the cortex. However, recent studies have indicated that the auditory brainstem is a site where experience-dependent plasticity does occur. The plastic nature of evoked human ABRs was first recognized in a study showing that attention modulates the properties of ABRs (Galbraith et al. 1998). Subsequent studies have shown that speakers of tonal languages have enhanced neural representation of pitch (Krishnan et al. 2005), thus illustrating that language experience shapes brainstem activity. More recent studies have shown the effect of musical experience in the plasticity of human evoked ABRs. Musical experience results in more robust brainstem encoding of speech sounds, linguistic pitch-patterns, and processing of vocal expressions of emotion (Fig. 9.1, middle and bottom panels; Musacchia et al. 2007; Wong et al. 2007). Musicians display greater "processing efficiency" of the fundamental frequency of vocal expressions of emotion, expressed selectively in response to complex portions of the stimulus (Strait et al. 2009). However, modification of auditory brainstem processing by language

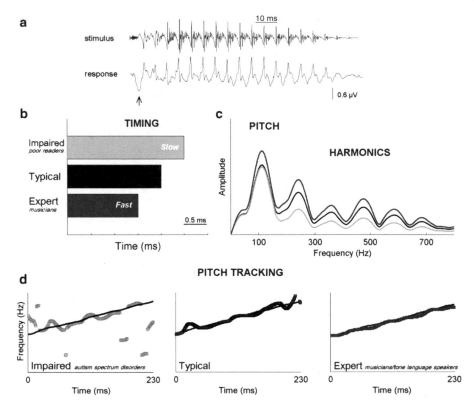

Fig. 9.1 Schematic representation of brainstem processing in impaired (*gray*), typical (*black*), and expert/specialized systems (*red*). This figure provides a schematization of the findings that have emerged from nearly a decade of research on impaired (poor readers, autism spectrum disorders or ASD), typical, and expert systems (musicians, tonal language speakers). (**a**) Time-amplitude stimulus "da" and brainstem response waveform. The stimulus has been shifted by ~8 ms (approximate neural travel time) to increase visual coherence with the response. Following a sharp onset response (demarcated with an arrow), the primary periodicity of the syllable – the fundamental frequency (F_0) – is clearly preserved in the response via phase locking. (**b**) A significant subset of children with reading problems (8–12 years old) have atypical subcortical timing resulting in later (i.e., slower) responses. In contrast, musicians have more precise subcortical timing leading to earlier (i.e., faster) responses than nonmusicians. These temporal disruptions and enhancements occur on the order of tenths of milliseconds (*x*-axis tic marks=0.5 ms) to selective components of the response. (**c**) A fast Fourier transform illustrates the frequency content of the response (the F_0 and its harmonics). Musicians represent the pitch and harmonics of the stimulus more robustly and efficiently than their nonmusician counterparts. A different pattern is seen in a subgroup of children with reading impairments who demonstrate reduced neural encoding of the harmonics, despite normal pitch representation. (**d**) By analyzing the brainstem response over small time bins, we can measure the precision with which brainstem nuclei phase-lock to the time-varying pitch of the stimulus, a phenomenon known as pitch tracking. In the three panels, the thicker lines (*gray, red*) represent the pitch contour extracted from the brainstem response and the thin *black line*, which is most apparent in the *left* panel, represents the pitch contour of the stimulus. Pitch contours were calculated using a running-window short-term Fourier analysis (40-ms time bins, 1-ms interval between the start of each consecutive bin). In these time-frequency graphs, the frequency with the largest magnitude for each given time bin is plotted. A subset of children with ASD showed poor pitch tracking relative to typically developing children, paralleling the prosodic deficit frequently occurring in autism. Musicians, on the other hand, show more accurate brainstem pitch tracking than nonmusicians (Reproduced with permission from Tzounopoulos and Kraus 2009)

and musical experience do not result in a nonspecific generalized gain effect. Instead, distinctive aspects of stimulus processing (e.g., high-frequency phase locking, onset synchrony) are impaired or enhanced depending on the behavioral relevance and relative complexity of the stimulus (Fig. 9.1). More clinically relevant studies have revealed that in a subset of children with language impairment (e.g., poor readers, children with autism) specific properties of brainstem response are disrupted, thus rendering ABRs reliable markers of disease states that involve subcortical impairments (Fig. 9.1, middle and bottom panels; Cunningham et al. 2001; Russo et al. 2008; Banai et al. 2009). While language, musical training, and subcortical impairments represent life-long experiences, brainstem processing can also be modified by shorter-term auditory training, such as over the course of weeks (Russo et al. 2005; de Boer and Thornton 2008; Song et al. 2008). Taken together, these results suggest that sound processing in the human brainstem is malleable and that experience-dependent plasticity results in specific alteration of receptive field properties.

2.2 Plasticity Mechanisms in the Auditory Brainstem

A key question that has emerged in the last few years is whether the auditory brainstem in mammals expresses cellular mechanisms that allow for activity-dependent modulation of neural circuits and that could support the learning phenomena observed in human studies. Activity-dependent changes in synaptic strength (LTP and LTD) represent the leading experimental model for the cellular changes that may underlie and support learning behavior (Malenka and Bear 2004). In addition to synaptic changes, many learning tasks and artificial patterns of activation in brain slices produce long-lasting changes in intrinsic neuronal excitability by changing the function of voltage-gated ion channels, a process called intrinsic plasticity (Zhang and Linden 2003). In agreement with activity-dependent changes in human auditory brainstem evoked responses and contrary to the traditional views that early sensory processing is largely hard-wired, recent studies have revealed that the auditory brainstem is a site where robust synaptic and intrinsic plasticity takes place.

2.2.1 Synaptic Plasticity

In the dorsal cochlear nucleus (DCN), an auditory brainstem nucleus bearing significant resemblance to the cerebellum, parallel fiber inputs to fusiform (principal neurons) and cartwheel (inhibitory interneurons) cells were first shown to exhibit synaptic plasticity in response to high- or low-frequency stimulation and various degrees of depolarization (Fujino and Oertel 2003). More recent work has demonstrated unique opposing forms of spike timing-dependent synaptic plasticity (STDP) at parallel fiber synapses onto fusiform and cartwheel cells (Fig. 9.2; Tzounopoulos et al. 2004, 2007). Over the last decade, STDP, a relatively new and more physiological way of

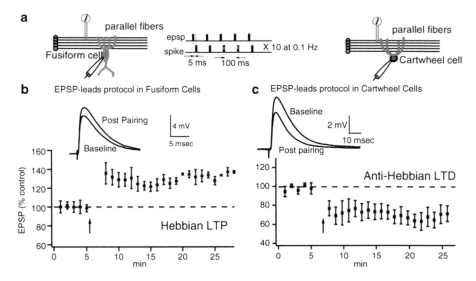

Fig. 9.2 Cell-specific synaptic plasticity in the DCN. Long-term cell-specific synaptic plasticity by time-dependent pairing of pre- and postsynaptic activity. (**a**) Plasticity was induced by a protocol of excitatory postsynaptic potentials (EPSP)-spike pairs. (**b**) Representative traces of averaged EPSPs before and 15–20 min after pairing and time course of induced synaptic plasticity for fusiform cells. (**c**) Representative traces and time course of induced synaptic plasticity for cartwheel cells. The same protocol induces LTP in fusiform cells and LTD at cartwheel cells. These studies have demonstrated unique opposing forms of spike timing-dependent synaptic plasticity (STDP) at parallel fiber synapses onto fusiform and cartwheel cells (Reproduced with permission from Tzounopoulos and Kraus 2009)

inducing synaptic plasticity, has emerged from the observation that the direction of synaptic change can be determined by the precise timing of pre- and postsynaptic action potentials (Levy and Steward 1983; Gustafsson et al. 1987; Bell et al. 1997; Magee and Johnston 1997; Markram et al. 1997). The STDP observed at parallel fiber-fusiform cell synapses resembles STDP observed in the cortex and hippocampus and is Hebbian, that is, presynaptic inputs that are successful in driving postsynaptic spikes are strengthened (LTP is observed when postsynaptic spike follows EPSP) (Bi and Poo 1998; Feldman 2000; Sjostrom et al. 2001; Froemke and Dan 2002). In contrast, parallel fiber-cartwheel cell synapses are characterized by an anti-Hebbian learning rule, where presynaptic inputs that reliably cause, or predict, a postsynaptic spike are weakened (LTD is observed when postsynaptic spike follows EPSP) (Bell et al. 1997). Thus, synapses in the DCN may undergo long-term increase or decrease in synaptic strength dependent on critical differences in the timing between pre- and postsynaptic activity and exhibit anti-Hebbian STDP.

Hebbian STDP has been linked with associative forms of learning such as the creation of a memory trace that sensitizes the circuit to particular profiles of subsequent sensory stimuli (Yao and Dan 2001). Anti-Hebbian STDP has been observed in the cerebellum and cerebellar-like structures and has been linked to cancelation of predictable sensory consequences of the organism's own motor actions (Bell et al. 2008; Sawtell 2010) or to the cancelation of redundant spatially diffused sensory input (Harvey-Girard et al. 2010) in the electrosensory system of the electric fish and with prediction and correction of motor errors and motor learning in the cerebellum (Ohyama et al. 2003). Therefore, Hebbian synaptic learning rules may create a memory trace (Yao and Dan 2001) that sensitizes the DCN to particular profiles of subsequent auditory and nonauditory stimuli, while anti-Hebbian synaptic learning may be responsible for constructing negative images of any ongoing auditory activity that is correlated with parallel fiber activity. Information about the position of the head and neck relayed by parallel fibers may provide the "raw material" that would be needed to cancel predictable consequences of movements on auditory input. In addition, anti-Hebbian plasticity may aid in responding to novel sounds by suppressing the response to self-generated or expected sounds. Thus, the DCN may act as an adaptive sensory processor in which the signals conveyed by parallel fibers in the molecular layer predict the patterns of sensory input to the deep layers through a process of associative synaptic plasticity.

Most studies of synaptic plasticity in the auditory brainstem have been performed in brain slices. This experimental approach has substantial differences from studying intact brains in vivo. For example, high spontaneous spike rates occur for some auditory nerve fibers under in vivo conditions, which have not been taken into account in brain slice recordings. Introduction of high spontaneous in vivo–like activity in brain slices prepared from the medial nucleus of the trapezoid body (MNTB) revealed synaptic failures during high-frequency activity (Hermann et al. 2007, but see Lorteije et al. 2009) not seen in previous studies. This is an important finding suggesting the MNTB is not a simple and faithful relay nucleus, and while it encodes precisely the onset of sound-like stimuli (as bursts of discharges), it may not reliably entrain to high-frequency stimuli for prolonged periods due to short-term plasticity. In addition, these results highlight that several plasticity protocols may be modified by in vivo–like conditions.

2.2.2 Intrinsic Plasticity

Recent findings have revealed that neurons in the MNTB and anteroventral cochlear nucleus (AVCN) change their firing pattern in an activity-dependent manner (Song et al. 2005; Steinert et al. 2008). The presence of rapidly activating and deactivating Kv3.1b potassium channels in these neurons allows for action potentials to be repolarized very rapidly without compromising the initiation or amplitude of a second action potential triggered by a stimulus closely following the first one (Rudy and McBain 2001). Recent studies indicate that changes in the acoustic environment alter the ability of auditory neurons to fire at high frequencies (Song et al. 2005). At low levels of sound intensity (quiet environment), phosphorylation of Kv3.1b by protein kinase C (PKC) reduces potassium current through this channel, thus allowing only low-frequency firing (<200 Hz). Conversely, in response to high-frequency auditory or synaptic stimulation, channel dephosphorylation and increased Kv3.1b channel current promote the ability of neurons to fire at high frequency (Song et al. 2005).

More recent studies have shown that nitric oxide (NO) is another activity-dependent modulator of Kv3 channels in MNTB neurons. Diffusion of NO from MNTB principal neurons provides modulatory control of excitability via direct suppression of postsynaptic Kv3 channels (Steinert et al. 2008). In these studies, activity in one MNTB neuron can modulate the excitability of adjacent neurons, suggesting that this modulation can serve as a homeostatic function for gain reduction during loud noise conditions.

2.3   Developmental Synaptic Plasticity

Synaptic plasticity has also been observed in the lateral superior olive (LSO). The LSO detects differences in the intensity of sounds at the two ears, cues that animals use to localize sources of high-frequency sounds. In adults, the LSO receives tonotopically organized excitatory inputs from the ipsilateral cochlear nucleus and tonotopically organized inhibitory inputs from the contralateral side via the MNTB. The topography of the MNTB–LSO projection becomes tonotopically refined during development and has served as a model system for studying the development of inhibitory circuits (Sanes and Siverls 1991; Sanes et al. 1992; Kim and Kandler 2003). Interestingly, the inhibitory GABA/glycinergic pathway from the MNTB to LSO is tonotopically refined before hearing onset by synaptic elimination and strengthening (Kim and Kandler 2003). Moreover, during this period of refinement, GABA/glycinergic MNTB axon terminals in the LSO contain vesicular glutamate transporter VGLUT3 and MNTB terminals co-release glutamate (Gillespie et al. 2005). This synaptically released glutamate activates postsynaptic AMPA- and NMDA-type glutamate receptors on LSO neurons and may be crucial in mediating synaptic plasticity that could be essential for the maturation and developmental refinement of this inhibitory sound localization pathway. However, these inhibitory synapses display a GABAb-mediated and NMDAR-independent age-dependent form of long-lasting depression, suggesting that this process could also support inhibitory synaptic refinement (Kotak et al. 2001; Chang et al. 2003). Future studies are expected to reveal the precise plasticity mechanisms underlying the developmental maturation and synaptic organization of nonglutamatergic circuits.

2.4   Brainstem Plasticity and Tinnitus

Although plasticity can lead to memory or learning or the compensation for loss of function and adaptation to changing demands, plasticity-induced changes can also cause signs and symptoms of disease, such as may occur with tinnitus. Tinnitus, the persistent perception of a subjective sound ("ringing of the ears"), is often a debilitating condition that ruthlessly reduces quality of life for those chronically affected. Estimates of the number of people experiencing tinnitus range from 13% to 18.6%

of the general population. Of those, approximately 10 million experience it with such severity that they seek medical attention and 2.5 million are considered disabled by it. Several lines of evidence, including clinical studies and animal models, indicate that the brain, rather than the inner ear, may in most cases be the site of maintenance of tinnitus. Tinnitus is most often the result of extreme sound exposure. How does tinnitus shift from a transient condition to a life-long disorder? Animal models of noise-induced tinnitus have shown that the output neuron of the DCN, the fusiform cell, exhibits increased spontaneous activity that could underlie the sensation of sound without external stimulation (Kaltenbach and Godfrey 2008). Therefore, one hypothesis on the mechanisms underlying the induction and expression of tinnitus predicts that activity-dependent mechanisms that change the balance of excitation and inhibition on fusiform cells could lead to hyperactivity of fusiform cells, via plasticity-like mechanisms discovered in the parallel fibers of the DCN (Tzounopoulos 2008). Previous studies have demonstrated that excitatory connections onto principal neurons and inhibitory interneurons exhibit opposing forms of synaptic plasticity (Tzounopoulos et al. 2004). Similar synaptic history leads to a potentiation of excitatory connections and to a depression of inhibitory connections (disinhibition). This is an important finding, suggesting that the net effect of these opposing forms of plasticity would bias the circuit toward excitation of the principal neuron. Although this type of synaptic plasticity may provide adaptation to changing sensory scenes, excessive plasticity-induced depolarization by high levels of potentiation and depression may also lead to abnormal hyperactivity similar to the one observed in the animal model of tinnitus. Experimental establishment of this hypothesis is expected to identify the molecular mechanisms of noise-induced tinnitus and suggest treatments.

Tinnitus, after becoming permanent, becomes a perceptual phenomenon, and its long-lasting nature involves higher-level structures (including the neocortex). Therefore, the auditory brainstem, which is the first stop of the auditory information in the central nervous system (CNS) and the earliest stage in the auditory CNS where synaptic plasticity occurs, represents the best target for preventing the conversion of transient tinnitus to a chronic disorder. Further understanding and potential manipulation of plasticity mechanisms observed in the auditory brainstem could perhaps be used to treat or manage tinnitus. Consistent with this view, signaling molecules involved in synaptic plasticity in the DCN, such as cannabinoid receptors (Tzounopoulos et al. 2007), are down-regulated in the cochlear nucleus in rats showing behavioral evidence of tinnitus (Zheng et al. 2007).

3 Plasticity in the Auditory Cortex

The neocortex performs sensory, motor, and integrative roles, all of which present a strong learning component. Primary neocortical sensory areas display diverse forms of plasticity, and the traditional mechanisms of synaptic plasticity have been extensively studied in sensory primary cortical areas (recently reviewed by Feldman 2009).

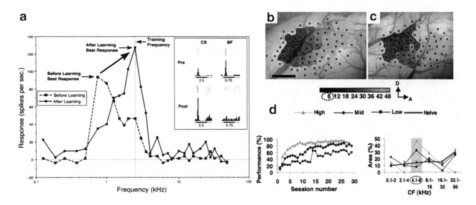

Fig. 9.3 Classical conditioning changes in frequency map. (**a**) Tuning shift of a single neuron in guinea pig's A1 produced by classical conditioning. Effect of 30 trials of tone-shock pairing with an acoustic CS with the frequency of 2.5 Hz. Note the shift from the pretraining best frequency (BF) (Reproduced with permission from Weinberger (2007a)). (**b–d**) Water-deprived rats were trained to press a lever for water in response to a CS of 6 Hz. Map of BFs in the A1 of a naïve (**b**) and a trained (**c**) rat. Each mark represents an electrode. Outside the map are nonresponsive neurons. (**d**) Increase in the representative area of the CF (*right*) was positively correlated with the performance in the task (Reproduced with permission from Rutkowski and Weinberger 2005)

Reports of learning-induced changes in the primary auditory cortex (A1) started in the 1950s with the observation that in adult cats classical conditioning (pairing an auditory stimulus with a puff of air to the face) was related to increased cortical-evoked potentials to the conditioned stimulus (sound) (Galambos et al. 1956). Plasticity in the auditory cortex in adult animals requires the combination of the auditory stimulus with a behaviorally significant paradigm, such as positive or negative reinforcements for producing alterations in the cortical auditory responses. This requirement illustrates the fundamental associative nature of the plasticity (for recent reviews on this subject, see Edeline (2003), Keuroghlian and Knudsen (2007), Ohl and Scheich (2005), and Weinberger (2007a, b)).

3.1   Learning-Induced Plasticity

Most studies of plasticity in the primary auditory cortex come from in vivo studies where the sound (conditioning or associative) stimulus is paired with a nonassociative (nonconditioning) stimulus like a paw shock or a food reward. These protocols lead to an increased cortical representation of the conditioned tone, as described later.

Paired tone-shock stimulation in guinea pigs showed that the animals developed conditioned responses to the specific tone. The frequency tuning of a single neuron in the primary auditory cortex (A1) was shifted from the original best frequency (BF) to match the frequency of the conditioning tone (Fig. 9.3a; Bakin and Weinberger 1990). The analysis of the frequency tuning curves shows not only that the conditioning tone became the BF, but also that the responses to the original BF decreased, producing a shift in the neuron's tuning curve (Fig. 9.3a). This response was dependent on the presence of a paired nonconditioned stimulus because presentations not associated with the nonconditioned stimulus (i.e., shock) produced similar increased responses across all frequencies (sensitization) (Bakin et al. 1992). However, repeated presentation of the same stimulus produced specific habituation to the response to the repeated tone (Ellinwood et al. 1968; Condon and Weinberger 1991). Thus, plastic alterations in the response of cortical auditory neurons to specific tones appear only when the tones are paired with a behaviorally relevant stimulus. In accordance, Rutkowski and Weinberger (2005) showed that water-deprived rats that learned to associate a specific tone as an opportunity to press a bar for a water reward, the representation of this specific frequency in A1 increased proportionally to the performance of the animal in the task (Fig. 9.3b–d). Interestingly, the cortical representation of the frequency of the sound produced by the water delivery machine (0.1–2 kHz) was also increased, reinforcing that the auditory cortex can specifically expand the representation of behavioral relevant stimuli. In owl monkeys, a similar expansion of the representational area of a specific tone in A1 was observed in animals subject to frequency discrimination training, and this parameter was also correlated with behavioral performance in the task (Recanzone et al. 1993).

Importantly, these learning-induced changes in A1 represent a specific alteration in the field map of A1 and not a general drift in the field map that could be occurring over time. The consistent specificity to the observed changes toward the BF provides good evidence for the specific nature of these phenomena. Furthermore, other studies have shown that drifts in the A1 field map that can occur spontaneously over time or due to other environmental changes are much smaller than the observed learning-induced changes (Galvan et al. 2001; Kisley and Gerstein 2001; Edeline 2003; Elhilali et al. 2007).

The critical importance of the behavioral relevance of the stimulus in producing plasticity was highlighted when the plasticity of representational field in A1 was tested in rats subjected to either frequency- or intensity-discrimination tasks (Polley et al. 2006). Animals submitted to frequency- or intensity-discriminatory tasks were rewarded for attending to a particular frequency or intensity. A double-dissociation in the map plasticity was observed: the tonotopic maps of A1 were affected in the frequency-trained animals, and intensity maps changed in the intensity-trained animals. Moreover these changes were correlated to the individual performance of the animal in each task. However, the increased representation of certain frequency is not enough to improve the ability to discriminate in that particular frequency. Increasing cortical representation of certain frequencies by local intracortical microstimulation in the rat A1 did not improve the animal's ability to discriminate that range of frequencies (Talwar and Gerstein 2001). This intriguing result suggests that the learning-induced reorganization in A1 per se is not enough to produce a behavioral improvement in auditory discrimination tasks and the correlation of the increased cortical tone representation with the animal's behaving performance strongly depends on the biological context that generated it. More complex sound features like multitone stimuli and temporal properties of sound (Fritz et al. 2007b) can also induce associative plasticity. Plastic changes were observed in neurons' responses to temporal properties of the sound when rats were trained to use sound repetition cues to localize a reward. These results suggest that auditory cortical plasticity can be related to more complex analysis of the auditory scene (Bao et al. 2004).

The induction of plasticity in mature cortical neurons requires training (Recanzone et al. 1993). Interestingly, behavioral training later in life can reverse response properties created by early auditory exposure. For example, infant rats that are exposed to broadband noise pulses display deficits as adults in the ability of their cortical neurons to be entrained by tonal stimuli presented at rates greater than 8 Hz (Zhou and Merzenich 2008). But this deficit can be reversed if the same animals are trained in a task in which they must discriminate between different rates of repetition (Zhou and Merzenich 2009).

It is important to note that despite the evidence for an increased representation of the conditioning frequency in A1 after the conditioning tasks, some groups reported different results. For instance, in cats trained to discriminate a specific frequency, although they presented an improvement on the frequency-discrimination thresholds for the trained tone, the cortical frequency map was unchanged (Brown et al. 2004). In gerbils, associative learning of a specific tone produced a decreased response at the conditioned frequency in A1, an effect proposed to represent contrast enhancement (Ohl and Scheich 1996, 1997). While these discrepancies may represent species-specific or task-specific differences, the majority of findings indicate that A1 cells gain responsiveness to some sensory stimuli as a result of associative learning tasks that render these stimuli biologically relevant.

3.2 Neuromodulation of Auditory Cortical Plasticity

It has been extensively shown that in adult animals, stimulus presentation per se is ineffective in inducing plasticity in A1 (Weinberger 2007a, b). This means that the animal must situate the stimulus within a biologically relevant context in order for the stimulus to produce biologically relevant cortical plasticity. Interaction with other brain regions is important to link the acoustic stimulus with the biological context with which it is associated. It has been proposed that this interaction could be via activation of neuromodulatory pathways projecting to the auditory cortex. The cortex receives robust dopaminergic and cholinergic inputs from the Ventral Tegmental Area (VTA) and the Nucleus Basalis (NB), respectively. These two regions could be activated during the conditioning protocol and taking part producing the learning-induced plasticity in A1.

Activation of the cholinergic system has been related to several instances of associative learning and memory (Hasselmo 2006). Activation of NB paired with tone presentation in guinea pigs and rats produced a long-lasting increase in the cortical representation of that tone (Bakin and Weinberger 1996; Bjordahl et al. 1998; Kilgard and Merzenich 1998; Dimyan and Weinberger 1999). This effect was dependent on activation of muscarinic receptors (Miasnikov et al. 2001, 2008), and it produced behavioral associative memory in rats (McLin et al. 2002). Increased acetylcholine release was also observed in the A1 of rats during associative auditory learning but not in the rats subjected to a nonassociative task (Butt et al. 2009).

The dopaminergic system is also involved in the processing of auditory information in the auditory cortex. Dopamine is released in the cortex in response to novel stimuli and is associated with reward behavior (Schultz 2001). VTA activation, along with tone stimulation, in rats increases the cortical representation of the tone in A1 and even recruits additional cortical areas not represented in the naïve animals (Bao et al. 2001). In gerbils, cortical application of an agonist of D1/D2 receptors improved the discrimination learning of frequency-modulated tones (Schicknick et al. 2008). Finally, it has been shown that experience-dependent learning in rats leading to improved discrimination of a specific sequence of sounds in response to a reward was abolished by pharmacological destruction of dopaminergic fibers and/ or by application of the D2 receptor antagonists (Kudoh and Shibuki 2006).

3.3 Developmental Plasticity and Critical Periods

Sensitive periods can be defined as specific developmental windows when the brain is more susceptible to environmental influences that can produce plasticity, which can lead to behavioral changes. These windows are species- and stimulus-specific, and their time span depends on the feature of interest. The auditory cortex presents fundamental differences in terms of plasticity in juvenile animals compared to adult animals (Knudsen 2004). A main difference in the auditory cortex of juvenile mammals is the ability to change their tonotopic map representation in response to exposure to a specific sound, but in this case without the need of pairing to an unconditioned stimulus. Persistent daily exposure of rat pups (from P9 to P30) to a specific tone increases the cortical representation of this tone at the expense of the representation of nonpresented tones (Zhang et al. 2001; de Villers-Sidani et al. 2007; however, see Norena et al. 2006, for an opposite effect observed in the cat A1). In rats, sensitive period was determined to be very short and only occurred between P11 and P13 (de Villers-Sidani et al. 2007) The duration of the sensitive period in A1 can be manipulated by environmental conditions. For instance rearing rat P7 pups in continuous white noise (which impairs proper stimulation with the "natural" acoustic environment) prevented the maturation of the tonotopic map in A1 (Chang and Merzenich 2003). Continuous presentation of a single tone to these rats as adults produced the augmented representation of this tone in A1, showing that the sensitive period was prolonged in these animals. Moreover, white noise cessation allowed for the establishment of mature tonotopic representation in A1, thus showing that continuous white-noise exposure prolonged the critical period instead of crystallizing A1 in an eternal immature state. Other experiments have revealed that exposing rat pups to continuous spectrally limited noise retarded the closing of the sensitive period of its specific representational region in A1, showing that the rules that affect the closure of the sensitive period in A1 are not global and are effective on local cortical areas (de Villers-Sidani et al. 2008). Multiple sensitive periods of different durations were also detected for other acoustical parameters (Insanally et al. 2009).

4   Cortical Synaptic Plasticity: LTP and LTD

The cerebral cortex expresses diverse types of plasticity such as LTP, LTD, STDP, non-Hebbian, intrinsic, and homeostatic plasticity (Feldman 2009). Despite the large array of information regarding the learning-induced plasticity in the auditory cortex it is not clear whether or how synaptic plasticity mechanisms mediate associative auditory cortical plasticity.

Early data obtained in monkeys have shown that the auditory cortex follows Hebbian-like rules (Ahissar et al. 1992). Simultaneous recording from two neurons showed that when they fired in synchrony the functional connection between them was strengthened. However, weakening of their connection was observed when simultaneous firing was prevented. These effects were strongly dependent on the behavioral context of the stimuli. These data pointed to the existence of Hebbian-like plasticity mechanisms in A1.

Frequency-tuning changes are induced in A1 via an STDP-like mechanism. These changes were induced in an A1 neuron when a tone of its preferred frequency (PF) is closely paired with another tone of a nonpreferred frequency (NPF) (Dahmen et al. 2008), shifting the frequency map of a PF A1 neuron to a frequency closer to the NPF tone. The effects were dependent on the time interval between two stimuli and on the presentation order. This effect may be related to STDP that occurs in other brain areas including the auditory brainstem (Tzounopoulos et al. 2004). In other cortical areas, STDP has been associated with altered sensory representation as a result of timing-sensitive pairing of stimuli (Meliza and Dan 2006; Mu and Poo 2006; Jacob et al. 2007).

LTP and LTD have been observed in the auditory cortex. LTP was first described in synapses in the supragranular layers II/III of the auditory cortex from adult rats (Kudoh and Shibuki 1994, 1996, 1997; Seki et al. 2003). LTP in the A1 layer II/II is NMDA-receptor-dependent and induced by both tetanic stimulation and theta-bursts (Kudoh and Shibuki 1997). Pyramidal neurons in layer II/III present strong horizontal excitation (Kubota et al. 1997) that can boost LTP in A1 (Kudoh and Shibuki 1997). NO-dependent LTP was also observed in layer V neurons (Wakatsuki et al. 1998).

LTD can also be induced in supragranular neurons (Kudoh et al. 2002; Watanabe et al. 2007) by low-frequency stimulation of layer IV. LTD mechanisms can vary depending on the stimulated synaptic pathway. Using lesions to the thalamic nucleus medial geniculate body (NGM) and slices obtained from different angles containing or not containing NGM fibers, Kudoh et al. (2002) determined that inputs from NGM produced mGluR5-dependent LTD, while inputs from cortico-cortical synapses showed NMDA receptor-dependent LTD. Potentiation of GABAergic transmission has also been suggested to produce an NMDA receptor-dependent LTD following focal stimulation of the supragranular layer (Watanabe et al. 2007).

Synaptic plasticity in A1 is also modulated by developmental factors. LTP is significantly stronger in P30–35 rats than in P100–110 and almost nonexpressed in rats older than P200 (Hogsden and Dringenberg 2009a, b). NR2B subunits of the NMDA receptor have been related to the increased expression of LTP in younger animals, and their expression in early ages seems to be important for expression of LTP in adulthood (Hogsden and Dringenberg 2009a, b; Mao et al. 2006). The finding that LTP is more pronounced in A1 from juvenile animals is consistent with previously mentioned findings showing that younger animals are more prone to plastic changes in A1 in response to environmental stimuli. Therefore, environmental manipulations may have more effects on synaptic plasticity mechanisms observed in juvenile animals.

Auditory input deprivation following cochlear ablation in immature gerbils eliminated LTP expression in layer V of A1 from adult animals but did not affect (and in fact increased) LTD expression (Kotak et al. 2007). On the other hand, hearing loss increased glutamatergic mEPSC amplitude and the expression of NMDA NR2B subunit in layer II/III neurons of the gerbil's A1 (Kotak et al. 2005). White noise exposure during the critical period, which impairs the developmental maturation of A1 in rats, also arrests plasticity maturation in A1 (Chang and Merzenich 2003). Thalamocortical synapses in the A1 of rats submitted to continuous white-noise exposure from birth until P50 expressed more robust LTP than in control animals, an effect not observed if white noise was presented in adult animals (Hogsden and Dringenberg 2009b; Speechley et al. 2007). This increased LTP was sensitive to NMDA receptor NR2B subunit antagonists, similar to what is observed in immature animals (Hogsden and Dringenberg 2009a, b; Mao et al. 2006).

Exposure of young animals to rich sensory environments has been shown to cause anatomical and physiological changes that persist during adult life (van Praag et al. 2000). Rats reared in such enriched auditory environments showed increased number of sound-activated neurons and stronger cortical responses to sound (Engineer et al. 2004; Percaccio et al. 2007). Consistent with in vivo recording experiments, in vitro recordings from these animals reveal larger glutamatergic currents in A1 (Nichols et al. 2007). Taken together, these studies suggest that the auditory cortex can alter its plasticity mechanisms in response to environmental conditions during the critical period in response to either degraded or enriched acoustic environment, thus providing a mechanistic basis for the enhanced plasticity observed in vivo.

Taken together these studies indicate that the classic mechanisms of synaptic plasticity, LTP and LTD, are widely expressed in the auditory cortex. However, it is unknown how these mechanisms are related to the plastic properties of A1 observed in vivo. In general, LTP is more pronounced in A1 from immature animals and is linked to the sensitive period of the auditory cortex, thus suggesting that LTP may mediate the enhanced behavioral plasticity observed in juvenile animals.

5   Mechanisms Underlying Cholinergic Neuromodulation

Acetylcholine can act both via activation of ionotropic nicotinic (nAChR) or metabotropic muscarinic (mAChR) receptors. Cholinergic muscarinic modulation by activation of NB has been shown to be a crucial step in developing auditory cortical plasticity as described previously. However, the precise circuitry and cellular and synaptic mechanisms of this effect are not known.

In vivo experiments have shown that stimulation of NB produced a depolarization of the resting membrane potential (Metherate et al. 1992) and an enhancement of EPSPs that was caused by potassium conductance inhibition (Metherate and Ashe 1993). On the other hand, in vitro experiments have shown that mAChR activation depresses glutamatergic EPSPs (Metherate and Ashe 1995), decreases the probability of release (Atzori et al. 2005; Salgado et al. 2007), and inhibits NMDA glutamatergic currents (Flores-Hernandez et al. 2009). Despite these contradictory effects between in vitro and in vivo studies, it has been concluded that the more prominent effect is an overall potentiation, as pyramidal neurons recorded in vivo are more depolarized than in vitro conditions and therefore more sensitive to the down-regulation of voltage-dependent potassium channels (Metherate et al. 2005).

Ionotropic cholinergic nicotinic receptors (nAChRs) are abundant in the sensory cortex and are relevant for thalamocortical glutamatergic neurotransmission and GABAergic interneuron excitability (Metherate 2004). Nicotinic receptors are found in terminals in layers 3 and 4 of the sensory cortex (Metherate et al. 2005). NAChRs control the excitability of thalamocortical axons both in vitro and in vivo by increasing the probability of firing axon spikes in thalamocortical axons thus increasing the amplitude of stimulus-evoked responses and decreasing their onset latency (Kawai et al. 2007). Furthermore, acetylcholine modulation of glutamatergic transmission via nicotinic α7 receptors seems to be especially relevant during the critical period in the development of the auditory cortex (Metherate and Hsieh 2003).

However, how these mechanisms contribute to the increased representation in A1 of the conditioned stimuli remains unknown. A recent study (Froemke et al. 2007) has started to unveil the mechanisms of cholinergic mechanisms on associative learning in A1. In vivo, whole-cell patch clamp experiments were used to measure the balance between excitation and inhibition in A1 neurons. These studies revealed that pairing a sound of specific frequency (tone) with NB stimulation produced a long-lasting increase in the amplitude of the EPSCs and a decrease of the IPSCs in response to the test tone (Fig. 9.4a, b), while the responses to unpaired tones were unaffected. These effects were inhibited by atropine, an mAChR antagonist. Interestingly, the effect on the inhibitory transmission preceded the effect on the excitatory transmission thus leading to a 3-h imbalance in the ratio of excitation over inhibition (Fig. 9.4c). These data strongly suggest that the rearrangement of the cortical map in response to tone-NB paired stimulation depends on a transient disinhibition. Initially, disinhibition leads to an increase of the cortical excitatory inputs to the conditioned tone and then a new representation of the NB-CS tone is established but now with balanced excitation and inhibition. This result might reflect a multistep process in the rearrangement of the cortical map, starting with transitory effect dependent on the unbalance of excitation and inhibition and culminating with a long-lasting effect no longer dependent on the disinhibition. These findings strongly support the role of LTP-like mechanisms in mediating receptive filed plasticity in A1 as a result of previous experience.

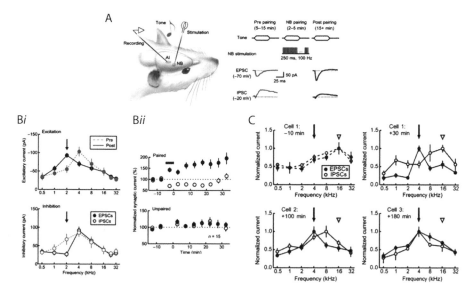

Fig. 9.4 Synaptic plasticity induced by pairing sensory stimulation with nucleus basalis activation (NB). (**A**) Scheme of the in vivo recording from the rat A1 and NB stimulation. On the right the EPSPs and IPSPs before and after the NB pairing. (**Bi**) Shift of the EPSP to the paired frequency (*top*) and the inhibition of the IPSP in response to the paired frequency (*bottom*). (**Bii**) Time course of the effect on the EPSP and IPSP in paired and unpaired animals. (**C**) Temporal course of the changes in synaptic field plasticity. Note the gradual shift of the peak IPSP to the paired tone after continuous stimulation (Reproduced with permission from Froemke et al. 2007)

6 Summary

Contrary to traditional views that early stages of sensory processing are not plastic, new studies discussed in this chapter have established that the auditory brainstem is dynamic. The plethora of intrinsic and synaptic plasticity mechanisms observed in the auditory brainstem in combination with the non-invasive methods of assessing auditory brainstem function in humans provide a platform for relating subcortical auditory processing to higher-order sensory and cognitive tasks involving speech and music. Therefore, it is suggested that the auditory brainstem offers an ideal model to study the mechanisms and functions of nontraditional aspects of sensory processing, such as synaptic and intrinsic plasticity and recurrent feedback from higher levels of processing. The past 10 years of research have revealed how timing information is fed through auditory brainstem pathways. This research has provided insight into how sounds are localized by vertebrates, but much less is known about how these pathways adapt to ongoing sensory activity and how they contribute to the perception and interpretation of environmental sounds, including speech, under normal and pathological conditions. Therein lies the exciting future of revealing the role of plasticity observed in the auditory brainstem.

Plasticity in the cortex has more history and has provided plenty of information about how the primary auditory cortex changes its responses to sound and its receptive fields during auditory associative learning. Moreover, extensive data have shown that the auditory cortex is extremely plastic during the critical or sensitive period of development, not needing the pairing with the conditioning stimulus to induce plasticity in response to repetitive auditory stimuli. The role of cholinergic and dopaminergic systems are also well established in inducing conditioned learning in A1. However, much less is known about the circuitry and synaptic mechanisms underlying these phenomena. It is well established that auditory cortical synapses present diverse forms of Hebbian plasticity like LTP, LTD, and STDP, but their role in producing the learning-induced plasticity in A1 is still unknown. The next challenge for the auditory cortical plasticity field is to understand how these and other forms of plasticity mechanisms interact in the different circuits that make up A1 in order to produce the observed changes in vivo, in both mature and immature animals. Combinations of diverse in vitro and in vivo approaches are expected to accomplish this task.

Acknowledgments This work was supported by NIDCD grant R01DC007905 to TT and by a grant from the American Tinnitus Foundation to TT.

References

Ahissar, E., Vaadia, E., Ahissar, M., Bergman, H., Arieli, A., & Abeles, M. (1992). Dependence of cortical plasticity on correlated activity of single neurons and on behavioral context. *Science, 257*(5075), 1412–1415.

Akhoun, I., Gallego, S., Moulin, A., Menard, M., Veuillet, E., Berger-Vachon, C.,Collet, L., & Thai-Van, H. (2008). The temporal relationship between speech auditory brainstem responses and the acoustic pattern of the phoneme/ba/in normal-hearing adults. *Clinical Neurophysiology, 119*(4), 922–933.

Aristotle (350.B.C.E.) On the soul. The Internet Classics Archive. http://classics.mit.edu/Aristotle/soul.html. Translated by J. A. Smith. Accessed 2 September 2010.

Atzori, M., Kanold, P. O., Pineda, J. C., Flores-Hernandez, J., & Paz, R. D. (2005). Dopamine prevents muscarinic-induced decrease of glutamate release in the auditory cortex. *Neuroscience, 134*(4), 1153–1165.

Bakin, J. S., & Weinberger, N. M. (1990). Classical conditioning induces CS-specific receptive field plasticity in the auditory cortex of the guinea pig. *Brain Research, 536*(1–2), 271–286.

Bakin, J. S., & Weinberger, N. M. (1996). Induction of a physiological memory in the cerebral cortex by stimulation of the nucleus basalis. *Proceedings of the National Academy of Sciences of the United States of America, 93*(20), 11219–11224.

Bakin, J. S., Lepan, B., & Weinberger, N. M. (1992). Sensitization induced receptive field plasticity in the auditory cortex is independent of CS-modality. *Brain Research, 577*(2), 226–235.

Banai, K., Hornickel, J., Skoe, E., Nicol, T., Zecker, S., & Kraus, N. (2009). Reading and subcortical auditory function. *Cerebral Cortex, 19*(11), 2699–2707.

Bao, S., Chan, V. T., & Merzenich, M. M. (2001). Cortical remodelling induced by activity of ventral tegmental dopamine neurons. *Nature, 412*(6842), 79–83.

Bao, S., Chang, E. F., Woods, J., & Merzenich, M. M. (2004). Temporal plasticity in the primary auditory cortex induced by operant perceptual learning. *Nature Neuroscience, 7*(9), 974–981.

Bell, C. C., Han, V. Z., Sugawara, Y., & Grant, K. (1997). Synaptic plasticity in a cerebellum-like structure depends on temporal order. *Nature, 387*(6630), 278–281.

Bell, C. C., Han, V., & Sawtell, N. B. (2008). Cerebellum-like structures and their implications for cerebellar function. *Annual Review of Neuroscience, 31*, 1–24.

Bi, G.Q. & Poo, M.M. (1998). Synaptic modifications in cultured hippocampal neurons: dependence on spike timing, synaptic strength, and postsynaptic cell type. *Journal of Neuroscience, 18*(24):10464–10472.

Bjordahl, T. S., Dimyan, M. A., & Weinberger, N. M. (1998). Induction of long-term receptive field plasticity in the auditory cortex of the waking guinea pig by stimulation of the nucleus basalis. *Behavioral Neuroscience, 112*(3), 467–479.

Bliss, T.V., & Gardner-Medwin, A.R. (1973). Long-lasting potentiation of synaptic transmission in the dentate area of the unanaestetized rabbit following stimulation of the perforant path. *Journal of Physiology, 232*(2), 357–374.

Bliss, T.V., & Lomo, T. (1973). Long-lasting potentiation of synaptic transmission in the dentate area of the anaesthetized rabbit following stimulation of the perforant path. *Journal of Physiology, 232*(2), 331–356.

Brown, M., Irvine, D. R., & Park, V. N. (2004). Perceptual learning on an auditory frequency discrimination task by cats: Association with changes in primary auditory cortex. *Cerebral Cortex, 14*(9), 952–965.

Butt, A. E., Chavez, C. M., Flesher, M. M., Kinney-Hurd, B. L., Araujo, G. C., Miasnikov, A. A., & Weinberger, N.M.. (2009). Association learning-dependent increases in acetylcholine release in the rat auditory cortex during auditory classical conditioning. *Neurobiology of Learning and Memory, 92*(3), 400–409.

Chang, E. F., & Merzenich, M. M. (2003). Environmental noise retards auditory cortical development. *Science, 300*(5618), 498–502.

Chang, E. H., Kotak, V. C., & Sanes, D. H. (2003). Long-term depression of synaptic inhibition is expressed postsynaptically in the developing auditory system. *Journal of Neurophysiology, 90*(3), 1479–1488.

Condon, C. D., & Weinberger, N. M. (1991). Habituation produces frequency-specific plasticity of receptive fields in the auditory cortex. *Behavioral Neuroscience, 105*(3), 416–430.

Cunningham, J., Nicol, T., Zecker, S. G., Bradlow, A., & Kraus, N. (2001). Neurobiologic responses to speech in noise in children with learning problems: Deficits and strategies for improvement. *Clinical Neurophysiology, 112*(5), 758–767.

Dahmen, J. C., Hartley, D. E., & King, A. J. (2008). Stimulus-timing-dependent plasticity of cortical frequency representation. *Journal of Neuroscience, 28*(50), 13629–13639.

de Boer, J., & Thornton, A. R. (2008). Neural correlates of perceptual learning in the auditory brainstem: efferent activity predicts and reflects improvement at a speech-in-noise discrimination task. *Journal of Neuroscience, 28*(19), 4929–4937.

de Villers-Sidani, E., Chang, E. F., Bao, S., & Merzenich, M. M. (2007). Critical period window for spectral tuning defined in the primary auditory cortex (A1) in the rat. *Journal of Neuroscience, 27*(1), 180–189.

de Villers-Sidani, E., Simpson, K. L., Lu, Y. F., Lin, R. C., & Merzenich, M. M. (2008). Manipulating critical period closure across different sectors of the primary auditory cortex. *Nature Neuroscience, 11*(8), 957–965.

Dimyan, M. A., & Weinberger, N. M. (1999). Basal forebrain stimulation induces discriminative receptive field plasticity in the auditory cortex. *Behavioral Neuroscience, 113*(4), 691–702.

Edeline, J. M. (2003). The thalamo-cortical auditory receptive fields: Regulation by the states of vigilance, learning and the neuromodulatory systems. *Experimental Brain Research, 153*(4), 554–572.

Elhilali, M., Fritz, J. B., Chi, T. S., & Shamma, S. A. (2007) Auditory cortical receptive fields: Stable entities with plastic abilities. *Journal of Neuroscience 27*(39), 10372–10382.

Ellinwood, E. H., Cook, J. D., & Wiilson, W. P. (1968). Habituation of evoked response to uniaural clicks. *Brain Research, 7*(2), 306–309.

Engineer, N. D., Percaccio, C. R., Pandya, P. K., Moucha, R., Rathbun, D. L., & Kilgard, M. P. (2004). Environmental enrichment improves response strength, threshold, selectivity, and latency of auditory cortex neurons. *Journal of Neurophysiology, 92*(1), 73–82.

Feldman, D.E. (2000). Timing-based LTP and LTD at vertical inputs to layer II/III pyramidal cells in the rat barrel cortex. *Neuron, 27*(1):45–56.

Feldman, D. E. (2009). Synaptic mechanisms for plasticity in neocortex. *Annual Review of Neuroscience, 32*, 33–55.

Flores-Hernandez, J., Salgado, H., De La Rosa, V., Avila-Ruiz, T., Torres-Ramirez, O., Lopez-Lopez, G., & Atzori, M. (2009). Cholinergic direct inhibition of N-methyl-D aspartate receptor-mediated currents in the rat neocortex. *Synapse, 63*(4), 308–318.

Fritz, J. B., Elhilali, M., David, S. V., & Shamma, S. A. (2007a). Auditory attention – focusing the searchlight on sound. *Current Opinion in Neurobiology, 17*(4), 437–455.

Fritz, J. B., Elhilali, M., & Shamma, S. A. (2007b). Adaptive changes in cortical receptive fields induced by attention to complex sounds. *Journal of Neurophysiology, 98*(4), 2337–2346.

Froemke, R.C. & Dan, Y. (2002). Spike-timing-dependent synaptic modification induced by natural spike trains. *Nature, 416*(6879):433–438.

Froemke, R. C., Merzenich, M. M., & Schreiner, C. E. (2007). A synaptic memory trace for cortical receptive field plasticity. *Nature, 450*(7168), 425–429.

Fujino, K., & Oertel, D. (2003). Bidirectional synaptic plasticity in the cerebellum-like mammalian dorsal cochlear nucleus. *Proceedings of the National Academy of Sciences of the United States of America, 100*(1), 265–270.

Galambos, R., Sheatz, G., & Vernier, V. G. (1956). Electrophysiological correlates of a conditioned response in cats. *Science, 123*(3192), 376–377.

Galbraith, G. C., Bhuta, S. M., Choate, A. K., Kitahara, J. M., & Mullen, T. A. Jr. (1998). Brain stem frequency-following response to dichotic vowels during attention. *Neuroreport, 9*(8), 1889–1893.The title is correct (Brains stem)

Galvan, V. V., Chen, J., & Weinberger, N. M. (2001). Long-term frequency tuning of local field potentials in the auditory cortex of the waking guinea pig. *Journal of the Association for Research in Otolaryngology, 2*(3), 199–215.

Gillespie, D. C., Kim, G., & Kandler, K. (2005). Inhibitory synapses in the developing auditory system are glutamatergic. *Nature Neuroscience, 8*(3), 332–338.

Gustafsson, B., Wigström, H., Abraham, W.C., & Huang, Y.Y. (1987) Long-term potentiation in the hippocampus using depolarizing current pulses as the conditioning stimulus to single volley synaptic potentials. *Journal of Neuroscience, 7*(3):774–80.

Harvey-Girard, E., Lewis, J., and Maler, L. (2010) Burst-induced anti-Hebbian depression acts through short-term synaptic dynamics to cancel redundant sensory signals. *Journal of Neuroscience, 30*(17), 6152–6169

Hasselmo, M. E. (2006). The role of acetylcholine in learning and memory. *Current Opinion in Neurobiology, 16*(6), 710–715.

Hermann, J., Pecka, M., von Gersdorff, H., Grothe, B., & Klug, A. (2007). Synaptic transmission at the calyx of Held under in vivo like activity levels. *Journal of Neurophysiology, 98*(2), 807–820.

Hogsden, J. L., & Dringenberg, H. C. (2009a). Decline of long-term potentiation (LTP) in the rat auditory cortex in vivo during postnatal life: Involvement of NR2B subunits. *Brain Research, 1283*, 25–33.

Hogsden, J. L., & Dringenberg, H. C. (2009b). NR2B subunit-dependent long-term potentiation enhancement in the rat cortical auditory system in vivo following masking of patterned auditory input by white noise exposure during early postnatal life. *European Journal of Neuroscience, 30*(3), 376–384.

Insanally, M. N., Kover, H., Kim, H., & Bao, S. (2009). Feature-dependent sensitive periods in the development of complex sound representation. *Journal of Neuroscience, 29*(17), 5456–5462.

Jacob, V., Brasier, D. J., Erchova, I., Feldman, D., & Shulz, D. E. (2007). Spike timing-dependent synaptic depression in the in vivo barrel cortex of the rat. *Journal of Neuroscience, 27*(6), 1271–1284.

Kaltenbach, J. A., & Godfrey, D. A. (2008). Dorsal cochlear nucleus hyperactivity and tinnitus: Are they related? *American Journal of Audiology, 17*(2), S148–161.

Kawai, H., Lazar, R., & Metherate, R. (2007). Nicotinic control of axon excitability regulates thalamocortical transmission. *Nature Neuroscience, 10*(9), 1168–1175.

Keuroghlian, A. S., & Knudsen, E. I. (2007). Adaptive auditory plasticity in developing and adult animals. *Progress in Neurobiology, 82*(3), 109–121.

Kilgard, M. P., & Merzenich, M. M. (1998). Cortical map reorganization enabled by nucleus basalis activity. *Science, 279*(5357), 1714–1718.

Kim, G., & Kandler, K. (2003). Elimination and strengthening of glycinergic/GABAergic connections during tonotopic map formation. *Nature Neuroscience, 6*(3), 282–290.

Kisley, M. A., & Gerstein, G. L. (2001). Daily variation and appetitive conditioning-induced plasticity of auditory cortex receptive fields. *European Journal of Neuroscience, 13*(10), 1993–2003.

Knudsen, E. I. (2004). Sensitive periods in the development of the brain and behavior. *Journal of Cognitive Neuroscience, 16*(8), 1412–1425.

Kotak, V. C., DiMattina, C., & Sanes, D. H. (2001). GABA(B) and Trk receptor signaling mediates long-lasting inhibitory synaptic depression. *Journal of Neurophysiology, 86*(1), 536–540.

Kotak, V. C., Fujisawa, S., Lee, F. A., Karthikeyan, O., Aoki, C., & Sanes, D. H. (2005). Hearing loss raises excitability in the auditory cortex. *Journal of Neuroscience, 25*(15), 3908–3918.

Kotak, V. C., Breithaupt, A. D., & Sanes, D. H. (2007). Developmental hearing loss eliminates long-term potentiation in the auditory cortex. *Proceedings of the National Academy of Sciences of the United States of America, 104*(9), 3550–3555.

Kraus, N., & Nicol, T. (2005). Brainstem origins for cortical "what" and "where" pathways in the auditory system. *Trends in Neurosciences, 28*(4), 176–181.

Krishnan, A., Xu, Y., Gandour, J., & Cariani, P. (2005). Encoding of pitch in the human brainstem is sensitive to language experience. *Brain Research Cognitive Brain Research, 25*(1), 161–168.

Kubota, M., Sugimoto, S., Horikawa, J., Nasu, M., Taniguchi, I. (1997) Optical imaging of dynamic horizontal spread of excitation in rat auditory cortex slices. *Neuroscience Letters, 237*(2–3):77–80.

Kudoh, M., & Shibuki, K. (1994). Long-term potentiation in the auditory cortex of adult rats. *Neuroscience Letters, 171*(1–2), 21–23.

Kudoh, M., & Shibuki, K. (1996). Long-term potentiation of supragranular pyramidal outputs in the rat auditory cortex. *Experimental Brain Research, 110*(1), 21–27.

Kudoh, M., & Shibuki, K. (1997). Importance of polysynaptic inputs and horizontal connectivity in the generation of tetanus-induced long-term potentiation in the rat auditory cortex. *Journal of Neuroscience, 17*(24), 9458–9465.

Kudoh, M., & Shibuki, K. (2006). Sound sequence discrimination learning motivated by reward requires dopaminergic D2 receptor activation in the rat auditory cortex. *Learning and Memory, 13*(6), 690–698.

Kudoh, M., Sakai, M., & Shibuki, K. (2002). Differential dependence of LTD on glutamate receptors in the auditory cortical synapses of cortical and thalamic inputs. *Journal of Neurophysiology, 88*(6), 3167–3174.

Levy, W.B, & Steward, O. (1983). Temporal contiguity requirements for long-term associative potentiation/depression in the hippocampus. *Neuroscience, 8*(4):791–7.

Locke, J. (1689). *An Essay Concerning Human Understanding.* Nidditch, P.H. (ed.) (1975). New York: Oxford University Press.

Lorteije, J.A., Rusu, S.I., Kushmerick, C., & Borst, J.G. (2009). Reliability and precision of the mouse calyx of Held synapse. *Journal of Neuroscience, 29*(44):13770–13784.

Magee, J. C., & Johnston, D. (1997). A synaptically controlled, associative signal for Hebbian plasticity in hippocampal neurons. *Science, 275*(5297), 209–213.

Malenka, R. C., & Bear, M. F. (2004). LTP and LTD: An embarrassment of riches. *Neuron, 44*(1), 5–21.

Mao, Y., Zang, S., Zhang, J., & Sun, X. (2006). Early chronic blockade of NR2B subunits and transient activation of NMDA receptors modulate LTP in mouse auditory cortex. *Brain Research, 1073–1074*, 131–138.

Markram, H., Lubke, J., Frotscher, M., & Sakmann, B. (1997). Regulation of synaptic efficacy by coincidence of postsynaptic APs and EPSPs. *Science, 275*(5297), 213–215.

McLin, D. E. III, Miasnikov, A. A., & Weinberger, N. M. (2002). Induction of behavioral associative memory by stimulation of the nucleus basalis. *Proceedings of the National Academy of Sciences of the United States of America, 99*(6), 4002–4007.

Meliza, C. D., & Dan, Y. (2006). Receptive-field modification in rat visual cortex induced by paired visual stimulation and single-cell spiking. *Neuron, 49*(2), 183–189.

Metherate, R. (2004). Nicotinic acetylcholine receptors in sensory cortex. *Learning and Memory, 11*(1), 50–59.

Metherate, R., & Ashe, J. H. (1993). Ionic flux contributions to neocortical slow waves and nucleus basalis-mediated activation: Whole-cell recordings in vivo. *Journal of Neuroscience, 13*(12), 5312–5323.

Metherate, R., & Ashe, J. H. (1995). Synaptic interactions involving acetylcholine, glutamate, and GABA in rat auditory cortex. *Experimental Brain Research, 107*(1), 59–72.

Metherate, R., & Hsieh, C.Y.(2003) Regulation of glutamate synapses by nicotinic acetylcholine receptors in auditory cortex. *Neurobiology of Learning and Memory, 80*(3):285–290

Metherate, R., Cox, C. L., & Ashe, J. H. (1992). Cellular bases of neocortical activation: Modulation of neural oscillations by the nucleus basalis and endogenous acetylcholine. *Journal of Neuroscience, 12*(12), 4701–4711.

Metherate, R., Kaur, S., Kawai, H., Lazar, R., Liang, K., & Rose, H. J. (2005). Spectral integration in auditory cortex: Mechanisms and modulation. *Hearing Research, 206*(1–2), 146–158.

Miasnikov, A. A., McLin, D. III, & Weinberger, N. M. (2001). Muscarinic dependence of nucleus basalis induced conditioned receptive field plasticity. *Neuroreport, 12*(7), 1537–1542.

Miasnikov, A. A., Chen, J. C., & Weinberger, N. M. (2008). Specific auditory memory induced by nucleus basalis stimulation depends on intrinsic acetylcholine. *Neurobiology of Learning and Memory, 90*(2), 443–454.

Mu, Y., & Poo, M. M. (2006). Spike timing-dependent LTP/LTD mediates visual experience. *Neuron, 50*(1):115–125.

Musacchia, G., Sams, M., Skoe, E., & Kraus, N. (2007). Musicians have enhanced subcortical auditory and audiovisual processing of speech and music. *Proceedings of the National Academy of Sciences of the United States of America, 104*(40), 15894–15898.

Nichols, J. A., Jakkamsetti, V. P., Salgado, H., Dinh, L., Kilgard, M. P., & Atzori, M. (2007). Environmental enrichment selectively increases glutamatergic responses in layer II/III of the auditory cortex of the rat. *Neuroscience, 145*(3), 832–840.

Norena, A. J., Gourevitch, B., Aizawa, N., & Eggermont, J. J. (2006). Spectrally enhanced acoustic environment disrupts frequency representation in cat auditory cortex. *Nature Neuroscience, 9*(7), 932–939.

Ohl, F. W., & Scheich, H. (1996). Differential frequency conditioning enhances spectral contrast sensitivity of units in auditory cortex (field AI) of the alert Mongolian gerbil. *European Journal of Neuroscience, 8*(5), 1001–1017.

Ohl, F. W., & Scheich, H. (1997). Learning-induced dynamic receptive field changes in primary auditory cortex of the unanaesthetized Mongolian gerbil. *Journal of Comparative Physiology[A], 181*(6), 685–696.

Ohl, F. W., & Scheich, H. (2005). Learning-induced plasticity in animal and human auditory cortex. *Current Opinion in Neurobiology, 15*(4), 470–477.

Ohyama, T., Nores, W. L., Murphy, M., and Mauk, M. D. (2003). What the cerebellum computes. *Trends in Neurosciences, 26*(4), 222–227.

Percaccio, C. R., Pruette, A. L., Mistry, S. T., Chen, Y. H., & Kilgard, M. P. (2007). Sensory experience determines enrichment-induced plasticity in rat auditory cortex. *Brain Research, 1174*, 76–91.

Polley, D. B., Steinberg, E. E., & Merzenich, M. M. (2006). Perceptual learning directs auditory cortical map reorganization through top-down influences. *Journal of Neuroscience, 26*(18), 4970–4982.

Recanzone, G. H., Schreiner, C. E., & Merzenich, M. M. (1993). Plasticity in the frequency representation of primary auditory cortex following discrimination training in adult owl monkeys. *Journal of Neuroscience, 13*(1), 87–103.

Rudy, B., & McBain, C. J. (2001). Kv3 channels: Voltage-gated K+ channels designed for high-frequency repetitive firing. *Trends in Neurosciences, 24*(9), 517–526.

Russo, N. M., Nicol, T. G., Zecker, S. G., Hayes, E. A., & Kraus, N. (2005). Auditory training improves neural timing in the human brainstem. *Behavioral Brain Research, 156*(1), 95–103.

Russo, N. M., Skoe, E., Trommer, B., Nicol, T., Zecker, S., Bradlow, A., & Kraus, N.. (2008). Deficient brainstem encoding of pitch in children with Autism Spectrum Disorders. *Clinical Neurophysiology, 119*(8), 1720–1731.

Rutkowski, R. G., & Weinberger, N. M. (2005). Encoding of learned importance of sound by magnitude of representational area in primary auditory cortex. *Proceedings of the National Academy of Sciences of the United States of America, 102*(38), 13664–13669.

Salgado, H., Bellay, T., Nichols, J. A., Bose, M., Martinolich, L., Perrotti, L., & Atzori, M.(2007). Muscarinic M2 and M1 receptors reduce GABA release by Ca^{2+} channel modulation through activation of PI3K/Ca^{2+}-independent and PLC/Ca^{2+}-dependent PKC. *Journal of Neurophysiology, 98*(2), 952–965.

Sanes, D. H., & Siverls, V. (1991). Development and specificity of inhibitory terminal arborizations in the central nervous system. *Journal of Neurobiology, 22*(8), 837–854.

Sanes, D. H., Song, J., & Tyson, J. (1992). Refinement of dendritic arbors along the tonotopic axis of the gerbil lateral superior olive. *Brain Research Developmental Brain Research, 67*(1), 47–55. <<au: journal title as meant?>>yes

Sawtell, N.B. (2010). Multimodal integration in granule cells as a basis for associative plasticity and sensory prediction in a cerebellum-like circuit. *Neuron, 66*(4):573–584.

Schicknick, H., Schott, B. H., Budinger, E., Smalla, K. H., Riedel, A., Seidenbecher, C. I., Scheich, H, Gundelfinger, E. D., & Tischmeyer, W. (2008). Dopaminergic modulation of auditory cortex–dependent memory consolidation through mTOR. *Cerebral Cortex, 18*(11), 2646–2658.

Schreiner, C. E., & Winer, J. A. (2007). Auditory cortex mapmaking: Principles, projections, and plasticity. *Neuron, 56*(2), 356–365.

Schultz, W. (2001). Reward signaling by dopamine neurons. *Neuroscientist, 7*(4), 293–302.

Seki, K., Kudoh, M., & Shibuki, K. (2003). Polysynaptic slow depolarization and spiking activity elicited after induction of long-term potentiation in rat auditory cortex. *Brain Research, 988*(1–2), 114–120.

Sjöström, P.J,, Turrigiano, G.G,, & Nelson, S.B.. (2001). Rate, timing, and cooperativity jointly determine cortical synaptic plasticity. *Neuron, 32*(6):1148–1164.

Song, P., Yang, Y., Barnes-Davies, M., Bhattacharjee, A., Hamann, M., Forsythe, I. D., Oliver, D.L., & Kaczmarek, L.K. (2005). Acoustic environment determines phosphorylation state of the Kv3.1 potassium channel in auditory neurons. *Nature Neuroscience, 8*(10), 1335–1342.

Song, J. H., Skoe, E., Wong, P. C., & Kraus, N. (2008). Plasticity in the adult human auditory brainstem following short-term linguistic training. *Journal of Cognitive Neuroscience, 20*(10), 1892–1902.

Speechley, W. J., Hogsden, J. L., & Dringenberg, H. C. (2007). Continuous white noise exposure during and after auditory critical period differentially alters bidirectional thalamocortical plasticity in rat auditory cortex in vivo. *European Journal of Neuroscience, 26*(9), 2576–2584.

Steinert, J. R., Kopp-Scheinpflug, C., Baker, C., Challiss, R. A., Mistry, R., Haustein, M. D., Griffin, S.J., Tong, H., Graham, B. P., & Forsythe, I. D. (2008). Nitric oxide is a volume transmitter regulating postsynaptic excitability at a glutamatergic synapse. *Neuron, 60*(4), 642–656.

Strait, D. L., Kraus, N., Skoe, E., & Ashley, R. (2009). Musical experience and neural efficiency: Effects of training on subcortical processing of vocal expressions of emotion. *European Journal of Neuroscience, 29*(3), 661–668.

Talwar, S. K., & Gerstein, G. L. (2001). Reorganization in awake rat auditory cortex by local microstimulation and its effect on frequency-discrimination behavior. *Journal of Neurophysiology, 86*(4), 1555–1572.

Tzounopoulos, T. (2008). Mechanisms of synaptic plasticity in the dorsal cochlear nucleus: Plasticity-induced changes that could underlie tinnitus. *American Journal of Audiology, 17*(2), S170–175.

Tzounopoulos, T., & Kraus, N. (2009). Learning to encode timing: mechanisms of plasticity in the auditory brainstem. *Neuron, 62*(4), 463–469.

Tzounopoulos, T., Kim, Y., Oertel, D., & Trussell, L. O. (2004). Cell-specific, spike timing-dependent plasticities in the dorsal cochlear nucleus. *Nature Neuroscience, 7*(7), 719–725.

Tzounopoulos, T., Rubio, M. E., Keen, J. E., & Trussell, L. O. (2007). Coactivation of pre- and postsynaptic signaling mechanisms determines cell-specific spike-timing-dependent plasticity. *Neuron, 54*(2), 291–301.

van Praag, H., Kempermann, G., & Gage, F. H. (2000). Neural consequences of environmental enrichment. *Nature Reviews Neuroscience, 1*(3), 191–198.

Wakatsuki, H., Gomi, H., Kudoh, M., Kimura, S., Takahashi, K., Takeda, M., & Shibuki, K. (1998). Layer-specific NO dependence of long-term potentiation and biased NO release in layer V in the rat auditory cortex. *Journal of Physiology, 513*(1), 71–81.

Watanabe, K., Kamatani, D., Hishida, R., Kudoh, M., & Shibuki, K. (2007). Long-term depression induced by local tetanic stimulation in the rat auditory cortex. *Brain Research, 1166*, 20–28.

Weinberger, N. M. (2007a). Associative representational plasticity in the auditory cortex: A synthesis of two disciplines. *Learning and Memory, 14*(1–2), 1–16.

Weinberger, N. M. (2007b). Auditory associative memory and representational plasticity in the primary auditory cortex. *Hearing Research, 229*(1–2), 54–68.

Wiesel, T.N. (1982). Postnatal development of the visual cortex and the influence of environment. *Nature, 299*(5884),583–591.

Wong, P. C., Skoe, E., Russo, N. M., Dees, T., & Kraus, N. (2007). Musical experience shapes human brainstem encoding of linguistic pitch patterns. *Nature Neuroscience, 10*(4), 420–422.

Yao, H., & Dan, Y. (2001). Stimulus timing-dependent plasticity in cortical processing of orientation. *Neuron 32*(2), 315–323.

Zhang, W., & Linden, D. J. (2003). The other side of the engram: Experience-driven changes in neuronal intrinsic excitability. *Nature Reviews Neuroscience, 4*(11), 885–900.

Zhang, L. I., Bao, S., & Merzenich, M. M. (2001). Persistent and specific influences of early acoustic environments on primary auditory cortex. *Nature Neuroscience, 4*(11), 1123–1130.

Zheng, Y., Baek, J. H., Smith, P. F., & Darlington, C. L. (2007). Cannabinoid receptor down-regulation in the ventral cochlear nucleus in a salicylate model of tinnitus. *Hearing Research, 228*(1–2), 105–111.

Zhou, X., & Merzenich, M. M. (2008). Enduring effects of early structured noise exposure on temporal modulation in the primary auditory cortex. *Proceedings of the National Academy of Sciences of the United States of America, 105*(11), 4423–4428.

Zhou, X., & Merzenich, M. M. (2009). Developmentally degraded cortical temporal processing restored by training. *Nature Neuroscience, 12*(1), 26–28.

Index

L.O. Trussell et al. (eds.), *Synaptic Mechanisms in the Auditory System*,
Springer Handbook of Auditory Research 41, DOI 10.1007/978-1-4419-9517-9,
© Springer Science+Business Media, LLC 2012